P9-CJR-100

SUN/EARTH

Richard L. Crowther
FAIA

SUN/EARTH

Alternative Energy Design for Architecture

UPDATED AND
ENLARGED EDITION

VAN NOSTRAND REINHOLD COMPANY
NEW YORK CINCINNATI TORONTO LONDON MELBOURNE

Library of Congress Catalog Card Number 82-7081
ISBN 0-442-21499-5c
ISBN 0-442-21498-7p

Printed in the United States of America
Designed by Loudan Enterprises

Published by Van Nostrand Reinhold Company Inc.
135 West 50th Street
New York, New York 10020

Van Nostrand Reinhold Publishers
1410 Birchmount Road
Scarborough, Ontario M1P 2E7, Canada

Van Nostrand Reinhold Australia Pty. Ltd.
480 Latrobe Street
Melbourne, Victoria 3000, Australia

Van Nostrand Reinhold Company Limited
Molly Millars Lane
Wokingham, Berkshire RG11 2PY, England

16 15 14 13 12 11 10 9 8 7 6 5 4 3 2 1

Library of Congress Cataloging in Publication Data

Crowther, Richard L.
Sun, Earth.

Bibliography: p. 243
Includes index.
1. Dwellings—Energy conservation. 2. Renewable energy sources. I. Title.
TJ163.5.D86C76 1983 720'.47 82-7081
ISBN 0-442-21499-5 AACR2
ISBN 0-442-21498-7 (pbk.)

Contents

Acknowledgments

Updated and Enlarged Edition
Robert F. Steimle, editing
Karin Spongberg, editing
James J. Day, delineation
Douglas Rosendale, delineation

Original Edition
Paul Karius, delineation, editing
Donald J. Frey, research, editing
Larry Atkinson, cover design, review

Appreciation is given to the following professionals who reviewed portions of *Sun/Earth* and offered their constructive criticism:

Updated and Enlarged Edition
John I. Yellott, School of Architecture, Arizona State University
DeVon Carlson, School of Architecture, University of Colorado
Dale Gibbs, School of Architecture, University of Nebraska

Original Edition
John I. Yellott, School of Architecture, Arizona State University
Benjamin T. (Buck) Rogers, Los Alamos Scientific Laboratory
DeVon Carlson, School of Architecture, University of Colorado
Jerry D. Plunkett, Montana Energy and MHD Research and Development Institute, Inc.
Frank R. Eldridge, Mitre Corporation
Elizabeth Kingman, Solar Energy Exhibition Program, University of Colorado at Denver

Foreword

Interesting, imaginative, and eminently readable—these are all adjectives that seek to describe adequately this remarkable survey of architecture and its relation to the natural environment. Architects are often gifted with the ability to illustrate their writings with strikingly effective graphics, and *Sun/Earth* shows this talent to excellent advantage. The many examples of the author's work that are depicted in this edition of *Sun/Earth* show how his concepts of energy conservation and use of passive and active techniques have been put to use in a wide variety of buildings.

Of primary significance is the large amount of information conveyed to the reader on the many ways natural energy resources can be used to produce comfortable living and working environments under climatic conditions that would normally require the expenditure of large amounts of nonrenewable energy. The author is keenly aware of the necessity of employing all our renewable energy resources to their best advantage. He knows that we must learn to live *with* our environments, and *Sun/Earth* will undoubtedly help us do so.

John I. Yellott, Professional Engineer
Professor Emeritus, College of Architecture
Arizona State University
Tempe, Arizona

Biographical Comment

To find Richard Crowther as the author of a book having this scope and content is no surprise to those of us who are familiar with his multifaceted career. He has concern for effective innovation, for assisting us to an improved state of living, and for stimulating us to think about the future—all of this is characteristic of him and his work.

Architecture and planning, to him, is an inventive science, the product of which is a composition of building and site that responds effectively to seasonal climatic conditions year round. Guided by these philosophic concerns, he early called for energy conservation and, in architecture and planning, a response to solar and other direct natural energies. He has prophetically led us through social and technological change to holistic creative design.

In the early thirties and forties, he was a pioneer in the development of contemporary architecture. This pioneering quality still guides his works and service, which encompass not only architecture but also interior design, research, and land and topographic planning for commercial, institutional, and residential purposes.

The effectiveness of this designer has been extended through consulting appointments by federal, state, and city agencies, as well as private organizations. His professional colleagues have sought his advice and assistance in programs promulgated by the American Institute of Architects (AIA) at both local and national levels. He served as a member of the Task Force for *Energy: AIA Energy Notebook* (Washington, D.C.: American Institute of Architects, 1975–1978) of the AIA Research Corporation.

photo by Malcolm Wells

He has been invited by environmental, political, business, and social organizations to give lectures, and these have stimulated others to awareness and action for a better environment and for more efficient use of resources. He has served as lecturer at the Universities of Colorado, Wisconsin, Kentucky, Missouri, Michigan, North Carolina, and Nebraska, as well as at the Smithsonian Institute, Midwest Research Institute, and at ISES (International Solar Energy Society), AIA, and other conferences. His contributions to various publications have reached those outside his professional circles; among these have been articles in *Progressive Architecture, The National Geographic, Journal of the American Society of Heating and Ventilating Engineers*, and numerous other domestic and foreign publications.

The theories espoused herein have been converted, over many years, into real aids to improved living through his service to clients and through projects that he has inaugurated. Currently more than fifty residential or commercial projects applying these principles are at the planning, construction, or completed stages. The latest example is a new Crowther research facility that employs more than twenty passive subsystems designed to function in a holistic relationship. All solar commercial and residential projects use optimized energy conservation, "passive" energy systems, and "active" solar-energy-collection systems.

This volume is his latest contribution to the increased understanding and more effective use of natural energies.

DeVon M. Carlson FAIA
Dean Emeritus
School of Architecture
University of Colorado
Boulder, Colorado

Introduction

The original *Sun/Earth* text and graphics remain timely. They espoused the anti-inflationary use of sun, earth, and water energies in homes and buildings. This updated and enlarged edition further explores the concepts of passive solar and natural-energy heating, cooling, and ventilating. In the interval of time between the original edition and now, numerous practical and technical books have been written about passive solar heating and passive cooling systems. They are good companions to *Sun/Earth*.

The original *Sun/Earth* laid the groundwork for holistic (whole) design in architecture. The challenge remains. A holistic energy design chart for architecture in the new edition addresses the realistic complexities of use and economics. All concepts and construction values must pass through the sieve of energy economics. The site, architecture, interior, and occupants are forms of energy. The design objective should be to appropriately join human metabolism and behavior with natural energy use. We need to call upon scientific research to fill in the gaps of knowledge about energy flows and transformations. With this and a holistic viewpoint on energy and design, we can free ourselves from the exorbitant waste of our energy-intensive society and minimize the use of nonrenewable energies.

A holistic design approach can open up employment in design and construction, rigorously reduce inflation, create a more healthful and vitalizing environment, deploy capital more effectively, increase savings in residential and commercial architecture and construction, and increase cash flow (by reducing money spent on utilities). Holistic design would also create a cohesive urban texture.

An opening statement in the preface to the first edition now becomes a conclusion:

> The forces of nature are regenerative, deriving their impetus from the sun and universe. To ensure man's survival, the fragmented, isolated systems and constructions of man need to acquire the elegant holistic interdependence of nature's resources and systems. Nature remains as the most sophisticated designer of the cosmic and world environment. Man's systems need to be made fully regenerative in concert with natural forces and nature's bounteous provisions.

1 Energies

Technological man has enjoyed the luxury of abundant energy and fuel supplies for many years. Now these reserves, which are finite or limited in quantity, have diminished to a point at which primary reliance upon them is no longer justified. Millions of years were necessary for their formation, and because they are being used at increasing rates, their quantity is rapidly decreasing. Most of this consumption has occurred during the two centuries following the Industrial Revolution. The fuel that is now consumed in one day, whether oil, natural gas, or coal, depletes the earth's present finite reserves. The technology does not exist to extract completely the energy potential from the fossil fuels used; their incomplete combustion is, in fact, a major contributor to air pollution. Developing countries are steadily increasing their demand for fossil fuels. These demands, added to present rates of consumption, will contribute to severe energy shortages in the future.

Man has traditionally attempted to dominate the environment instead of harmonizing with it. He has used his intelligence to try to subdue nature and resist the natural forces of the universe. By playing the role of a superior being, he is destroying the earth as he extends his influence and accelerates his technology. Since the Stone Age, man has modified his physical surroundings for protection and comfort and sometimes merely as a demonstration of his control

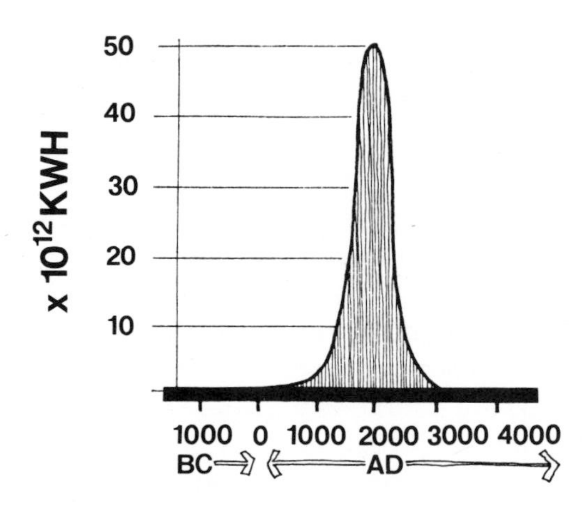

fossil fuel usage

over nature. As he developed the use of tools, he decreased the proportion of direct physical energy needed to complete a task. Early man sought to amplify his own energy by means of simple machines such as levers, pulleys, and various forms of inclined planes. Fuel (wood) combustion, meanwhile, was used only to produce heat. As his development continued, man harnessed the power of animals, slaves, and the more complex machines they drove, minimizing his own energy input. This "biological" power source was ultimately superseded by the combustion of fuels to perform mechanical work. Our contemporary use of fossil fuels now includes a third service, the creation of synthetic materials. Among competing applications, these fuels should be allocated to products of long-term durability, to the production of medicines, and to other uses of immediate importance that cannot be met in other ways.

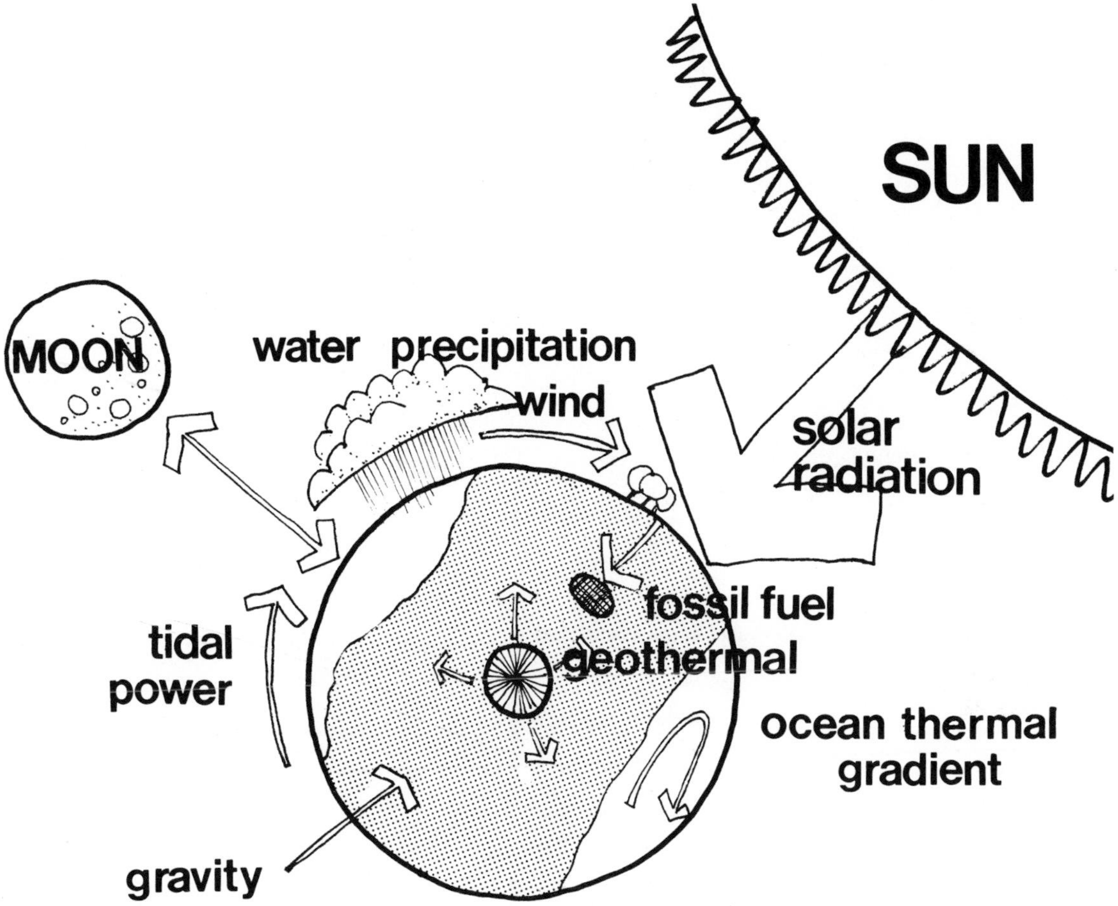

Contributing to the present "energy crisis" is society's dependence on combustion heat derived from fossil fuels instead of the almost limitless energy available from the sun. Man must realize that he can benefit directly and indirectly from the natural energy that is available to him.

Most of the energy on the earth comes or has come indirectly from the sun. The direct solar energy that strikes this planet (*in*cident *sol*ar radi*ation* has evolved into the term *insolation*) provides and sustains the biosphere in which we live. The radiation from the sun heats our atmosphere and causes air movement or wind. It also heats the oceans and water reservoirs of the earth, and in combination with wind creates weather and sustains the continued water cycle of the planet. The sun provides the energy for the metabolic activities of plants. Energy is stored by them in the form of carbohydrates, wood, or as fossil fuel. Fossil fuels are created with energy stored by the forces of heat and pressure acting upon decaying plant material.

The amount of energy consumed by technological man is phenomenally large in comparison to that used by primitive man, who consumed only the energy he received from his food. His heat was a product of his own metabolic activity. As man advanced he began to cook food and warm his dwelling with the heat from a fire. Agricultural man initiated the use of tools for farming and the use of animals for labor. These were the first industrial uses of energy. Industrial and technological man increased the use of energy in all categories. This is especially true in the area of transportation, as a result of man's dependence on the automobile.

Many inputs and outputs from man's biosphere affect the amount of energy he uses. He has direct uses for such things as heating, cooking, lighting, and transportation and indirect uses such as the manufacture of goods, use of resources, production of food, and handling of wastes.

Inputs to the biosphere are converted by many processes to outputs. The outputs from the consumption of fuel and energy are solid wastes (particulates and pollution in the air), carbon dioxide (a by-product of combustion), and waste heat (heat produced by combustion exhausted to the environment). Oxygen that is used for respiration is converted to carbon dioxide. Plants take in carbon dioxide and water and produce oxygen during photosynthesis, thus maintaining the oxygen content in the atmosphere. The input of food and water results in an output of solid wastes, human wastes, and heat wastes. While at rest, the body of an average adult human male gives off a minimum of 400 Btu per hour

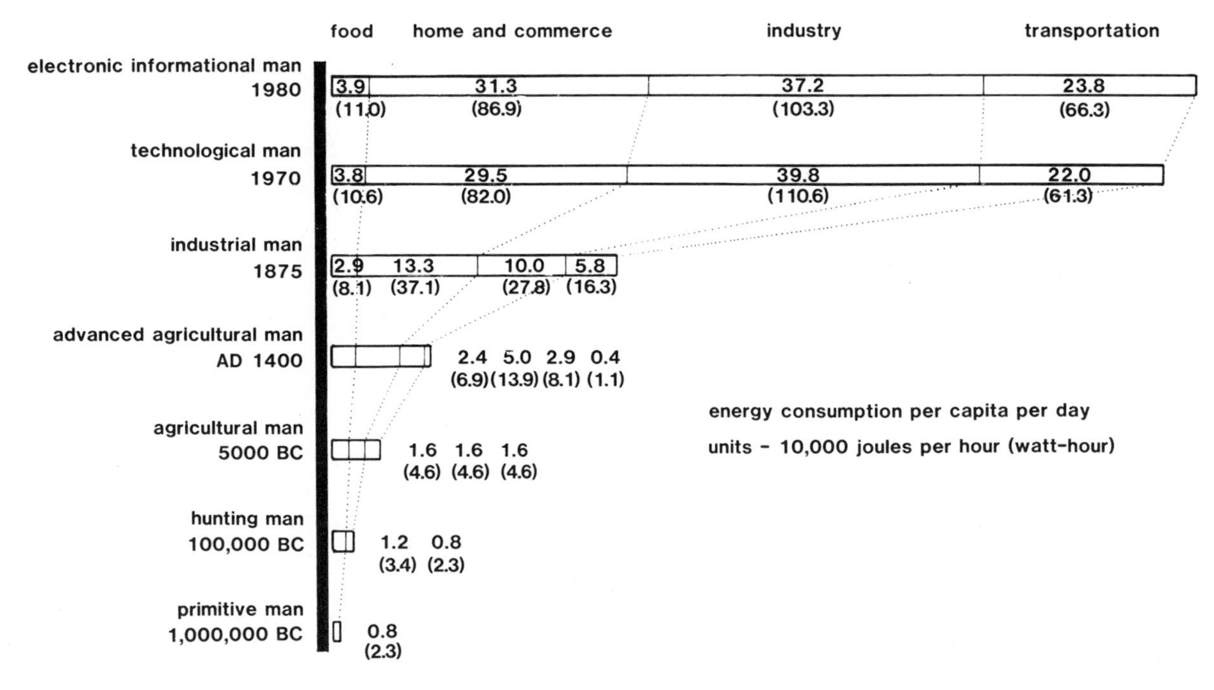

From "The Flow of Energy in an Industrial Society," by Earl Cook.

(Btu/hr), or 117 watts, by convection, radiation, and evaporation.

The processes involved in the production and distribution of consumer goods require large amounts of energy. This energy is consumed during the extraction and transportation of resources and during the manufacturing processes. It is needed for packaging the final products and transporting them to their destinations. Additional energy is needed to purchase any product and transport it to its place of use. The numerous and complicated processes along the way produce waste heat, solid wastes (disposing of these is a major problem in many large cities), air pollution, and carbon dioxide.

The United States, with only 6 percent of the world's population, uses 33 percent of the world's energy and produces more than 50 percent of its solid wastes.

As leaders in world technology and in per capita general resource consumption, we in the United States should be developing systems that use the sun and other natural energies to a much greater extent. We should examine all of our individual energy-use patterns to attempt an immediate reduction in the massive energy waste of our society.

When we consume fossil fuels to produce energy, we must recognize the source of the energy and the by-products of its use.

Plants store the sun's energy as they grow. Certain plants are used as a source of energy by releasing sun-stored energy in a useful form. Burning is a means of releasing the energy of wood, dried grass, or any other combustible material. As a log is burned, the energy that a tree received from the sun and stored while growing is released. The burning process does not convert all the stored energy to usable heat. Some of the wood is not completely burned, and it escapes as soot and smoke. When wood is completely burned, it releases only carbon dioxide and water vapor (wood is a carbon compound).

Some plant materials and other organic matter are converted to fossil fuels. This process takes millions of years and requires proper conditions of heat and pressure.

Deposits of the fossil fuels (oil, natural gas, and coal) are formed only by unusual geological circumstances. In general, the fossil fuels are derived from the remains of land-dwelling plants and other organic matter. Younger sedimentary strata cover the plant remains, and

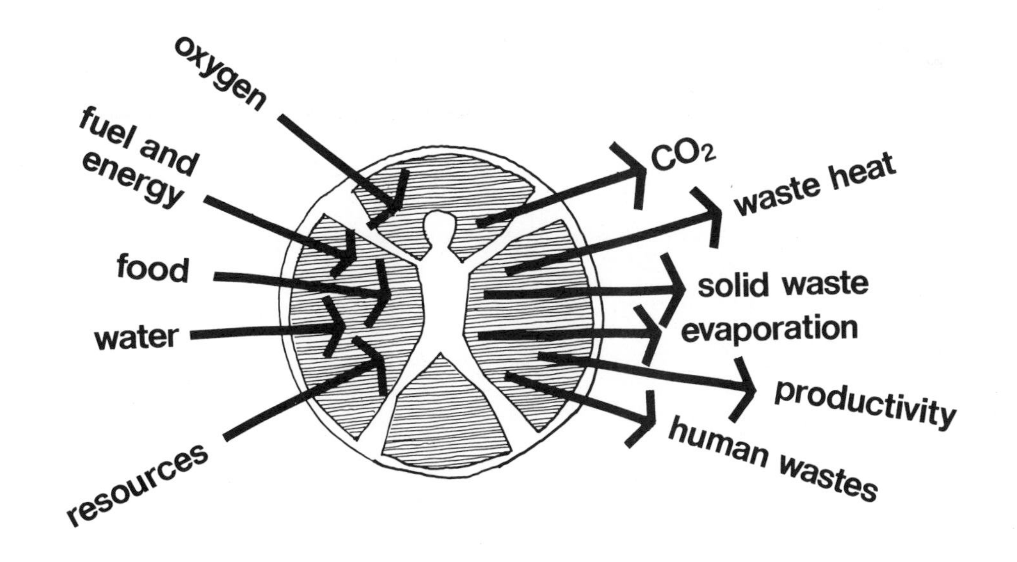

man's inputs and outputs

the extreme pressure and temperature produced by succeeding layers are responsible for the transition from plant compounds to the simpler higher-energy-level hydrocarbons.

Fossil fuels are rarely found in their purest form. Minerals such as sulfur are considered impurities in fossil fuels and are released as oxides during combustion. The introduction of clay and sand, which can occur during coal formation, will remain as ash after the coal is burned, incurring a substantial cost for its removal.

The combustion of fossil fuels adds the following to the atmosphere:

- carbon dioxide (natural product of combustion)
- carbon monoxide (product of incomplete combustion)
- water vapor (natural product of combustion)
- solid particles (noncombusted matter carried as smoke and soot)
- hydrocarbons (products of incomplete combustion)
- sulfur, nitrogen, and mineral oxides (resulting from the combustion process)
- waste heat (from burning and mechanical sources)

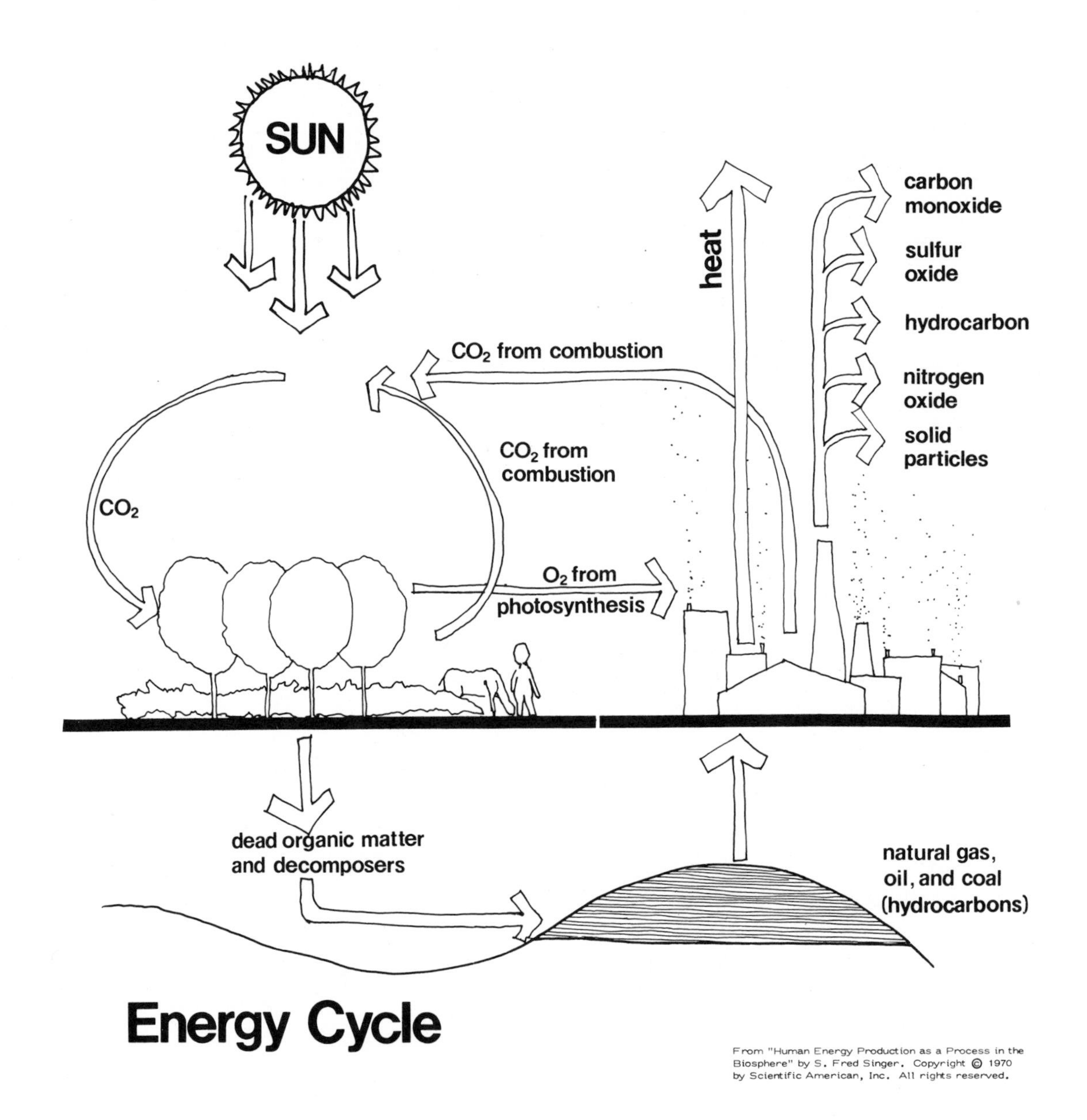

Energy Cycle

Vehicular and industrial emissions of sulfur dioxide and nitrogen oxides combine with airborne moisture to produce acid rain, which is seriously damaging forest and lake ecosystems as well as buildings and other structures. Although the most serious concentrations are in the Northeast and Northwest sections of the United States, the fragile ecosystems at high altitudes are particularly threatened. Only decreased reliance on fossil fuels and improved pollution controls can reduce the risk of acid rain.

Technological man has ignored the sun as a major energy source. Instead he has built an energy-intensive fossil-fuel technology that is difficult to modify. (An oil company in its TV ad states: "A country that runs on oil cannot afford to run short.") As we begin to witness the dwindling supply of fossil fuel on our earth and realize the tremendous time it took for that fuel to form, we see that the sun and nature can be providers of our energy.

The sun provides energy to the earth for the maintenance of the planet and the support of the life forms that inhabit it.

As we begin to move away from the fossil-fuel age, we must analyze the forms of energy that will take its place and weigh the advantages and disadvantages of each. A determination must be made about how to best utilize the remaining supplies of fossil fuels. We have to analyze the motivation behind our worldwide energy consumption.

All life processes depend on the ability of plants to capture the sun's energy. Green plants use sunlight to build sugar and other compounds from carbon dioxide and water. All other forms of life use this primary energy source in some form.

Animals and man derive all or part of their energy from plants by breaking down and restructuring plant compounds into forms useful to their own metabolism for the extraction of energy. This energy is then converted into biomechanical energy and heat as man and animals go about their daily activities.

Some animals, including man, also utilize the energy locked into animal compounds. Again the process of restructuring takes place with reduced efficiency, such as when vegetable proteins are subsumed into meat.

It becomes apparent when viewed in this cyclic pattern that the most efficient energy use is the one most directly linked to the source of energy—the sun—and that as we proceed through the secondary and tertiary recycling of this energy base, the efficiency is diminished accordingly.

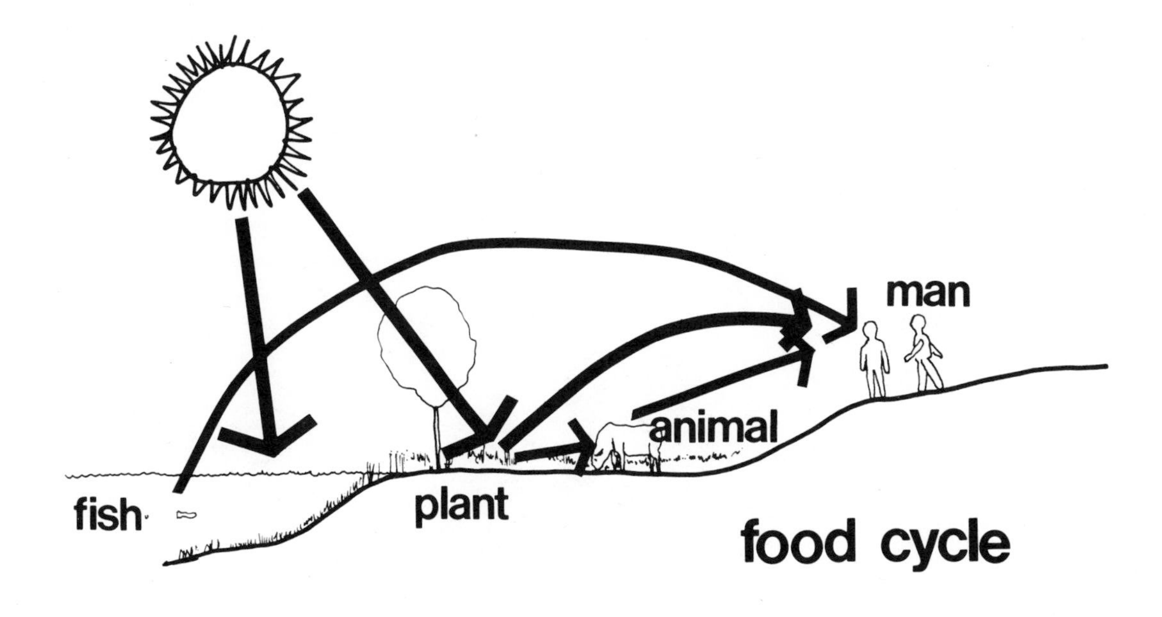

Fossil Fuels

The three most universally used fossil fuels are hydrocarbons: coal, oil, and natural gas. Each contains mainly two elements—hydrogen and carbon. Statistics about world reserves and consumption of these fuels must be in terms of their conversion to other forms, such as mechanical power, electricity, and synthetic materials, in addition to their familiar use in direct combustion for heat.

Because public utility companies are appropriating these three fossil fuels to generate electricity, they are becoming less available for direct use in buildings. Unfortunately, thermal energy is "lost" in the form of waste heat at each stage of the conversion from one form of energy to another. Natural gas burned in a furnace can deliver less up to 96 percent of its stored energy as heat to a home. If the same natural gas is first converted to electricity at a power generation station, and then the electricity is converted to heat in the home, only 30 percent of the energy stored in the gas is delivered as heat. During this process, up to 70 percent of the stored energy is lost as waste heat. Other fossil fuels have even lower conversion efficiency.

The above facts indicate why solar energy, wind energy, and other natural-energy forms are being more seriously considered as power sources. Most of these direct sources involve very short, efficient chains of conversion from initial form to final use.

More important, the fossil fuels extracted from the earth are not renewable and are limited in supply, whereas the supply of energy from the sun is practically unlimited.

Coal

Coal is used to supply the United States with approximately one-fifth of its energy needs. However, the reserves of coal in terms of energy potential are actually far more abundant in the United States than the reserves of oil or natural gas. Coal is an advanced-stage hydrocarbon and is the highest-density naturally occurring solid fuel. This indicates a lengthy formation period, probably hundreds of millions of years. The duration of this process clearly shows why "mines bear no second crop." Although coal is not renewable, it has become prominent as a fuel because:

- It burns at a higher temperature and is more compact than wood.
- Its initial abundance has, so far, outweighed its nonrenewability. Ninety percent of all proven U.S. fuel reserves are coal.
- It can be recovered using several different methods and can be either burned directly or converted to other forms prior to combustion.

The disadvantages inherent in the use of coal have recently become more evident, partly because of increased concern about coal miners' health and safety and about the effects of air pollution. Problems facing us in the continued use of coal include:

- **Hazards to mining personnel**
  Technical difficulties in deep-mining coal and accumulations of explosive gases ensure that the possibility of mine cave-ins will always be a certain risk. Mining of coal can cause several serious respiratory ailments for miners exposed to coal dust.

- **Negative environmental impact**
  The alternative method of extracting coal, strip mining, is beset with public disapproval for environmental reasons.

- **Coal substituting for oil**
  Coal is being used more frequently for the production of synthetic materials as well as for combustible fuel. This is partly to compensate for oil shortages in the same areas of use. Such trends are expected to accelerate.

- **The demand on other resources**
  Elaborate processing of coal involves vast quantities of water, which may be needed elsewhere.

- **Disruption of arable land**
  Much of the coal reserves are located under arable land that must either be sacrificed or restored at great cost after coal extraction.

- **Unrecoverable reserves**
  The above factors may render large percentages of U.S. coal reserves unrecoverable.

- **Air pollution**
  Coal varies widely in heating value per unit weight and in ash and sulfur content. The products of coal combustion require intensive capital investment to reduce atmospheric pollution to acceptable levels.

- **Low grade coal is less efficient**
  Coal that has a high level of ash and moisture content offers very low heating value per unit weight.

- **High transportation costs**
  Coal of low heating value cannot be shipped over a wide radius without transportation cost exceeding the worth of the fuel itself. Certain public-utility companies now pay shipping costs of five times their coal material costs.

Despite the negative and unhealthful aspects of burning coal to produce power, it remains the most immediate, logical, abundant source for electrical

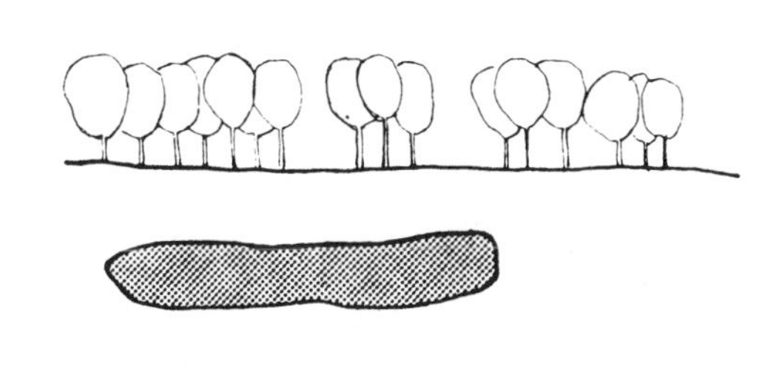

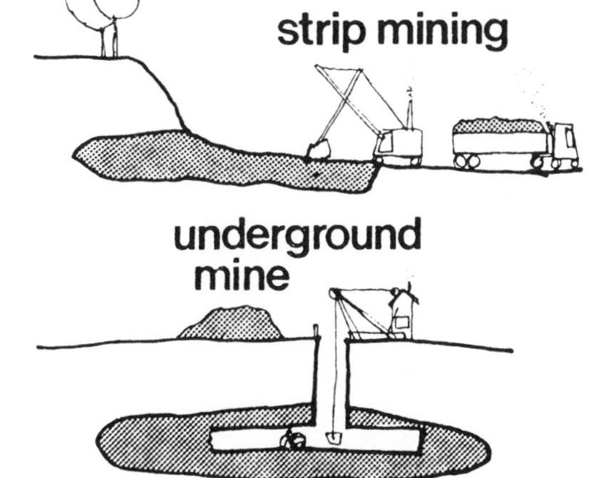

generation. Every effort is needed to decrease our escalation of national energy demand and to supplant coal and other environmentally degrading fuels with clean, natural-energy alternatives.

Oil

Oil is a highly versatile fuel, supplying more than 36 percent of the total energy requirements of the United States. Because of the extremely slow creation cycle for this hydrocarbon, each year's worldwide oil use represents millions of years in original "production" time. This indicates a very rapid depletion of reserves. Increasing problems in its recovery, refining, political influence, and distribution include:

- Sudden oil "blowouts," originally termed "gushers," are now controllable during land-based drilling but are more dangerous in offshore operations.

- Routine refinery operations have produced a greater number of river and ocean "spills" each year as a natural consequence of increased volume and cumbersome methods of transport. An average of four to six million metric tons per year are spilled into the world's oceans.

- Although much of the United States land area is leased for oil and gas exploration, yields from new discoveries began decreasing in 1970, and the United States' share of known world petroleum reserves had already dropped from 40 percent in 1937 to 5 percent in 1979. As little as twenty years of recoverable domestic reserves now remain.

- As the western world's dependence on offshore oil begins to overtake land-drilled oil use, the net energy value obtained will decrease drastically. The amount of energy required to extract oil will inevitably approach the level of the energy extracted.

The use of oil obtained from unconventional sources such as tar sands, oil shale, and coal liquefaction appears to involve net energy disadvantages similar to those noted above.

Despite the negative and polluting effects of oil as a fuel, it is the only present abundant energy source that can adequately serve transportation needs. Its use for heating should be curtailed, and its reserves should be extended, possibly by adding ethyl alcohol to gasoline.

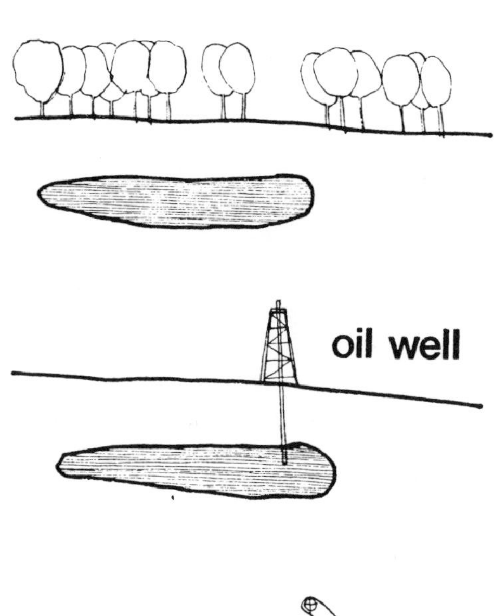

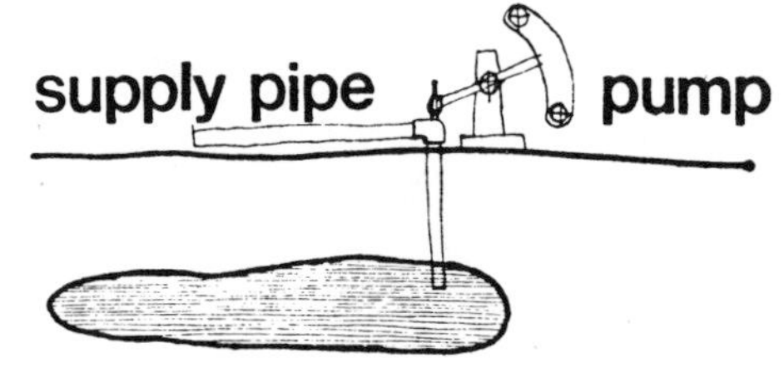

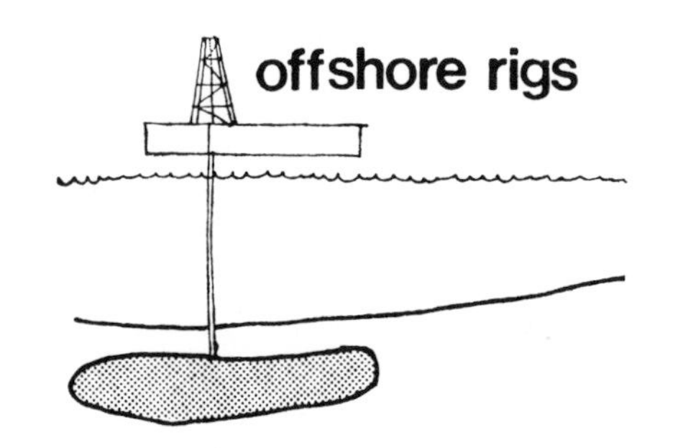

Natural Gas

Natural gas now fulfills approximately one-third of U.S. energy demand. Considering the economics of pipeline distribution, gas is the most efficient direct-combustion hydrocarbon fuel now in general use. It is also the cleanest burning and the least polluting of the fossil fuels.

The fundamental drawback to the use of this excellent fuel is its limited quantity. This situation and artificial price controls have led to various problems that are becoming more critical.

- Natural-gas prices in interstate commerce have been held low by federal controls. The deregulation process will continue to escalate natural-gas cost to the consumer.

- After intensive exploration and drilling are resumed, the United States will be facing total depletion of natural gas in twenty to fifty years. Imported and synthetic gas will then be the only fuel available in this clean form. In the meantime, prices will rise rapidly to levels approaching those of electricity.

- Drilling beyond 4,572 meters (15,000 feet) is not economical in most cases. Most authorities agree that the cost of drilling doubles for each 3,600 feet of depth.

- As a consequence of the above trends, some gas is now being transported from other continents, primarily Africa. The gas must be in a compact, liquid form during transportation, a task accomplished by refrigeration. Liquid natural gas (LNG) ocean tankers carry more potentially destructive power aboard than any other energy "container" except nuclear reactors.

The probability of tanker casualties increases closer to land. And as tankers increase in size and number, the magnitude of danger grows.

The use of natural gas as a combustible fuel, or in any other form, is slowly declining. In many locations no new gas connections are available, and industrial and institutional users face the probability of reductions or termination of supplies. Natural gas is an exceptionally valuable raw material from which a host of products is made, ranging from refrigerants to fabrics. These factors should accelerate a shift in emphasis toward various forms of solar and natural energy. Methane produced from biomass and hydrogen produced by direct solar energy are two renewable gases that could replace natural gas in the future. It should be noted that hydrogen could be difficult to transport in existing utility pipelines.

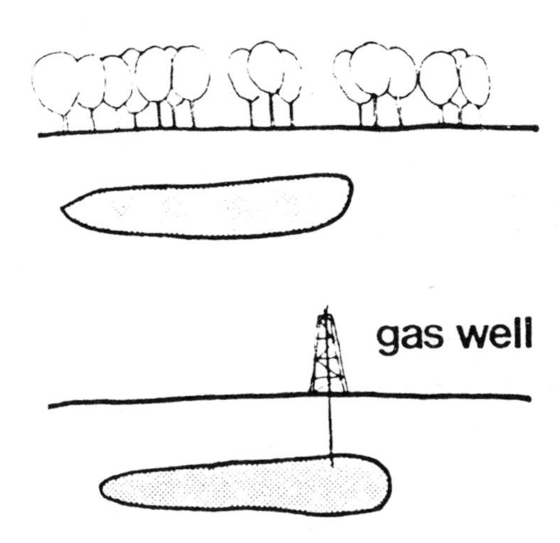

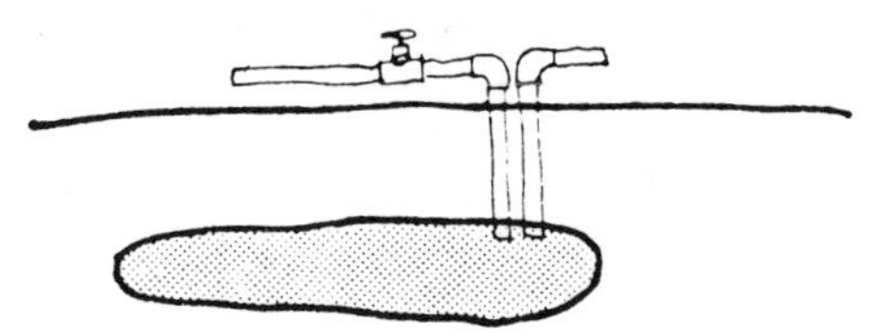

Nuclear Fission

All nuclear-reactor power stations currently in operation use fission processes in which heavy uranium (235) atoms are split to produce heat for steam-generated electricity. In 1980, it represented less than 12 percent of the nation's electrical generation and 3.8 percent of its overall energy supply.

Despite heavy government funding of research and the fact that one "fission event" (with a uranium-235 atom) releases 50 million times as much energy as the burning of a hydrocarbon (such as coal) atom, the development of nuclear energy has not yet approached its projected potential.

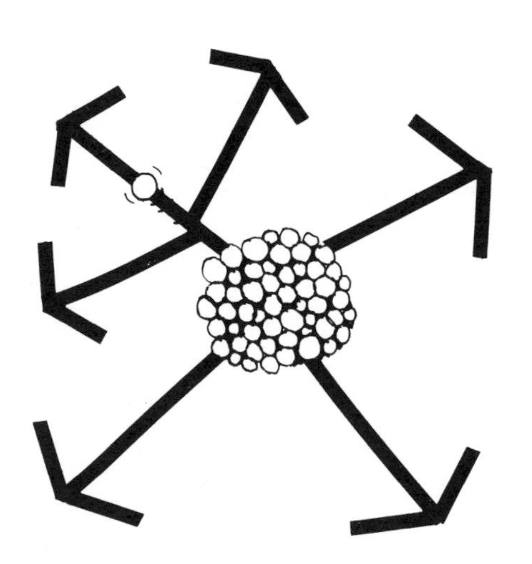

Drawbacks to the expansion of the nuclear industry are numerous, major, and often cited but little understood. The most publicly debated aspect of such energy is its safety. Indeed, most of the reactor stations operative in the United States have at one time or another been shut down for a range of reasons from physical failures to design flaws. The following are factors contributing to the drawbacks of nuclear energy.

- Reactors (except for high-temperature, gas-cooled reactors) must be kept cooler than fossil-fuel-powered plants in order to protect uranium fuel rods. Consequently, their waste heat has been as much as 70 percent of all the energy released, and thermal pollution is far greater than for any fossil-fuel plant.

- Radioactive waste products, especially plutonium, are potentially lethal for thousands of years. The complexity and cost of a waste-disposal system that ensures the safety of the public and the environment are major factors that the industry may not be able to overcome.

- Uranium fuel comprises less than two parts per million of the earth's crust. Unless fission technology shifts to fast-breeder reactors from conventional uranium reactors, uranium supplies may be exhausted within the century. The technology of liquid-metal fast-breeder reactors is still more complex and creates still more dangerous waste products than do the more conventional fission processes.

- Conventional industrial "economies of scale" have so far worked in reverse with nuclear energy. Maintenance problems are compounded by safety precautions with every increase in size. Thus, while the fission reactors powering submarines and aircraft carriers operate faithfully and with impressive efficiency, central utility plants for urban electricity are plagued by repairs taking up to one year apiece. These plants must be completely inoperative for any major maintenance or repair; most operate on line only 60 percent of the time and then function at only 50 percent to 60 percent of their rated capacity.

- The initial capital cost for a nuclear/electric power station is considerably higher than for any other equivalent-scale fossil-fuel facility.

- The time required for licensing, planning, and construction of nuclear plants now in service has ranged from seven to seventeen years, averaging ten years. This greatly exceeds construction time for all other existing energy-production plants, except perhaps hydroelectric dams.

- The quantities of fossil fuels expended in the mining of uranium, its processing, and its ultimate disposal in other forms of by-products even more lethal are impressive. It could lead to the irony that the amount of hydrocarbon fuels consumed to support all nuclear activities would be greater than the electrical power produced by nuclear facilities.

- There is more energy from the sun striking the land area required for a nuclear power plant and its surrounding area than is produced by the plant itself.

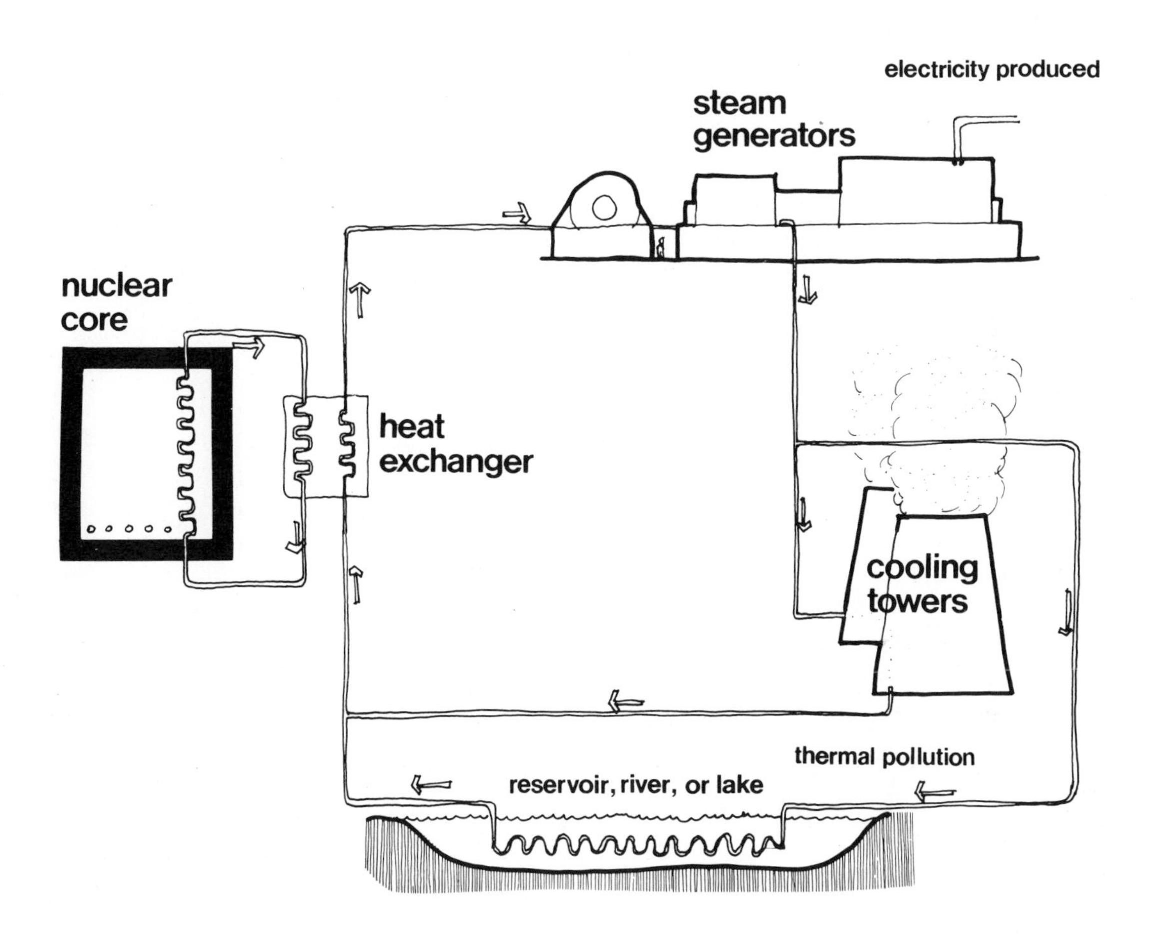

Nuclear Fusion

Thermonuclear fusion as potential energy is theoretically far more abundant than any form of fission and is possibly less dangerous. It has, however, shown itself even more difficult to master than fission and may induce radioactivity in the construction materials of the generating plant.

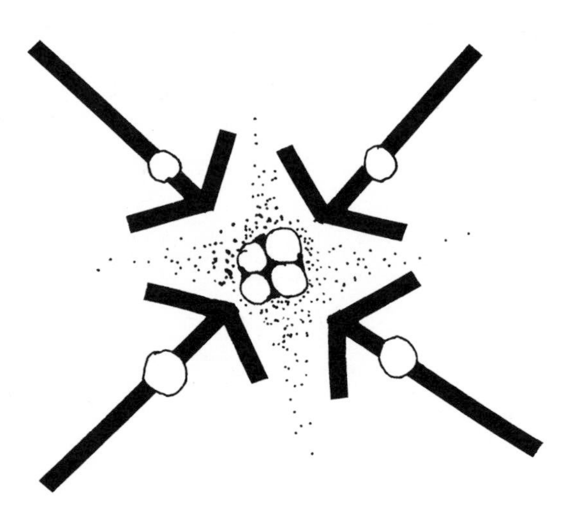

In principle, fusion proceeds by the forceful combining or fusing of very light (hydrogen) atoms. Since hydrogen is the most abundant element in the universe and the sun continues successful operation as a fusion reactor, such application on earth has become most attractive to scientists. The sun, however, relies on intense gravitational forces as well as high temperatures for hydrogen fusion to take place deep in its core. With such gravity or its simulation unattainable on earth, temperature for fusion must compensate by being still higher than at the sun's core. Magnetic-confinement fusion reactors and, more recently, laser-fusion reactors have each approached for an instant the required temperature range between 60 to 100 million degrees Celsius (108 million to 1.8 billion degrees Fahrenheit); but after thirty years of research in several countries (at least on magnetic-confinement fusion) such reactions remain highly undependable. Potential radiation hazards as a result of using heavy hydrogen (deuterium and tritium) are much smaller than those that accompany the use of nuclear fuels; nevertheless they must be dealt with.

One little-known problem presented by fusion technology is its present requirement for exotic metals, such as vanadium, niobium, and lithium, in processing and for heat resistance.

Generally, both fission and fusion nuclear energy have met deepening doubt in the public mind, if not in the minds of the scientific and political communities. Fission is looked upon with due apprehension for its dangers, which have overshadowed an equally severe cost-efficiency disadvantage and a limited reserve of fissionable materials.

Breeder reactors raise the problem of the dangers inherent in their principal product, plutonium. Fusion has yet to be accomplished, and there is no certainty that it will ever be practically achieved on earth. There is a certain futility in attempts to duplicate the fusion process of the sun, which already delivers its products to earth in a worthwhile form.

Energy Forms

There are two basic states of energy—potential and kinetic. Potential energy is stored energy that is available for release. A rock at an elevated point has stored energy. As it is released from that point, its potential energy is liberated. Kinetic energy is moving or dynamic energy. As the rock falls, it converts its potential energy into kinetic energy.

The forms that energy takes can be divided into five categories: radiant, electrical, chemical, mechanical, and heat. Energy cannot be created or destroyed, but one form of energy can be transformed or converted into another energy form. Heat is both the first and the final form of the energy used in many processes. The conversion from one form to another influences the cost of energy.

The more difficult it is to convert energy to useful work or heat, the more expensive and less practical it becomes. An example of this is the automobile. In its internal-combustion engine, the potential chemical energy stored in gasoline is converted to mechanical energy (kinetic energy) and heat. The mechanical energy is also converted into heat by the friction of air, tires, and bearings. There are many energy losses before a fraction of the potential energy stored in the gasoline is converted to the kinetic energy (velocity) of the automobile.

Below are some examples of each energy form.

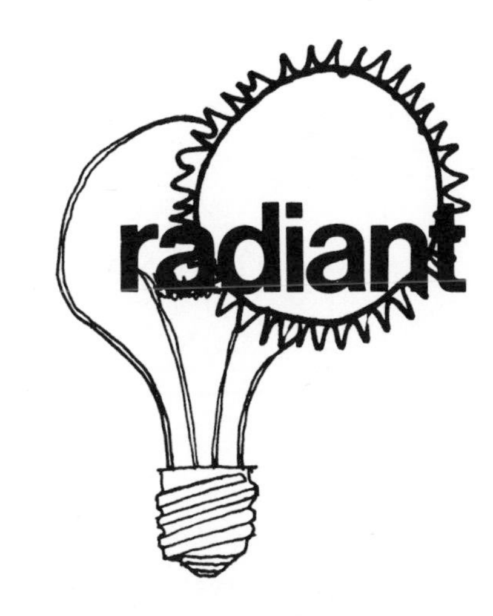

Radiant energy
- sunlight
- fire (conversion from chemical energy)
- radiant heaters
- light bulbs (conversion from electrical energy)
- fireflies (conversion from chemical energy)

Chemical energy
- plants, food, wood (conversion from solar energy by photosynthesis)
- man's energy (conversion from chemical energy)
- fossil fuels (long-term conversion and storage of radiant energy)
- battery (conversion to and from electrical energy)

Electrical energy
- electricity (can be readily converted to any of the other forms)
- static electricity (conversion from mechanical energy)

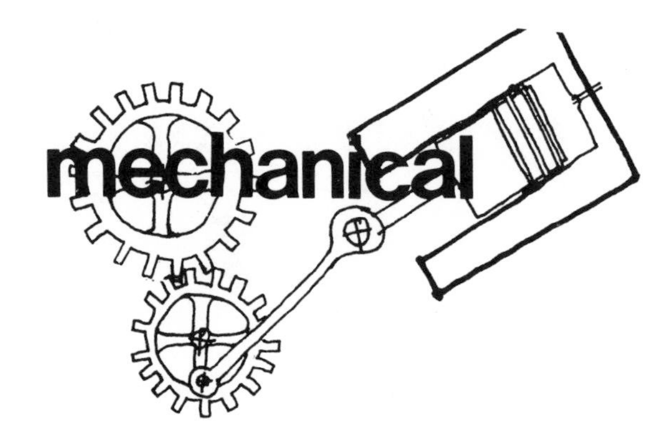

Mechanical energy
internal- and external-combustion engine (conversion from chemical energy)

Heat energy
fire (conversion from chemical energy)
heating coils (conversion from electricity or geothermal energy)
waste heat (by-product of conversion of all forms of energy)
solar heat (conversion from light when light strikes an absorbing surface)

Energy Transmission

There are three ways of transmitting heat energy: conduction, radiation, and convection.

Conduction is the transmission of energy between two bodies that are in direct contact. As a common example, a teapot when placed on an electric stove receives energy by conduction.

Radiation is the transmission of heat by electromagnetic rays. Only these rays can travel through a vacuum such as outer space and heat the object that intercepts them. The sun warms the earth by radiation that travels in this manner.

Convection is the transmission of heat through a fluid. An object heats the air (or liquid) in contact with it, and the warmed air or liquid travels to nearby objects and warms them. The use of convection for heating is exemplified by a home forced-air furnace. The furnace warms the air, and then the hot air moves to heat occupants and the house.

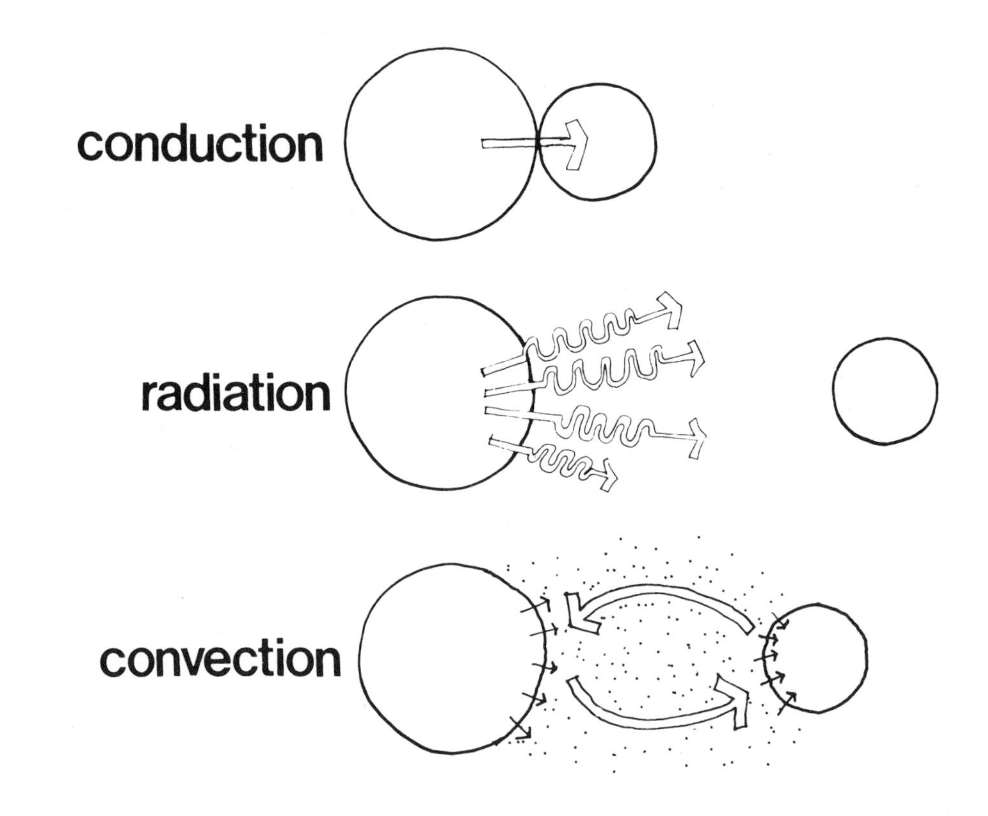

A change in temperature of a body indicates a change in the energy content of that body. This energy is usually in the form of heat or light, but it can also be in a chemical form. An increase in temperature signals an increase in energy content. A decrease in temperature indicates a decrease in energy content. Some of the more important units for measuring energy are presented below.

energy content

hot

cold

The three units that are used for measurement are the calorie, the Btu (British thermal unit),* and the joule. A calorie is the amount of energy required to raise one gram of water one degree Celsius. A Btu is the amount of energy required to raise one pound of water one degree Fahrenheit. The joule is the unit of energy, hence heat or work, in the système international d'unités (metric), also known as the S.I.

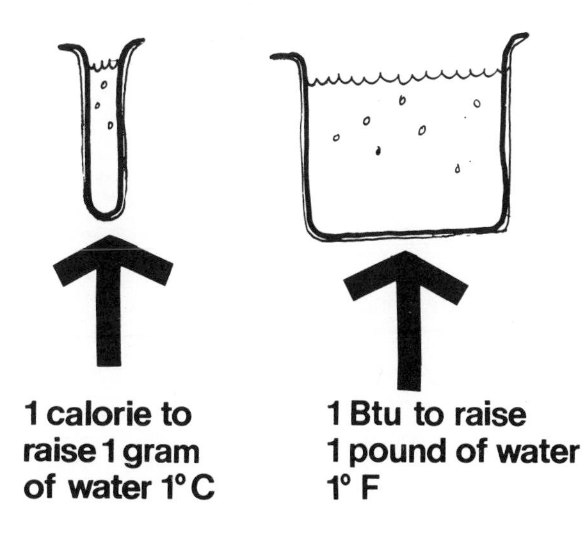

1 kilowatt-hr. = 3414 Btu

1 calorie = .003968 Btu

1 Btu = 1055.87 joules

The units of temperature are either degrees Celsius (° C) or degrees Fahrenheit (F). The Celsius scale will be implemented in the United States as part of a phased program to adopt the S.I. (metric) system of measurements. S.I. units are used in this book, with the more familiar units in parentheses. Also, throughout the book the degree symbol (°) is used with Celsius temperatures, but not with Fahrenheit temperatures, as is consistent with accepted engineering practice.

*The British no longer use this unit, but it is still used in the United States.

Energy Phase Change

Heat can be thought of as being sensible heat or latent heat. Sensible heat is the heat a body gives off without changing its state, or phase, and is measured as a change in temperature. Latent heat is the heat required by a body to change its state, or phase, without changing its temperature.

A change of state, or phase, of a substance is a change in its physical condition. A substance may exist in any one of three states:

- Gas (water vapor, steam, air)
- Liquid (water, fluid, mercury)
- Solid (ice, steel, wood)

Many substances occur naturally in only one state because of the range of temperatures and pressures that exist on the earth. One example of this is hydrogen, which normally occurs in the gaseous state. If the temperature is lowered to −157° C (−250 F), hydrogen changes from a gas to a liquid.

Latent heat is required to change a substance from a solid to a liquid or from a liquid to a gas. It is given off when the processes are reversed. Under standard atmospheric conditions, water will increase in temperature as heat is added to it. Once a temperature of 100° C (212 F) is reached, additional heat does not raise the temperature but provides the impetus for a change of phase. Steam may exist at the same temperature of 100° C (212 F) at sea level. Per unit weight it contains a great deal more energy than water at this temperature.

Water gives off only sensible heat when the temperature range is between boiling and freezing. So if water at one atmosphere of pressure is cooled from 88° C to 10° C (190 F to 50 F), the heat removed or given off is sensible heat. If water is cooled from 1.7° C to .6° C (35 F to 31 F), the heat removed is a combination of sensible and latent heat, since water freezes at 0° C (32 F). Thus, much more heat is removed from water in this range of cooling than, for example, −2.8° C to −.6° C (27 F to 31 F).

All substances have the ability to store heat. Some can store more heat per unit weight than others; the term for this capacity is specific heat. Temperature is a measure of the kinetic energy of the molecules of a substance and is an indicator of the amount of heat stored in it.

The land, sea, and air share heat in thermal interaction, and each influences the temperature of the others. The temperature of the earth's surface determines the temperature of the atmosphere above it. The atmosphere in turn helps maintain and equalize the temperature of the earth. A coastal city experiences less extreme seasonal and day/night temperature differences than its inland counterpart. This is due to the relative thermal stability of water compared to air and land. Such stability is a result of the high specific heat of water.

The moon has vast temperature differentials and is an example of a planet without an atmosphere. The bright side of the moon (the side that is in direct solar radiation) heats up to several hundred degrees Celsius; the dark side is hundreds of degrees below zero Celsius.

The atmospheric temperature on the

latent heat

sensible heat

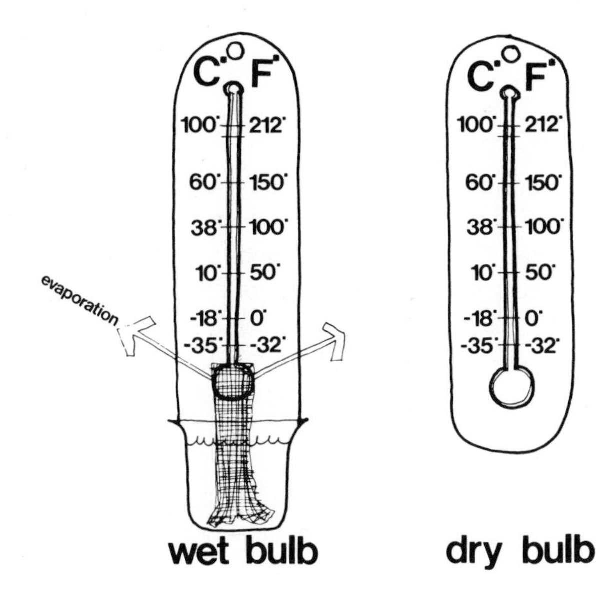

earth varies seasonally and with latitude, but it doesn't reach the extremes experienced by the moon. The warmth of the atmosphere varies in proportion to the amount of direct solar radiation (insolation) reaching the earth's surface. The water content, or humidity, greatly affects the air temperature that is perceived. Heat is required to evaporate water and add humidity to the air.

The influence of humidity on temperature can be quantified by comparing temperatures measured in two different ways. These are called dry-bulb and wet-bulb temperatures. Dry-bulb temperature is a measure that doesn't take into consideration the humidity. This is the method of measuring temperature with which the majority of people are most familiar.

The temperature measured by a wet bulb takes into account the humidity in the air by measuring the rate of cooling through evaporation at the bulb of the thermometer. When the humidity is low, the wet-bulb temperature will differ by many degrees from the dry-bulb temperature. As an example, in Arizona during the summer, the dry-bulb temperature (the temperature that is announced in a weather report) may be 42° C (108 F) while the wet-bulb temperature may be 21° C (70 F). When the humidity is high the wet-bulb temperature will tend to differ only slightly from the dry-bulb temperature. St. Louis on a hot muggy day may have a 21° C (70 F) dry-bulb temperature and a 20° C (68 F) wet-bulb temperature, indicating that the humidity is very high.

The difference between wet-bulb and dry-bulb temperatures will help explain the use of natural cooling and heating systems in different climates.

Human Thermal Response

The energy relationship between the sun and man's biosphere, and other heat sources outside the biosphere, includes many factors that help to maintain the heat balance of the human organism.

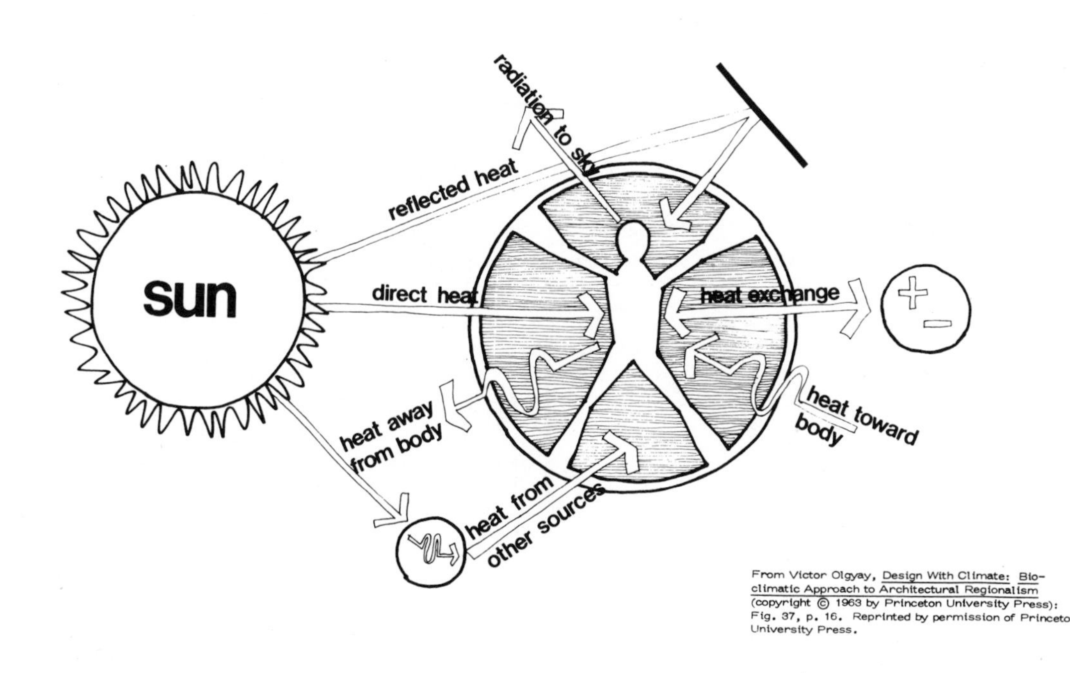

From Victor Olgyay, Design With Climate: Bioclimatic Approach to Architectural Regionalism (copyright © 1963 by Princeton University Press): Fig. 37, p. 16. Reprinted by permission of Princeton University Press.

Man can be a receptor of direct or reflected radiation from the sun. He also generates heat from his body processes. The amount of heat he generates depends on his activity level (whether he is sleeping, sitting, or running), the amount of food energy he takes in, and his metabolic rate.

Heat can be lost by the human body as a result of convective transfers or by radiation to the sky. It can also be lost or gained through conduction by contact with objects that are warmer or colder than he is. Heat can also be gained by convection (from warm air) or from sources that received radiant energy from the sun and then reradiate that energy.

All of these affect the perceived temperature in a space. People in different cultures have different ranges of acceptable temperatures for comfort. Many of these are psychocultural influences of society. In the United States the suggested comfort zone is from 20° to 24° C (68 to 75 F), while in England it lies between 17° and 20° C (63 to 68 F). Adaptation of the human to the environment is well illustrated by the contrast between an Alaskan igloo environment, where indoor temperatures of 5° to 16° C (41 to 61 F) are tolerated, and a tropical habitat, where the comfort zone ranges from temperatures of 22° to 30° C (72 to 86 F).

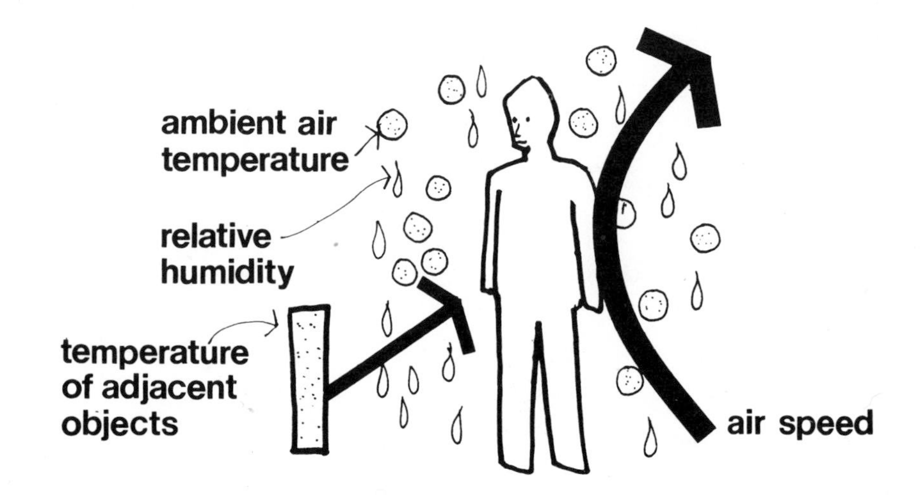

The comfort zone of tolerable temperatures for the average United States citizen has historically narrowed in range as a result of increasing dependence upon mechanical systems for comfort. This reveals a psychophysical conditioning of man in proportion to a technical ability to narrowly control temperatures.

The human comfort zone depends on many factors. A few are listed below.

ambient air temperature: the temperature of the air surrounding the body (the dry-bulb temperature).

relative humidity: the percentage of water vapor in the air in relation to the maximum amount of water vapor it can hold at a given temperature.

air movement or speed: how fast the air is moving adjacent to the body.

temperature of adjacent objects: the air temperature can be 24° C (75 F), but if the wall and floor temperatures are low, the perceived temperature is lower. Sometimes this is called the mean radiant temperature of the environment.

Air Movement

An understanding of the ways in which air moves when subjected to various conditions is necessary for designing efficient energy-conserving buildings and systems. The following are simplified air-movement principles.

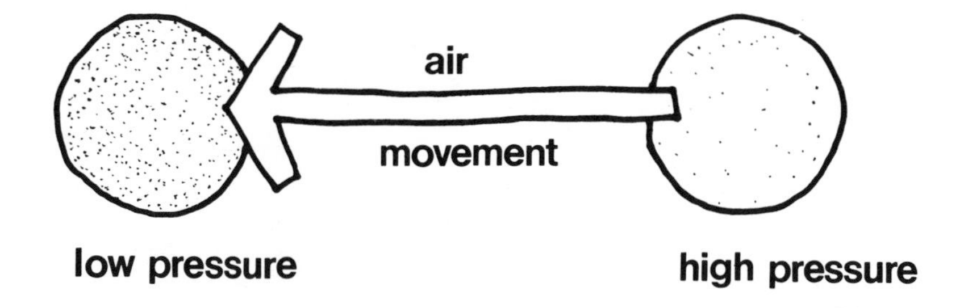

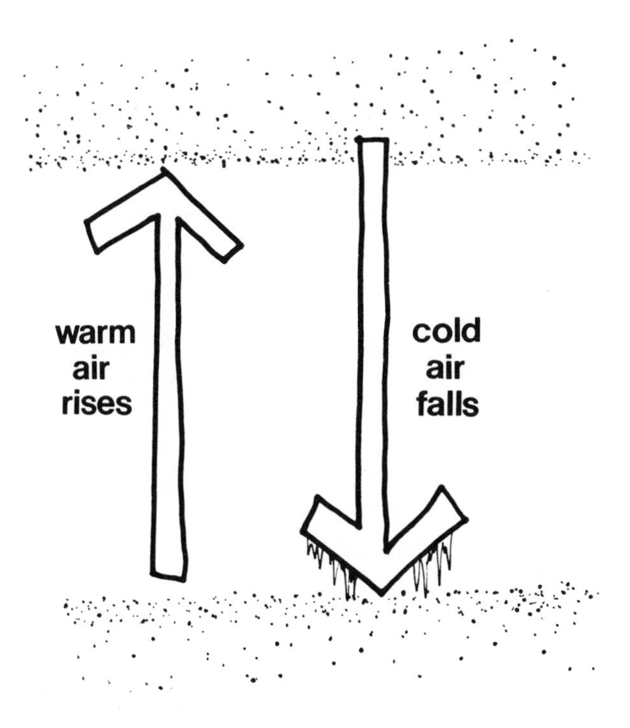

When air is warmed, heat energy is converted to molecular kinetic energy, measured as an increase in temperature. Air-pressure differences tend to be equalized by the flow of air from a region of high pressure to one of relatively lower pressure. Wind and weather "fronts" are a result of this equalization.

Air will tend to stratify into horizontal thermal layers. As the air is warmed, its molecules will move farther apart, decreasing its density. With the decreases in density, the weight of air per unit volume will decrease, causing it to rise above the cooler, denser air.

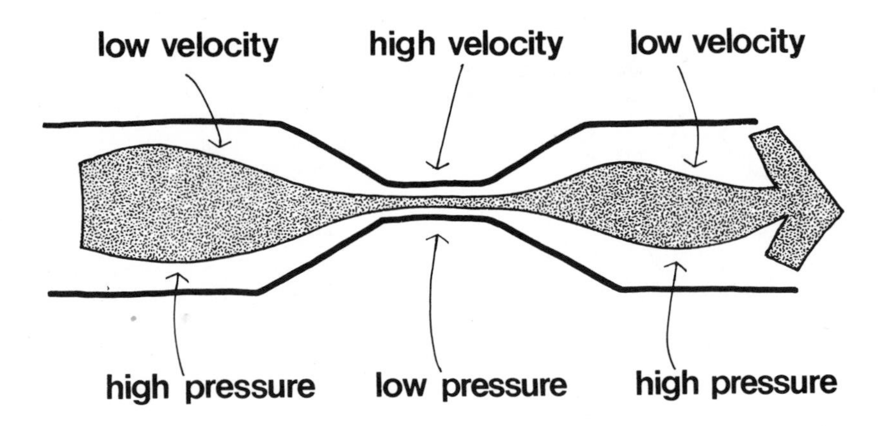

venturi effect

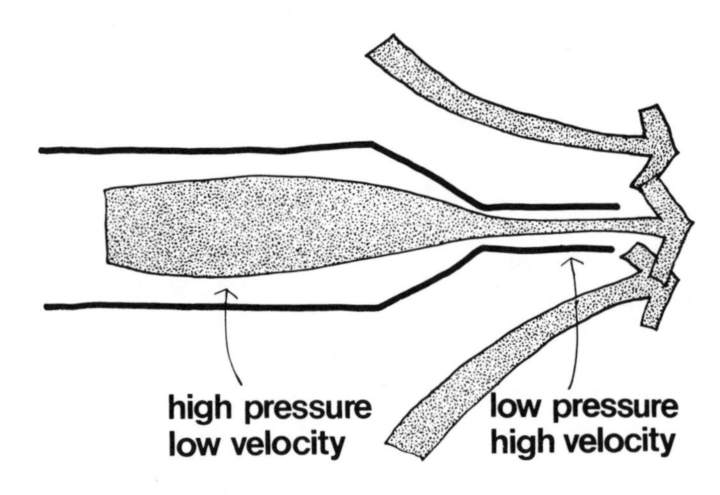

This difference in pressure is the basis for many aerodynamic and thermodynamic principles. One of these is the venturi effect, which results from the conversion of pressure to velocity when a fluid flows through a constriction.

Another useful principle is that as air passes through a reduced opening, its velocity increases in a "bottleneck" effect and upon exiting through the small aperture draws adjacent air along with it by induction.

Light

Electromagnetic radiation that is visible to the eye is called light. Energy from the sun is transmitted in forms other than visible light, which occupies a comparatively narrow band of the electromagnetic spectrum between the short wavelength ultraviolet and the long wavelength infrared. Because the source of the earth's natural light, the

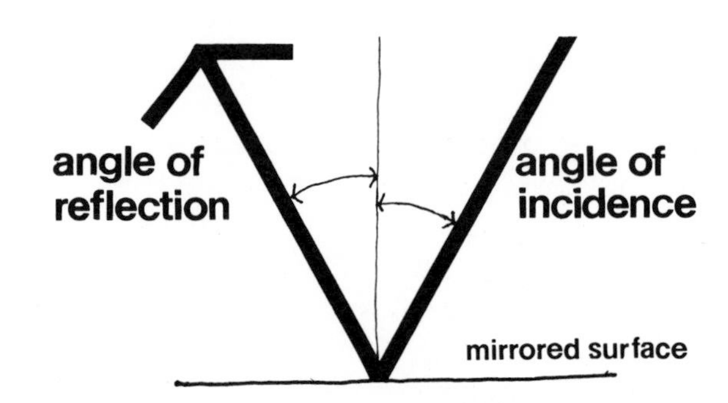

sun, is approximately 148,990,000 kilometers (92,600,000 miles) away, the light arriving here is thought of as rays that are parallel to each other. These rays can be manipulated with the aid of lenses to intensify or diffuse the light energy. They can be bent or reflected, using a suitable surface. When light is reflected from a polished surface, the angle at which the light hits this reflecting surface (called the angle of incidence) is equal to the angle at which it rebounds from the surface (called the angle of reflection).

Materials may either absorb, reflect, refract, diffract, diffuse, or transmit light. Colors are a result of certain wavelengths being absorbed and others being reflected or transmitted. To reflect all the colors contained in "white" light a surface should have a coating or be of a material that does not absorb any of the light rays. White rocks have a surface color that reflects light, but because of the irregularities of the surface, they do not reflect the rays in a parallel or predictable manner. Scattering (diffusion) of the light is the result. A mirror, by contrast, produces predictable (specular) reflection, as shown by the illustration.

reflected-light wavelengths

absorbed-light wavelengths

If an object absorbs all visible incoming light energy and reflects none, it appears black. If it reflects all the light and absorbs none, it appears white. A leaf reflects green, while absorbing all the other colors; hence it appears green. The visible light spectrum (or range) is made up of different wavelengths, each wavelength producing a different color within our perception.

Light and other radiant energy can be controlled and collected. It can be reflected from one surface onto an absorbing surface, thereby intensifying the density of the energy. Parallel rays of radiation can all be focused onto one point or line of points with a parabolic reflector.

Radiation can also be bent or refracted with the use of lenses. Lenses can be used for concentrating or diffusing the radiant energy.

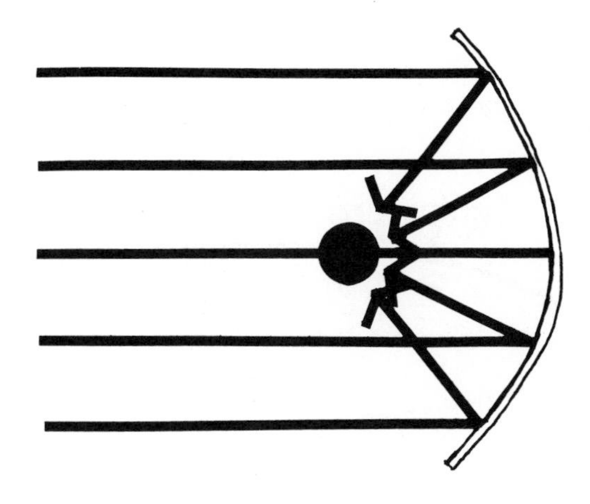

reflection of light onto a point

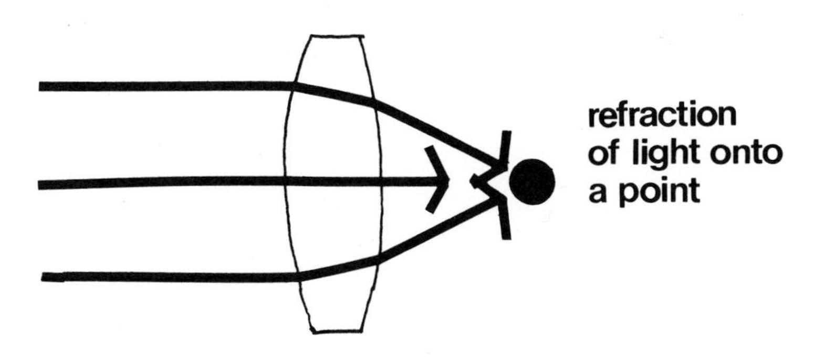

refraction of light onto a point

A thin-section lens used in solar-energy collection is the Fresnel. The Fresnel lens duplicates the curvature of a much thicker lens on a flat sheet. The curvature of the large lens is divided into segments and duplicated exactly by the Fresnel lens.

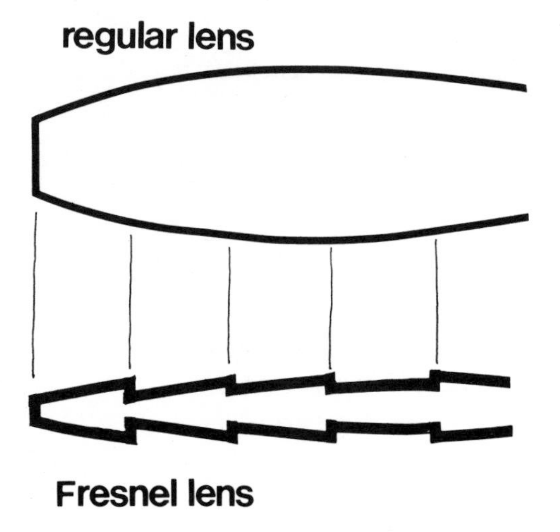

An important need in the design of habitat is a comprehension of all forms of energy in regard to time-related human response and interaction. Planning and design for homes and buildings should be based upon the reality of our human encounter with all energy forms. Cultural, sociologic, and economic considerations always modify our human structuring and use of energy. Economics is the fundamental sieve that should guide our determinations of the most appropriate and effective long-term uses of finite and renewable energies.

The question is how to make all of our natural, regenerative resources and energies most productive and least damaging to our biosphere. Each choice that we make in planning and design irretrievably affects the total world environment.

2 Natural Energies

All ecological systems on the earth are interrelated and interdependent. Changes in one system have influence on others. Man has manipulated and changed his natural environment to suit the requirements of his own physical-comfort zone.

As our technological dependence changes from exploitation of the earth and its biosphere to a belief in living in harmony with nature and utilizing the natural forms of energy continuously available to us, we will use those forms that effect minimum disruption to natural cycles of the earth. In our buildings we can employ such forces to minimize the pollution and degradation of the environment.

The approach to energy in building advocated throughout this book may be summarized in the analogy of civilization to nature, in which the greatest complexity offers the greatest survival potential. As an example, consider the sophisticated interdependence of all living organisms inhabiting a rain forest or jungle. This is nature in its most intricate and advanced state: There are safeguards against all forms of damaging weather, insect blights, and intervention of predators. In fact, there are many duplicate systems that allow such a forest/jungle to survive the extinction of any one species. Nowhere else is the term "balance of nature" more appropriate.

By contrast, the Valley of the Nile River

rain forest

in Egypt shows very elementary life forms and terrain in which one major storm can threaten an entire year's crop and the human population that depends on it. Civilizations surrounding the Nile have never escaped the cyclic extremes of flooding and drought. No long-term equilibrium exists there.

The analogy to mankind's energy needs is clear: The most complex power-source network with the greatest number of backup systems offers the greatest continuity and survival value to the daily business of any civilization. An oversimplified fuel economy is as vulnerable to changing conditions as a single-crop land area.

In the case of fossil-fuel consumption, the increased demand and consequent depletion are particularly serious because these commodities are not renewable. The customary reliance on one fuel source must not simply shift from oil to, say, uranium or coal, for these are finite also. Such renewable resources as timber are more valuable as building materials than as fuels, and even if these were diverted to be used only as fuels, they would fail to keep pace with accelerating (or even present) demands for combustion fuel.

A proper concern, therefore, is to optimize a balance of intersupportive, redundant natural-energy systems and energy-conservation features in the process of total urban and community planning.

The most promising survival-community network includes many dependable fuels and materials similar to the redundant systems of rain forests and jungles.

On a large scale, this means the development of every fuel type known to be safe, with long-range priority given to those that are least finite. Various forms of solar energy and lunar gravitation (creating ocean tides) will remain available for millions, perhaps billions, of years hence.

On a smaller scale, in individual dwellings and other buildings, survival entails the use of as many conservation features as practicable to allow a smooth transition from one fuel source to another. The thermal efficiency of a building is directly related to implementing the most intereffective balance of redundant systems.

The health and well-being of man depends largely on responsive architecture and the utilization of nature in buildings. The natural-energy systems of the earth can be divided into four major areas:

- Sun: energy derived directly from the sun
- Wind: dynamic movement of air caused by temperature differences across the planet's surface (indirect solar energy)
- Water: movement and cycles of water influenced by sun and weather (indirect solar energy)
- Earth: energy from the natural mass and heat of the planet (gravity, natural radioactivity, magnetic fields)

Natural-energy forms can be used passively and actively. In active systems, motorized mechanical devices are necessary to utilize the energy source, whereas in passive systems, the need for motors is eliminated by various applications of physical principles, such as the vertical stratification of air temperatures.

A solar system that uses motors to pump fluid through a collector is an example of an active energy-collection system. A solar collector that uses the natural thermosiphoned movement of air or water in its heat collection is considered a passive system. Some energy collection and storage systems use both passive and active devices and are known as hybrid systems.

Sun

The direct energy from the sun has many different architectural applications. Some of these are:

Heating of Buildings
Both passive and active solar energy can be used to heat internal and external areas in our built environment. Using the energy that falls naturally upon a building can bring about less dependence on fossil-fuel systems.

Cooling of Buildings
Solar cooling of buildings can be achieved mechanically through the use of absorption-cooling methods. Evaporative-cooling systems depend on water vaporization to lower the temperature of the air and are most effective in dry climates.

Hot-Water Heating
Collection systems can utilize the sun's heat to supply domestic hot water, to supplement existing hot-water systems, or to heat swimming pools.

Ventilation
The use of the sun's energy to induce air movement can be an economical method of passively ventilating a building. Depending on the relative elevations of intake and outlet, summer cooling may also be induced.

Ventilation Air Tempering
Incoming cold air can be tempered (preheated or precooled) before it enters the building.

Humidification
The sun's energy can be used to evaporate water to provide humidification of buildings.

Dehumidification
With the use of rechargeable desiccants, solar energy can dehumidify building spaces.

Desalination and Distillation
Through the use of a solar still, salts and minerals can be removed from water.

Drying or Dehydration
When dried, food and grains are preserved and lumber prevented from warping.

Solar-Responsive Architecture
Buildings themselves, by virtue of their design, can act as efficient solar collectors during cold winter months and as cooling chambers during periods of warm and hot weather.

Phototropic Change
Changing the color or value of buildings, such as darkening in winter and lightening in summer, could make them responsive to thermal need through solar radiation.

Landscaping and Site Planning
Designing with nature provides an effective response for all seasons of the year

by coordination between the building and its site. Every tree, bush, and ground cover affects the thermal response of the building.

Natural Lighting

The use of natural lighting from the sun can considerably reduce electrical energy usage by buildings. Daylight can be effectively introduced through roofs as well as outer sidewalls of buildings. Reflective external surfaces can be used to increase interior daylighting.

Solar Furnaces and Ovens

Concentrated solar energy can be used to cook food or provide heat for industrial purposes.

Photochemical Reactions

Energy from the sun causes chemical reactions in plants, animals and materials, bringing about changes in color, form, and growth. Greenhouses can effectively extend the growing season for various crops, lessen transportation distance for distribution, and increase food supplies.

Germicidal Reactions

The sun's ultraviolet radiation is germicidal by nature. With the proper compass orientation, this can be used to great advantage in certain rooms in all buildings, an obvious example being hospitals. Glass or plastic windows that pass full solar-spectral radiation are needed for this process.

Electrical Power Generation

Use of photovoltaic cells to convert the sun's energy directly to electricity has so far been confined to space vehicles and geographically remote locations on earth. Increasingly broad feasibility is likely, however. Indirect electrical conversion uses heat concentrators in developing steam to generate electricity and is presently more economical for large-scale applications than photovoltaic conversion.

Switching

Sunlight can be used to activate photocells controlling switching devices. This application is already in use with lighting for city streets, shopping-center parking lots, and building exteriors.

Wind/Air

The movements of air and wind can be used in many active and passive energy systems. The air moves to equalize any temperature or pressure differentials caused by the sun. The wind can be used for:

Water Pumping

Wind-driven pumps for wells and irrigation were familiar on the rural American landscape in the early 1900s. Because of expanding rural electrification, windmills were later supplanted by electric pumps.

Grain Milling

Many of the old European windmills were used for grain milling. The slow rotary action of the windmill provided high torque for the turning of the large milling stones.

Mechanical Power

The turning force of a wind machine can provide direct drive power for factories and industries.

Electrical Generation

Wind machines can generate electrical power for homes and buildings. Large-scale generating plants, using wind machines or vortex generators, could provide power for towns or cities.

Wind-Responsive Architecture

The shapes and forms of architecture can be arranged to control external and internal wind-induced ventilation and to use prevailing wind patterns to power wind turbines.

Combustion Air

Oxygen in the air allows the combustion by fire of various substances or their reduction by heat.

Fresh Air

One of the most immediate, continual needs for the sustenance of humans, plants, and oxygen-breathing life forms is fresh air. For these organisms, air pollution has serious long-range health implications.

Cooling Media

Air that is cooler than an object in the airstream will carry away the heat being given off by the object. The rate at which heat is carried away increases as the velocity of the air increases.

Dehumidification

Air can be used for cooling by first passing it over a desiccant, a moisture-removing substance, or by allowing its moisture to condense on a surface.

Humidity

The addition or subtraction of humidity in the air can provide temperature changes necessary for natural heating and cooling. The sensation of comfort is greatly influenced by wind speed and humidity level.

Air Support

Structural support by inflatable and floating structures can use the compressive potential of air in confined spaces.

Movement

Wind provides the movement of air to induce the propulsion of ships, gliders, or other vehicles. Pneumatic air can also be employed to move objects, actuate control systems, and propel devices through tubes.

Compression
Energy can be stored in air by means of compression. Containers can be filled with air, which can later be released to power turbines and generators during peak electrical-demand periods.

Insulation
Still air can be used as insulation. To keep the air from moving, pockets or chambers can be designed to minimize air currents and thereby minimize convection-heat losses. Closed-cell plastic foams are widely used as insulation because of their ability to entrap air.

Purification
Landscaping, using the texture of trees, bushes, and ground cover, has a purifying effect on air. Plants withdraw carbon dioxide and provide oxygen in the process of photosynthesis. Air can also be purified by fiber, mechanical and electrostatic filters, and ionization devices.

Water/Precipitation

The natural cycle of water and precipitation is a force needed to sustain life on earth. About three-fourths of the planet's surface is water. Many forms of energy can be developed from the movement of water in its natural cycle.

Life Support
Approximately two-thirds of the human body is water. Water is a vitally needed life-sustaining substance for all organisms.

Plant Support
The circulation of water through plants and over the land maintains the food supply for the planet and, through photosynthesis, provides our oxygen. Water can be retained in ponds or cisterns for landscaping and greenhouses.

Cooling
Water can be used as a natural cooling medium. As cool water runs over a warm surface, it draws off heat, cooling the surface. Water can be sprayed over surfaces to remove heat by evaporative cooling. Groundwater or cold lake water can be used to cool building spaces. The circulation of cold water can be used to reduce air-conditioning costs. Ponds on building roofs or on land near buildings can be used for cooling by evaporation.

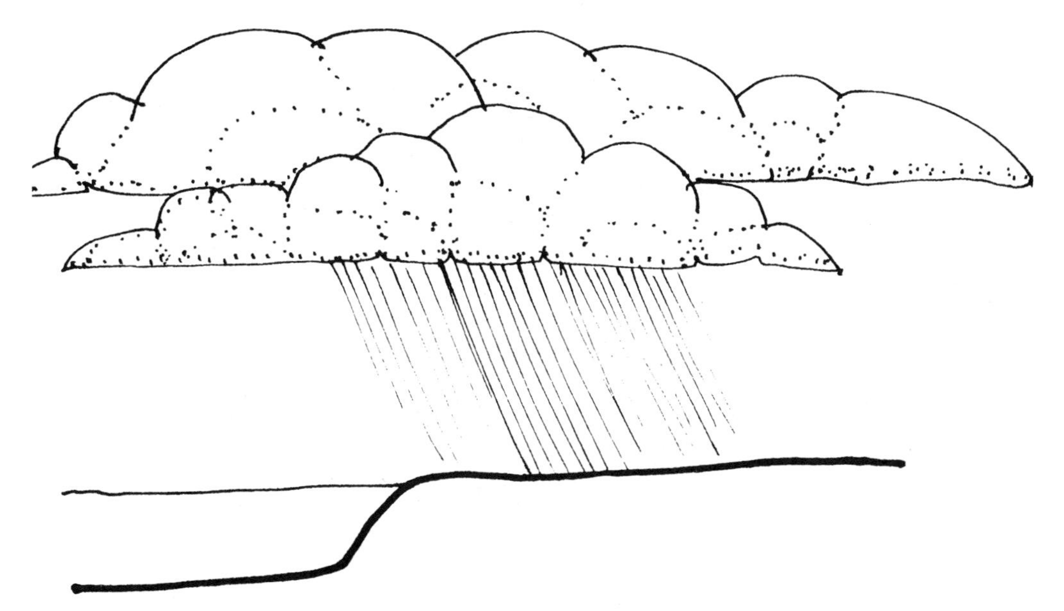

Humidification

Water can be evaporated to provide humidification of buildings.

Thermal Storage

Each gallon of water can hold 69.7×10^4 joules (660 Btu) of heat energy, more than four times the heat of an equal volume of 5 centimeters (2 inches) gravel. Nevertheless, this property for solar-energy water storage is not always a decided advantage because of its thermal diffusion. Gravel storage with equal energy can usually maintain higher concentrations of temperature.

Air Purification

Water in the form of precipitation removes pollutants from the air. Rain and snow in some areas are suitable for consumption, while in major cities air pollutants in rain cause impurities that make such water nonpotable. Fountains and waterfalls provide airborne beneficial negative ionization of moisture vapor.

Cleaning

The movement of water acts as a natural cleaning mechanism, dropping heavier impurities as it moves across stream- and riverbeds. Algae and microorganisms living in the water can help clean it. Purification of water can also occur by osmosis through a suitable membrane.

Thermal Inertia

Large bodies of water such as oceans, lakes, and broad rivers naturally moderate day-to-night temperature changes internally and on adjacent land areas. Coastal and riverfront cities are known for their mild climates and refreshing breezes. Covered swimming pools can be used as heat sinks for solar-energy thermal storage.

Recycling

There is a fixed supply of water on and surrounding the earth that is continually recycling in one of its three states: liquid, vapor, or ice. More concern with water and how it recycles in the earth's biosphere will help to curtail wasteful uses of the water supply. Waste water can be filtered and recycled for buildings.

Electrical Generation

The generation of electrical energy is facilitated economically by hydroelectric dams and their turbine generators. It should be noted, however, that dams can have the negative effects of damaging the ecology and of causing oversalinization of irrigated land.

Hydrogen

Water can be reduced to its components of hydrogen and oxygen by electrolytic action. Hydrogen can be stored or directly used as a clean, gaseous, powerful fuel. Oxygen can be similarly used as an aid to the combustion of other substances. Hydrogen fuel cells are an advanced technology needed for energy conversion.

Mechanical Power Generation
Waterwheels can be used for rotary power generation. The force from turning waterwheels was used in the 1700s and 1800s to run weaving and other equipment. Also, the force derived from wheels was used in early industrial plants and for grinding ore in mines.

Ocean Thermal Gradients
Because the oceans cover approximately 75 percent of the planet's surface, they are the largest receptors of solar energy. The thermal gradients of the ocean (temperature decreases with depth) can be used to power heat engines that could produce great quantities of electrical power.

Tidal Power
Electrical power can be produced by harnessing the movement of tides with dams and generators. Tidal power stations are most efficient when using reversible turbines, which take advantage of the opposite directions of movement of ebb and flow. The gravitational force of the moon causes tides, and this type of application is an example of "lunar energy." Very few sites in the contiguous United States are suitable for electrical generation using tidal power. Worldwide, however, a number of locations appear promising.

Earth

The earth can be thought of as an energy source, but its quantity of energy is very limited in comparison to that of the sun. Some forms of energy from the earth and methods by which the earth may be beneficially used are:

Life Support
Countless organisms and microorganisms find a home within the earth. Nutrients in the soil, formed and released by nature's regenerative processes, nurture concentrated life forces. The habitat constructed by these creatures in response to the natural environment offers a remarkable example for the design of homes and buildings.

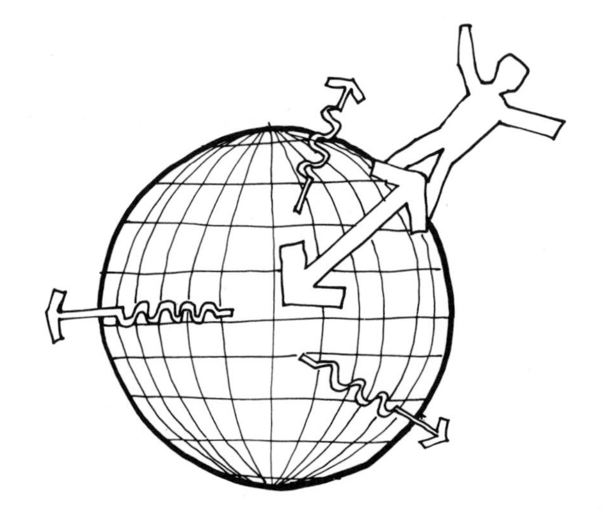

Resource Container
The earth provides minerals and materials that can be used as energy sources. Fossil fuels are contained and formed under the earth. These are renewable only on an extremely protracted time scale, measured in millions of years.

Filtering
Sands and gravel on the earth act as a filter by purifying water. Many sewage-treatment plants use varying sizes of stone and gravel as filter systems.

Purification
As water percolates down through the soil, impurities are removed. The process is aided by aerobic and anaerobic bacteria that reduce wastes to a harmless state.

Geothermal
The earth is a natural heat generator producing intense heat at its core, which is conducted out to the surface and into subterranean water.

Gravity
The force of attraction between two masses can be used to store energy. Off-peak electrical power, as an example, can be stored as potential energy by pumping water to a high-level retaining area (water tank or dam). The gravitational force acting on the water after it is released will convert the potential energy to kinetic energy, which will ulti-

mately power a turbine for the generation of electricity.

Interior Thermal Mass

In buildings, large interior masses of earth, concrete, or masonry that are insulated from ground temperatures and from changes in external climatic conditions will stabilize internal building temperatures.

Thermal Cooling

At depths of 1.2 meters (4 feet) or more, earth temperatures usually remain at about 10° C to 16° C (50 F to 60 F). During warm and hot weather this can be used as an effective medium for cooling buildings.

Decomposition

The waste products of living earth organisms and their activities are decomposed into combustible gases such as methane that can be used as an energy source.

Thermal Inertia

The planet as a whole exhibits strong thermal stability from the ground level downward. In most temperate climatic zones, a "frostline" is found at about one meter depth, below which the soil never freezes.

Solar-Heat Absorption

The earth, in varying degrees, is a natural absorber of solar energy. Although water accounts for 75 percent of its surface and retains solar heat effectively, certain land masses are more efficient absorbers. This is particularly true in equatorial regions, where the earth's surface is most nearly perpendicular to incident radiation. Rain forests and jungle growth also trap heat.

Building Material

Stabilized earth, adobe, rammed earth, earth floors, sod roofs, and mud surfacing are useful as low-cost building materials. The energy expended for their manufacture and transportation is reduced by using excavated on-site materials for such structures.

Insulation

Earth can act as an insulator by tempering heat losses and heat gains through roofs, walls, and floors. The earth can be manipulated into forms to protect structures from wind and harsh weather conditions. A mass of earth can also act to protect plumbing pipes from freezing and reduce excessive heat losses from building ducts.

Earth Concrete

Dry cement mixed with earth and sprinkled with water can provide a durable paving surface.

Magnetic Fields

The earth's magnetic fields and electrodynamic forces could become possible energy sources. It appears that energy extracted from these sources would be greater than the energy invested.

3 Sun

A star in the universe, the sun is a giant nuclear-fusion reactor. At a speed of 300,000 kilometers per second (186,000 miles per second), the energy from the sun takes 8.3 minutes to reach the earth. The sun is extremely large, with a diameter of 1,390,000 kilometers (865,000 miles) and a volume a million times greater than that of the earth. The temperature of its photosphere, or surface, is approximately 5760° C (10400 F). Nuclear transformations at the core produce intense heat and consume about 3.6 billion kilograms (4 million tons) of mass per second. The sun acts as a fusion reactor combining four nuclei of hydrogen into one helium nucleus. This results in a decrease in mass. There is no need for alarm because the supply of

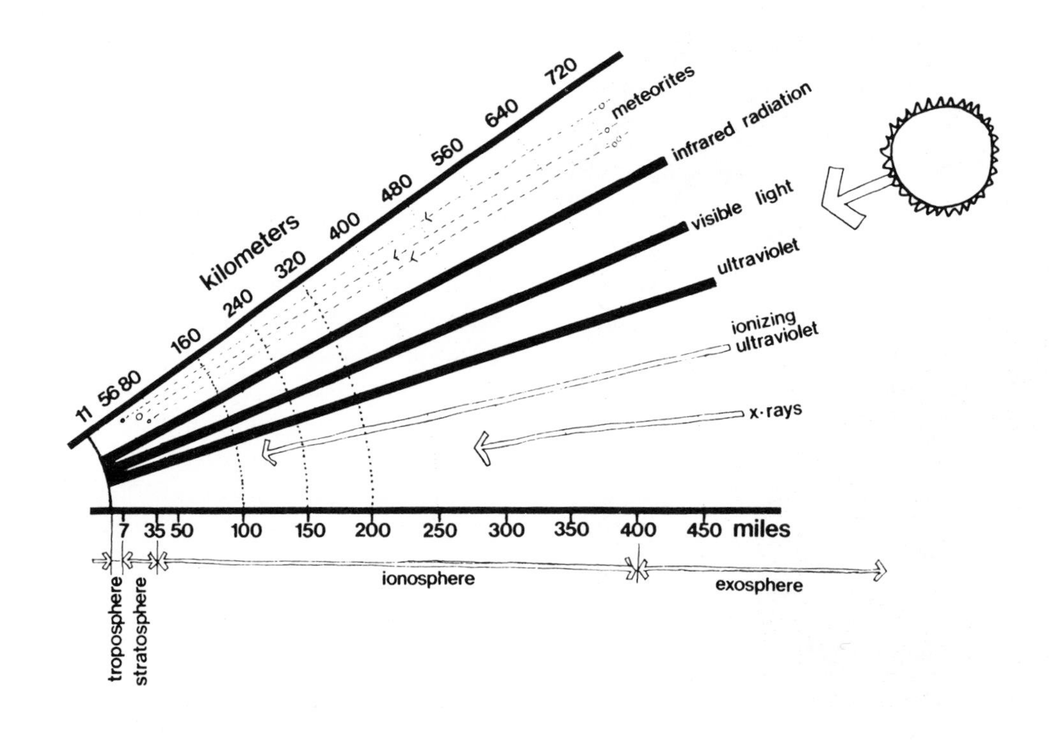

hydrogen will probably last another five billion years.

The sun's core radiates heat as a result of its nuclear reactions. This is transmitted by radiation in areas closest to the core; heat is then transmitted by convection up to its outer layers, where it is radiated outward. Since the sun is extremely far from the earth (150 million kilometers, or 93 million miles), and the earth is small in comparison to the sun, solar radiation is considered as traveling to earth in parallel rays.

Radiation from the sun has catalyzed or directly supplied most of the natural-energy systems on the earth. Many different levels of energy from the sun continuously flow toward the earth. Not all of these different radiative frequencies penetrate the atmosphere and reach the earth's surface. This is because the atmosphere is composed of many layers, each with its own composition of oxygen, nitrogen, hydrogen, and other matter. These layers protect the earth and help sustain the life forms on its surface. Visible light, which makes up a portion of the sun's energy, is able to penetrate all the different levels of the atmosphere, whereas much of the ultraviolet and infrared energy (which we cannot see) is absorbed or reflected by it. The upper levels of the atmosphere contain ozone, which absorbs dangerous ultraviolet radiation and X-rays. Most meteors and other debris from space burn up, never striking the earth, because the atmosphere is denser near the earth's surface than it is at higher altitudes.

solar radiation composition

ultraviolet
visible light
infrared
4.6%
46%
49%

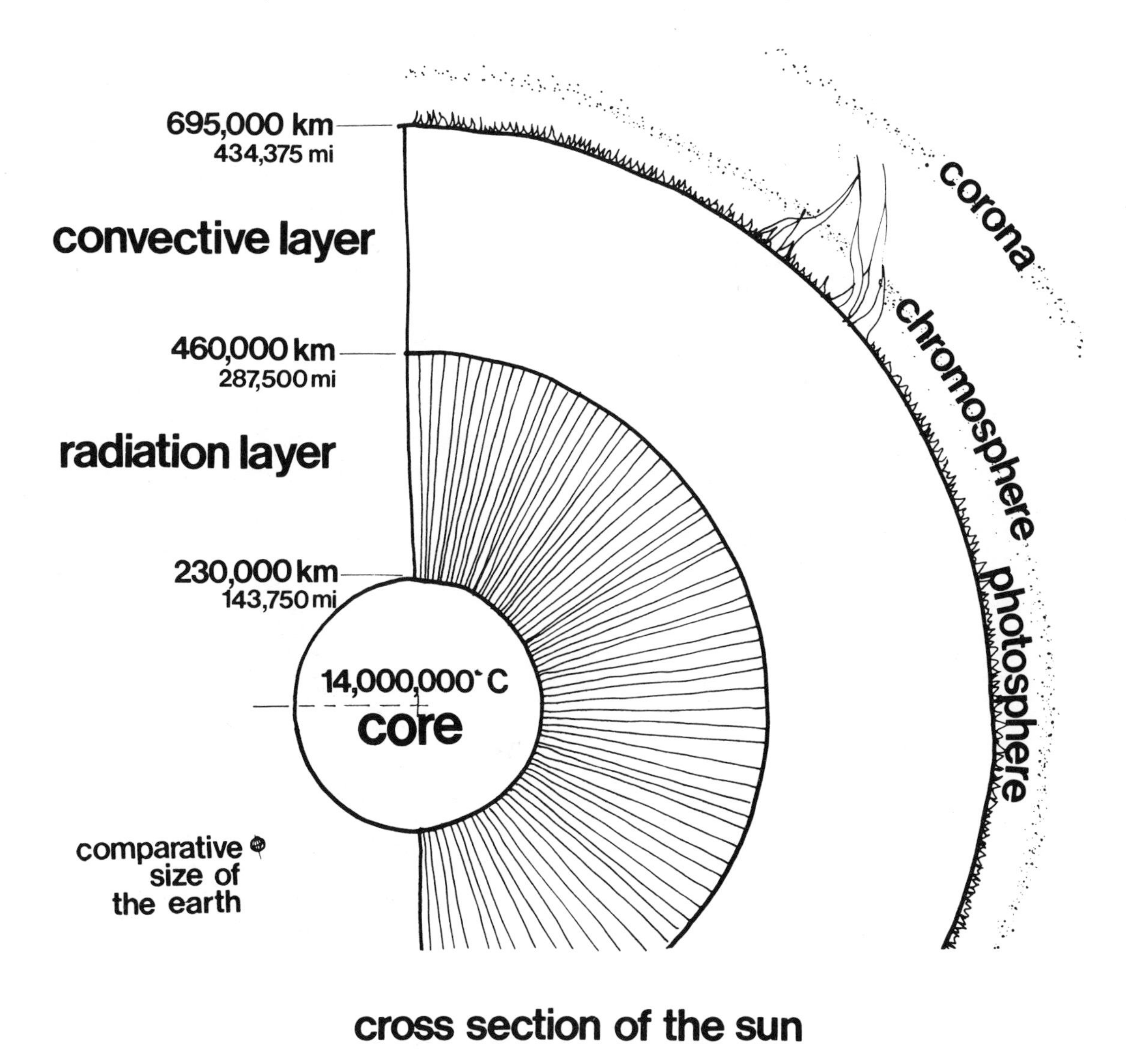

cross section of the sun

Some of the solar radiation that pierces the atmosphere is reflected by the earth's surface or by cloud cover. Some of it is absorbed by the lower levels of the atmosphere, causing warming of the air. This sets up a temperature differential that generates air movement or wind. Major currents or patterns of the wind are a result of the earth's rotation about its axis and day-night cycles of solar heating and subsequent cooling.

A great amount of energy is absorbed by the earth's surface itself, as well as by oceans, plants, and buildings. The energy that plants absorb is potentially convertible into food energy and material for combustion (logs, etc.) or fossil fuel (long-term decay). The energy absorbed by the oceans causes water evaporation and motivates the complete water (hydrologic) conversion cycle.

In combination with the earth's motion around the sun, the water cycle and the wind provide weather and the change of seasons.

Some of the physical considerations of the earth and the sun are their distance apart, their masses, and the earth's speed of rotation and its orbital velocity. The earth is not a static, stable object, but a moving, dynamic body protected only by a thin atmosphere. It is a spaceship planet, traveling through an infinite void and powered by the sun's energy, which provides it with food, water, oxygen, and fuel.

The earth is a giant spacecraft, supporting life as it orbits the sun. The sun provides the earth with the energy it needs to sustain and maintain life. The earth with its thin atmosphere (a layer only 160 kilometers, or 100 miles, deep) supports more than four billion people. Smaller spacecraft in many ways duplicate the conditions and functions of the earth by utilizing the sun's energy and shielding their occupants as they travel through space. Unable to shield themselves with atmosphere as does earth, however, these craft must rely on so-

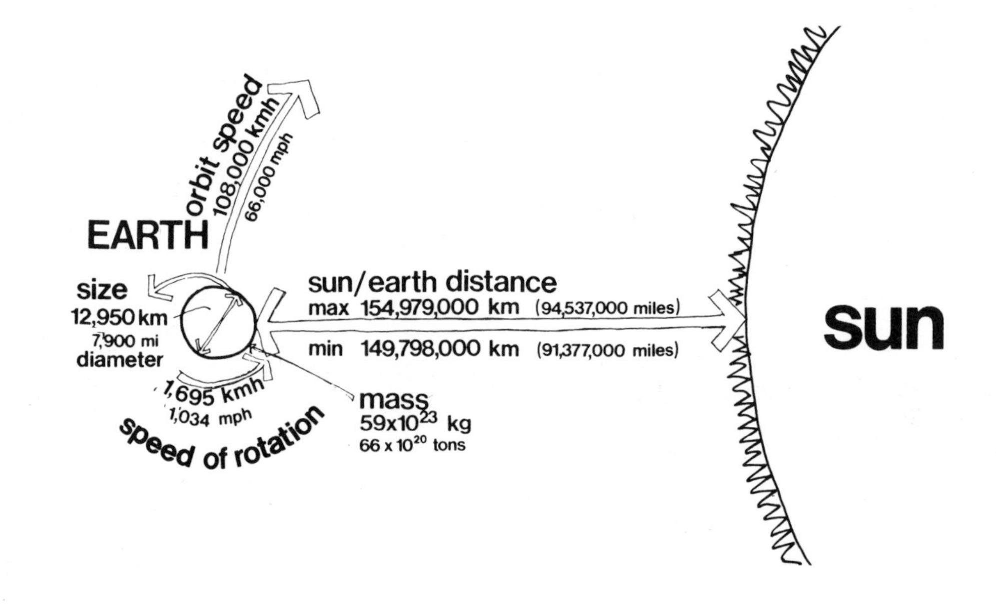

phisticated metallic-alloy skins and sensitive thermal-conditioning systems. Many use photocells to collect the sun's energy and convert it to electricity. The energy received on the earth's surface in one day exceeds the human consumption of energy in that period of time by a thousandfold and will continue to do so throughout the forseeable future.

Both the position of the earth and locations on the earth greatly affect the feasibility of using natural-energy systems powered by the sun. The earth rotates and in so doing appears to cause the sun to "rise" and "set." The angle of the sun and the intensity of its radiation are affected by our location on the earth. The earth's axis of rotation is tilted at 23½ degrees perpendicular to the plane of its orbit. Geographic latitude (the angular distance measured from the equator) affects the angle at which solar rays strike an object.

The earth also travels around the sun, completing one orbit every 365.24 days. The tilt of the earth in combination with its orbital motion around the sun brings about seasonal changes in the northern and southern hemispheres.

The apparent path of the sun changes every day. For example, at the winter solstice position (December 21 is the shortest, or solstice, day) at 40 degrees north latitude, the sun travels through only a 120 degree azimuth, or plan arc,

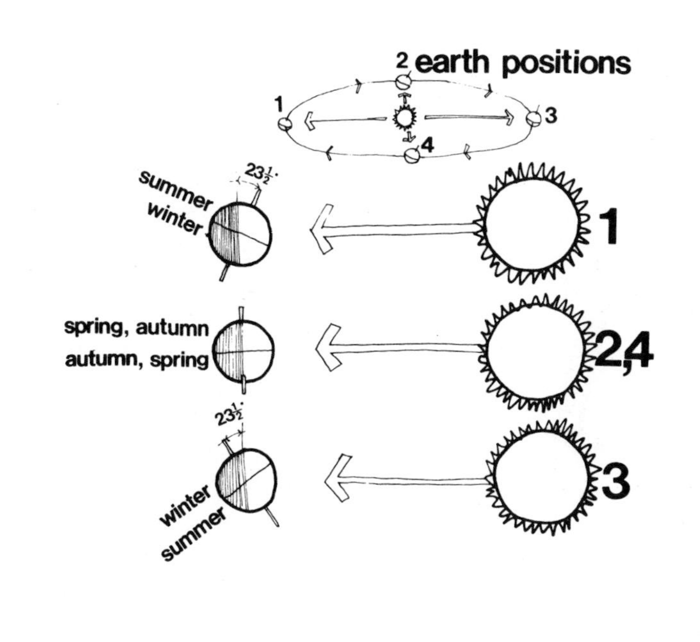

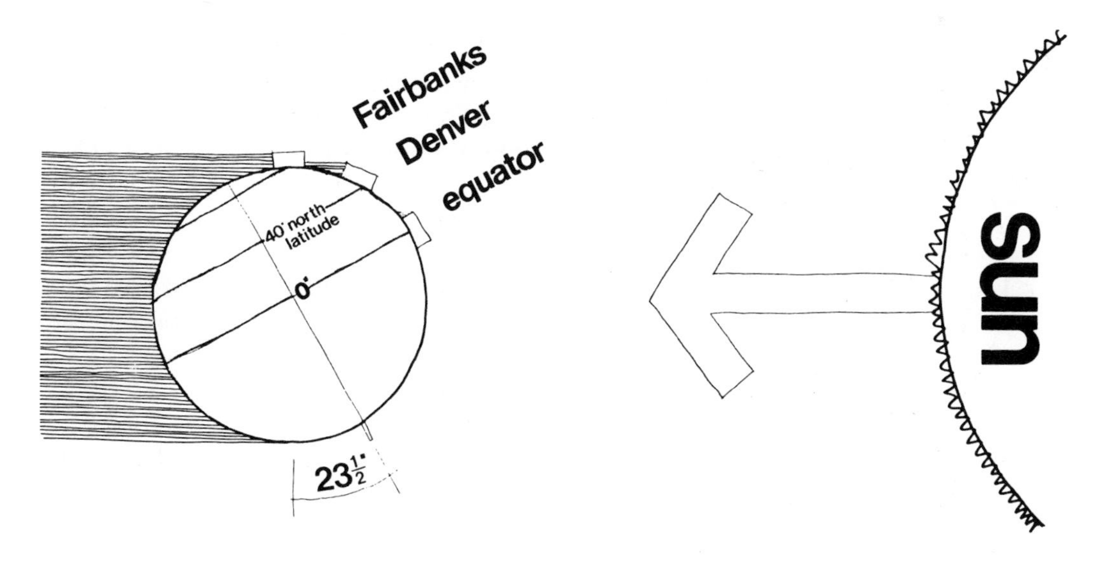

from dawn to dusk. On June 21, the summer solstice day, the sun's azimuth arc totals 240 degrees. Occurring between solstices are spring and autumnal equinoxes, in which daytime-elapsed time equals nighttime-elapsed time. The sun appears to travel on a curved path from sunrise to sunset, with the zenith occurring at noon. The difference between the maximum solar altitude in the summer and the minimum solar altitude in the winter is 47 degrees.

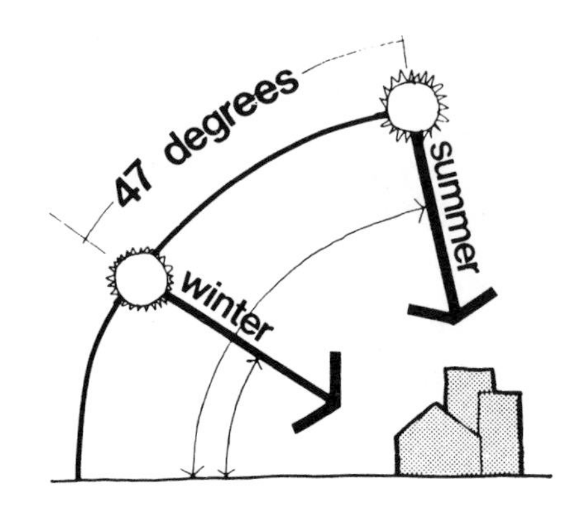

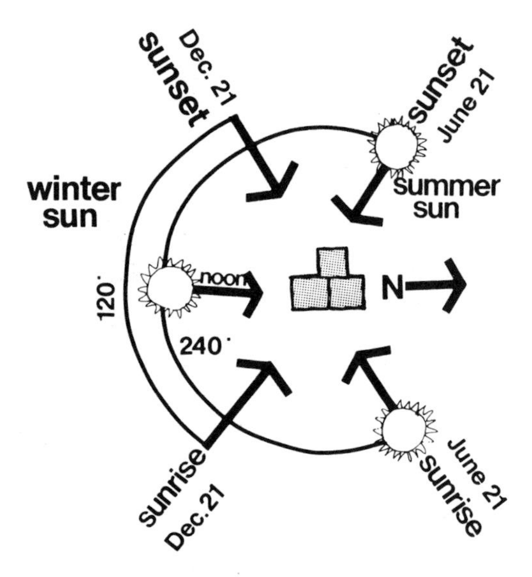

typical for 40 degrees north latitude

Orientation in respect to the sun affects the amount of insolation a surface will receive. Considerations when designing solar-heat-collection systems are: the tilt angle of a fixed solar collector, the plan axis on which the building is oriented, and any opaque obstructions that interfere with solar radiation, especially during the hours from 9:00 A.M. to 3:00 P.M.

The tilt angle for maximum solar collection by a collector in a fixed position depends on the time of the year when the greatest amount of the sun's energy is needed. For maximum winter collection (for heating purposes), this angle should equal the geographic latitude plus about 15 degrees. For maximum summer collection (to power air-conditioning or ventilation equipment), the angle should equal the latitude minus about 10 degrees. A compromise angle for both summer and winter lies between these two angles, the ideal being dependent on climatic conditions. Variations from these figures of up to 20 degrees can be allowed with only a minor loss in collector output.

Collecting the Sun's Energy

A solar collector is a device used to absorb heat that comes from the sun in the form of radiation and to transfer it to a circulating fluid (air or liquid).

The flat-plate solar collector consists of a black metal plate covered with glass or plastic and backed with insulation. The metal plate is known as an absorber and may have tubing built in to contain the circulating fluid. An air space separates the cover from the absorber.

The glass or plastic is transparent to incoming solar radiation, yet glass in particular is opaque to radiation of longer wavelengths (heat). Solar radiation passes through the cover and is absorbed by the black surface, increasing the temperature of the metal. Longer wavelength infrared radiation is emitted from the absorber, but most of it cannot pass back through the glass. This produces what is known as the greenhouse effect. (It commonly occurs in automobiles on a hot summer afternoon.)

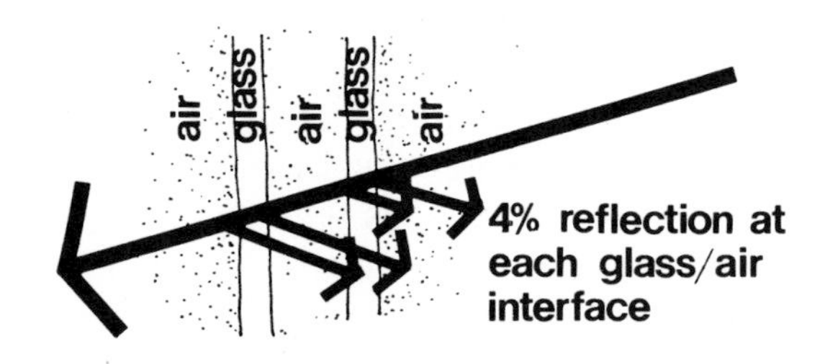

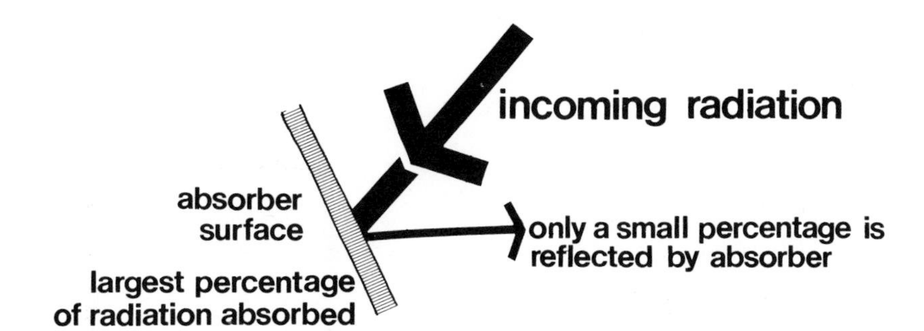

A liquid is circulated through the tubes in the absorber, or air is blown behind it. If these fluids are at a lower temperature than the absorber, heat will be transferred to the fluid. Insulation decreases heat loss from the back of the collector.

If the maximum potential of the sun's energy is to be utilized, as much heat energy as possible must be absorbed and retained. To illustrate the basic workings of a flat-plate collector it is necessary to examine the way in which energy is transported into and out of the collector.

A typical flat-plate collector is an absorbing unit. It cannot yield more energy output than the maximum solar radiation striking its surface.

The glass or plastic cover of a solar collector can either reflect, absorb, refract, diffuse, or transmit incoming solar radiation. If the glazing material is clear, only a small percentage of the incoming radiation is reflected or absorbed while most is transmitted through to the absorber surface. As the angle of inci-

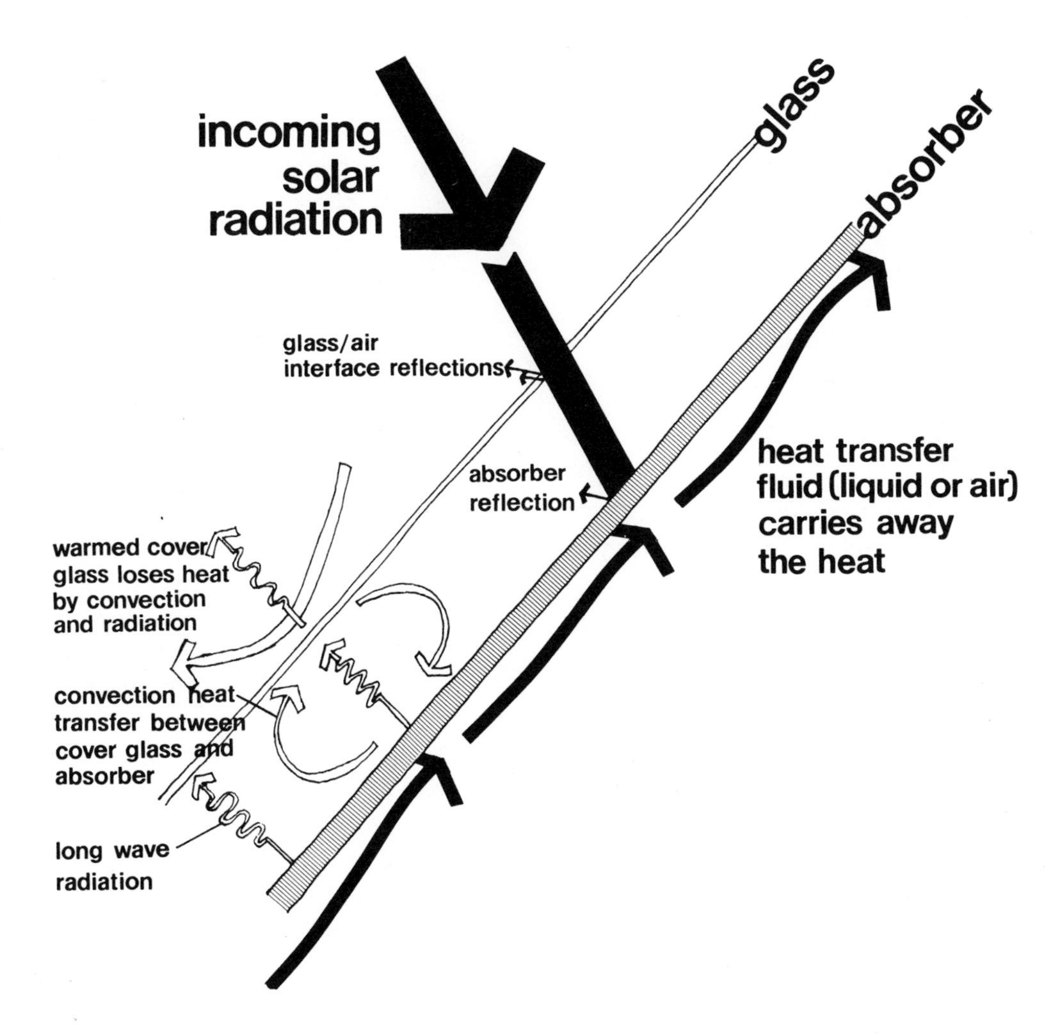

dence increases, a greater percentage of the radiation is reflected from the surface of the glazing. Boundaries created by media with different optical densities reflect radiation. At each interface between glass and air approximately 4 percent of the radiant energy is reflected.

Most of the incident radiation remains at the absorber surface while a small fraction is reflected. Glass is much more opaque to long-wavelength heat than the plastics available today. Hence, a plastic-encased collector will lose considerably more heat than the same one protected by glass. It is desirable for the absorber surface to collect as much energy as possible.

Since no light is reflected by the absorber surface, it appears black. Plants reject the color green as a means of tempering the intense heat they receive from the sun. In this way they receive only energy within their most beneficial spectral range.

Selective surfaces have been developed to absorb as much of the incoming solar-heat radiation as possible (90 percent) and emit the least possible amount of long-wavelength radiation (10 percent). Some paints offer as much as 99 percent absorptance. However, these paints and other nonselective coatings allow greater heat losses by reradiation.

As a collector warms up it transfers heat by both radiation and convection. It is desirable to transfer the maximum amount of heat possible to the collector fluid and reradiate or emit as little as possible to the outer surfaces of the collector. The fluid (liquid or air) behind the absorber plate absorbs heat and then transports it into the heating system. Some of the heat is carried off the absorber surface by convection and eventually warms the inner surface of the cover glass. It is then conducted through the glass to the outer surface and is lost to the atmosphere by radiation and convection.

Collector temperatures are regulated by controlling the flow rate of their heat-transfer fluid (air, water, or other). The faster the fluid flows through the absorber, the cooler its surfaces stay. The heat losses (both by radiation and convection) in proportion to the heat obtained for utilization when operating at very high temperatures are much greater than at low temperatures. Because of these losses, flat-plate collectors should operate at low rather than high temperatures. Generally, the range between 40°C (104 F) and 94°C (201 F) is best. Parabolic and other concentrating collectors, by contrast, operate at much higher temperatures (sometimes as high as 1093°C (2000 F). However, they are designed to operate at these high temperatures and have much less

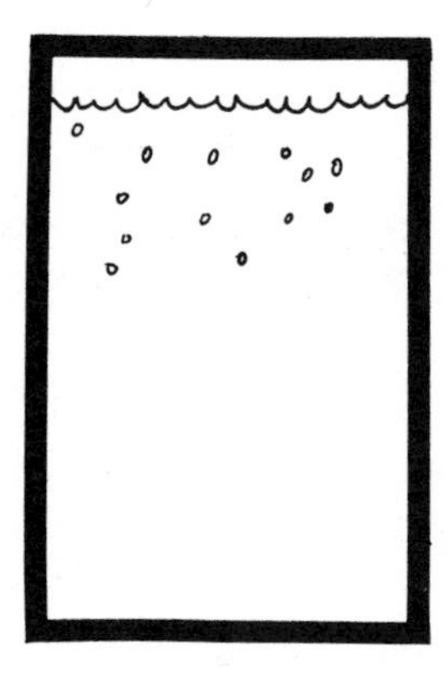

liquid energy storage

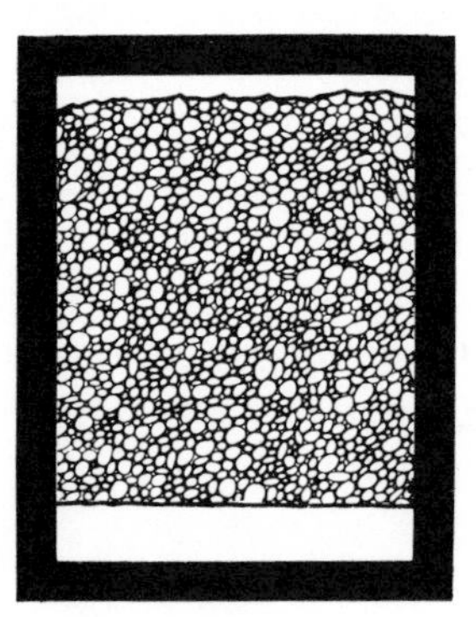

rock bed energy storage

absorber surface area than do flat-plate collectors of equal projected area.

One thing to consider when designing a collector is the number of glazing sheets to use. Some units have one, but most units have two panes of glass. Some are insulating glass. Water-white glass with a low iron content allows more solar radiation to pass through it. The number of sheets of glass or plastic used affects the amount of incoming radiation reaching the absorber surface, as well as the amount of heat lost to the exterior. As the thickness and the number of sheets of glass are increased, the amount of solar radiation that is able to penetrate this barrier is reduced.

Certain configurations work best in certain climates. For example, the decrease in heat gain through two panes of glass might have to be weighed against the heat-loss increase through only one pane of glass. The most critical climatic feature to consider when making these decisions is the relationship between the ambient outdoor air temperature and the amount of insolation available. Problems can develop when plastic is used for glazing the collector. Plastic is susceptible to attack by ultraviolet radiation, which may deteriorate the plastic and turn it yellow, thus decreasing its efficiency. Some plastics cannot withstand the high temperatures that collectors can attain when the flow of cooling fluid is interrupted. Those that can suffer from extremely high coefficients of thermal expansion compared to glass. This factor in turn modifies the design requirements of the collector frame.

Other considerations in the design include the spacing between the glass plates, the spacing between the absorber and the glass, the thickness of backing insulation, placement of condensation barriers, and the control of collector-edge heat loss.

Illustrated are some examples of general types of flat-plate collectors, tube collectors, and parabolic collectors. Also included are examples of collection systems and solar-powered cooling cycles.

Liquid-Type Flat-Plate Collector

Some flat-plate collectors use a liquid as the heat-absorbing fluid. Water and other liquids have been used for this purpose. Some of the considerations in choosing a fluid for a liquid flat-plate collector are:

Are the materials of which the collector surface and piping are made susceptible to fluid corrosion?

What is the quantity of fluid required and what is its cost?

At what temperatures does the fluid freeze and boil?

Is there homogeneity among all the materials with which the fluid comes in contact? Electrolysis will occur if two different metals are bridged by a liquid containing acid or dissolved inorganic salts. This will cause chemical corrosion.

Liquid flat-plate collectors may have a variety of collector-tube configurations. Some of these configurations are:

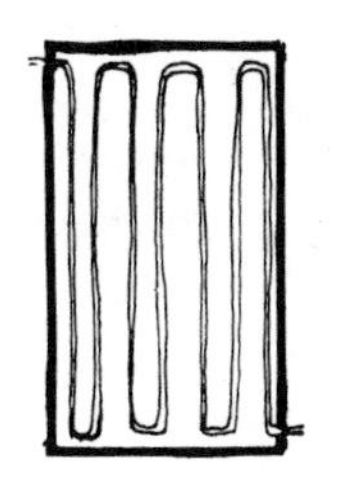

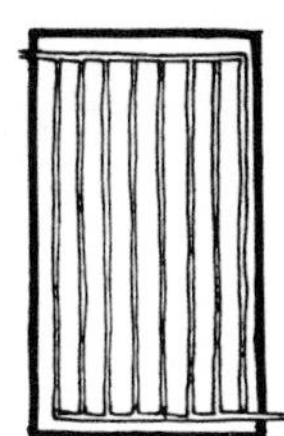

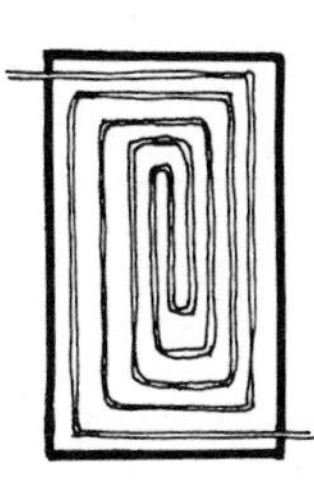

The liquid is usually pumped into the low side or bottom of the collector. This is done to distribute the liquid evenly in the system and decrease the trapped air. The sun heats the collector surface. As the liquid travels through the tubes, it picks up heat from the surface by conduction. Heated liquid then leaves the collector at the top and is pumped to storage or is used in the heating system.

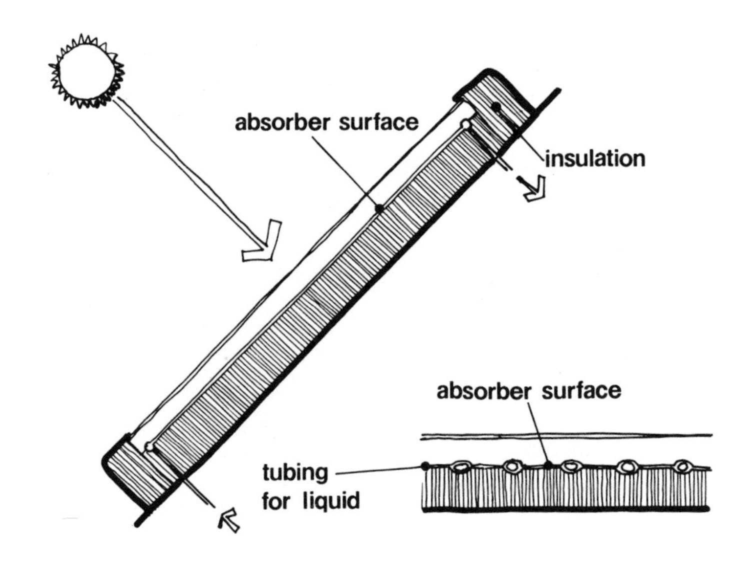

Fluid heat-storage volumes tend toward thermal equilibrium (equal temperatures throughout), as a consequence of their natural convection currents, which cause mixing. Temperatures in a tank full of fluid cannot be changed as rapidly as the temperature of one layer of rocks in a rock bin.

The Thomason "trickle" type collector uses water flowing in the troughs of a corrugated surface to collect heat. The surface is painted black to absorb the maximum amount of the incident radiant

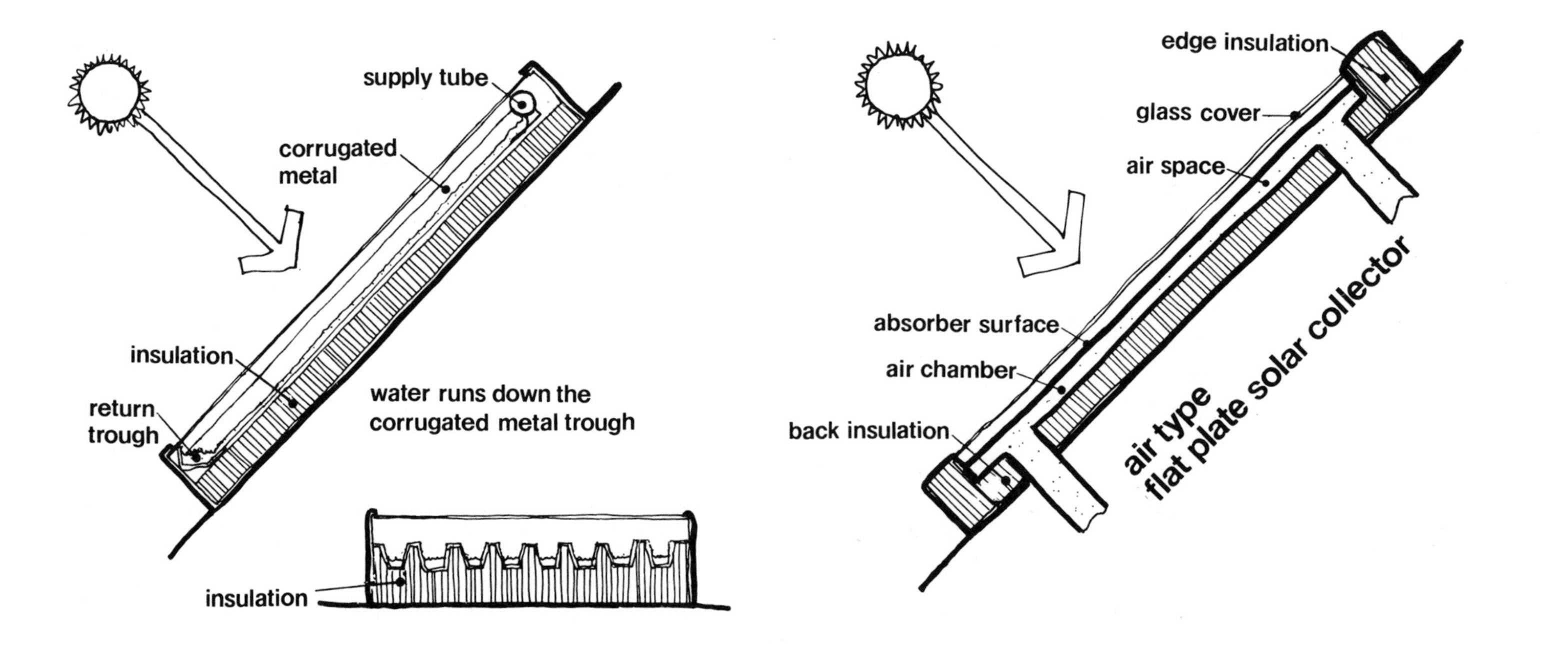

energy. Water is supplied through a tube located at the top of the collector and then runs down the corrugated surface (aluminum or galvanized steel). As the water flows over the collector surface it picks up heat. The warmed water is collected by a gutter and is directed into a hot-water storage tank.

The air-type flat-plate collector works similarly to the liquid-type flat-plate collector. Air is pumped or blown by means of a circulating fan through the manifold for distribution into the collector. The collector surface gains heat from the sun and raises the temperature of the air in contact with it. As the air heats, it rises in the collector. The air exits at the top of the collector, normally at a peak temperature of 54°C (130 F) to 66°C (151 F), and then is supplied to the house-heating system or is pumped to the storage bin where its heat is transferred to rocks.

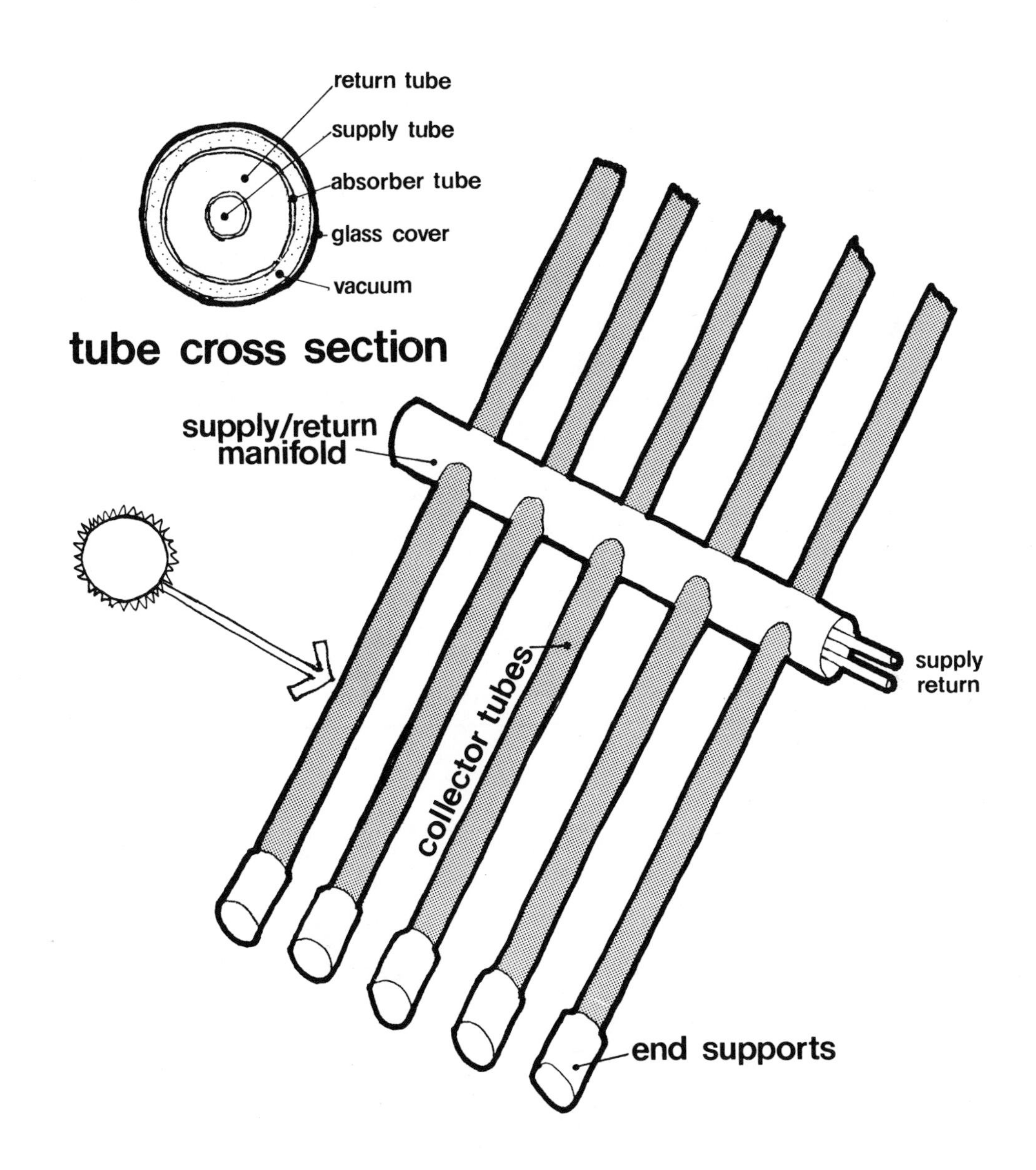

Evacuated-Tube Collector

One type of evacuated-tube collector is a recent development that utilizes a series of tubes with special absorptive coatings to collect solar energy. The tubes have the advantage that almost the same amount of surface area is exposed and perpendicular to the sun at any time during the day.

The evacuated-tube collector is composed of a series of three concentric cylindrical tubes, with a vacuum (nearly zero air pressure) between the outer and middle tubes and a black selective coating on the outer surface of the middle tube. Water or air (usually water because of its higher specific-heat capacity) is circulated from supply pipes through the inner tube. As it travels through this space it picks up heat. Upon reaching one end of a collector module, it enters the volume between the inner and middle tubes where it reverses flow direction and continues to build up heat content. It is then drawn off by the return tube and circulated into the heating system.

An evacuated-tube system has the advantage of higher collection efficiency at standard operating temperature and the utilization of high temperatures in its collection process without excessive heat loss. The vacuum between the outer tubes of glass helps attenuate conductive and convective heat loss, but it can do nothing for radiation-heat loss. This latter loss can be decreased by applying a "selective surface" interior coating to the outer tube.

Parabolic Collectors

The paraboloidal collector concentrates large amounts of solar energy on a small area. This concentration allows high temperatures to be attained. One use for these temperatures is in the absorption-cooling system used in air-conditioning. For effective collection, the parabolic concentrator must track the sun so that the sun's rays are perpendicular to the frontal plane of the paraboloid. In the altazimuth system, the unit moves vertically to match the sun's altitude and tracks horizontally to follow the solar azimuth during the day. Since the sun is in different positions every day, the tracking mechanism must be very sensitive and exact. This is likely to limit such devices to use in large buildings, where capital costs for mechanical systems may be offset by large fuel savings.

Another system to concentrate the solar radiation has a parabolic trough oriented along an east-west axis. The angle of the trough with respect to the horizontal is adjusted weekly to accommodate changes in solar altitude. The reflective trough concentrates energy on an absorber tube through which a heat-absorbing fluid is circulated.

The paraboloidal surface of the collector

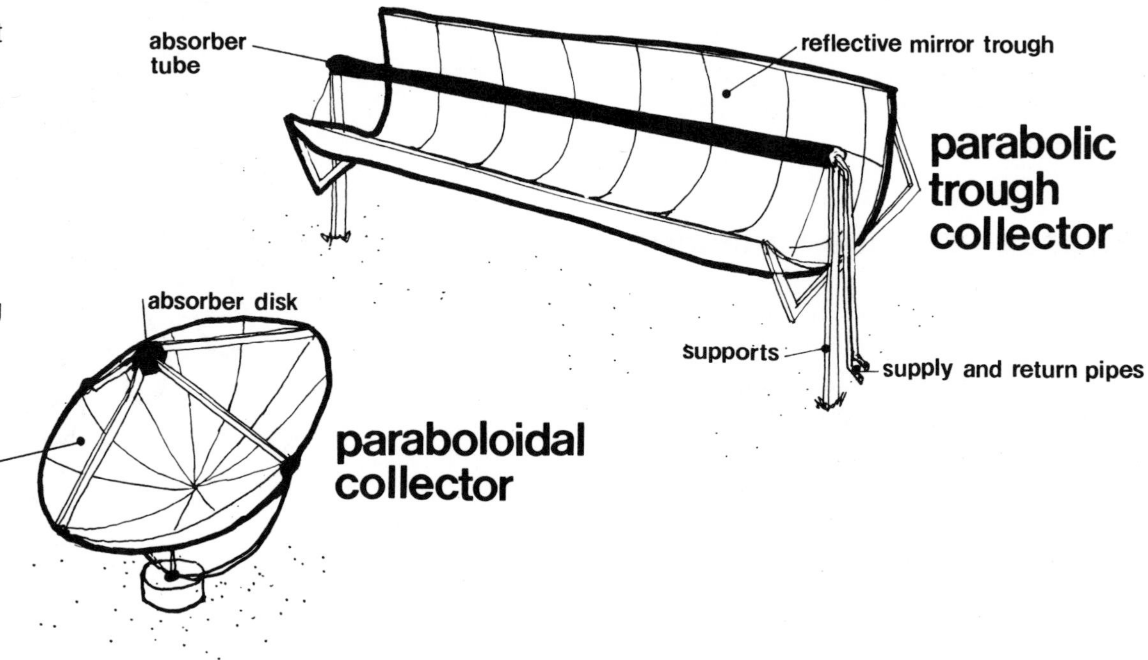

is mirrored and reflective. Radiant energy is reflected onto an absorber area, which is usually glass covered to minimize heat loss by convection and radiation. The absorber has a liquid circulating inside it to carry away the heat energy. It is essential to keep reflective surfaces very clean to avoid reflective loss.

SRTA Collector

The major drawbacks to the parabolic collector are the high costs of optically precise surfaces, the need for expensive tracking mechanisms, and the difficulty in maintaining a clean optical surface. The SRTA (stationary-reflector tracking absorber) collector has been developed to eliminate the need for tracking the sun with the mirror reflector. With this collector the hemispherical reflector is stationary and the absorber tracks the sun. The light energy is reflected off the mirror and onto the absorber.

The SRTA absorber is made up of tubing covered with glass. Water or other fluid circulates in the tubing to transfer the heat. Whatever the fluid, it must nearly always be confined under high pressure to prevent boiling.

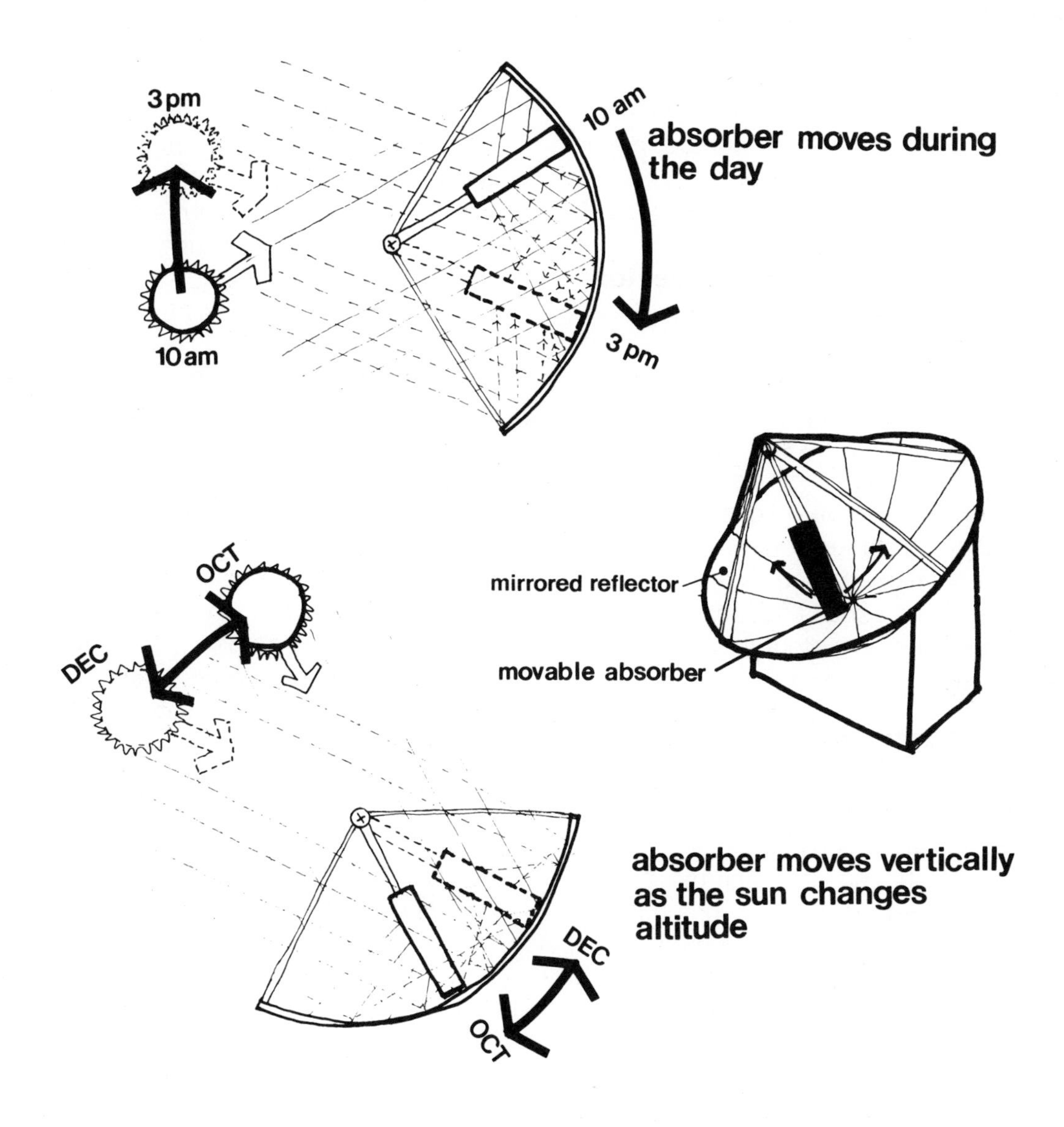

Air-Type Collector System

An air-type flat-plate collector system can operate in several modes, some of which may be in operation simultaneously:

Heating with the Solar Collector

Air in the return duct from a room is routed to the base of the collector. It is then passed through a manifold to distribute the air evenly in the collector. The air is warmed by the sun in the flat-plate collector (see preceding section on collectors), then delivered directly to supply ducts and distributed in rooms. When any room has attained its desired temperature, excess heat energy is transported to the storage system.

Charging the Heat Storage

The heat captured by a flat-plate air-collector system is stored in a rock bin. The rock bin is sealed in an insulated container that has inlet and outlet connections. When the rooms of a building are sufficiently warm and there is still adequate sunshine for the collectors to operate, heat can be stored for use at a later time.

To increase the amount of heat stored in the rocks, air is drawn out of the rock bed, routed through the collector, and then returned to the rock bed. Cold air leaves the rock bed near the bottom, and hot air enters near the top. As warm air enters the top of the rock bed, it transfers or loses its heat to the rocks. The warm air flows down through the rock bed until it is drawn out and returned to the collector. The rocks gain heat each time newly warmed air is passed over them. Because each rock in the bin has a very large surface area (smooth, graded pebbles of 2.5 to 5 centimeters [1 to 2 inches] in diameter are used), they are capable of removing heat from flowing air at an extremely fast rate. Air entering such a bin at a temperature 20° C (68 F) above that of the bin, and at low velocity (less than a meter per second), will lose most of its temperature difference and reach thermal equilibrium within the first 30-centimeters (1 foot)-thick layer of rock. This means that full room-heating temperature may be obtained from a rock bin that is depleted to its last 30-centimeters-thick rock layer. The storage process continues until each horizontal 30-centimeters "layer" of rock has reached the incoming air temperature and the bin has been fully charged. When the sun has set or has been obscured by clouds, and the house requires heat, air is drawn from the hottest air layers at the top of the rock bin.

In contrast to the rock-heat storage system, the entire volume of liquid in a liquid-heat storage system must be maintained near its use temperature for it to be effective. Such performance differences, plus the difference between the sizes of ducts and pipes and the power needed by fans and pumps, would lead to a preference for heat storage in a solid medium rather than a liquid for most building applications involving solar collection in hot air. Moreover, air-to-liquid heat exchange is less efficient than air-to-air, liquid-to-air, or liquid-to-liquid.

Heating with Rock Storage

Air drawn out of a room via the return duct and then routed to the storage bin (which was previously solar heated) increases in temperature as it passes through the rocks and is then supplied back to the room. As the air is drawn through the rock bed (from bottom to top since heat will stratify in rocks), it gains the heat that the rocks have stored. This heat is then supplied to the room.

Hot Water Preheat

In most climates domestic hot water for homes and other buildings can be preheated using solar energy year-round. Warmed air from the collector can be drawn across a heat exchanger to warm the water to a temperature requiring little or no fuel to reach final-use temperature. The warmer the air from the solar collector, the higher the preheated water temperature. In the summer, when solar collectors are unnecessary for room heating, they may be fully applied to domestic water heating and in most cases will raise water to final-use temperature.

Auxiliary Heating

An auxiliary heating system is required during times when the solar system cannot supply all the heating requirements of a building. This can be during periods of limited sunshine or abnormally cold weather. If the supply air to a room from the collector or storage system is not warm enough, an auxiliary heating system should automatically activate. Most storage systems are designed to meet a normal midwinter demand for about half a day (essentially overnight).

Summer Venting

In the summer when warm air from the collector is not needed to heat a building, the system should be vented. This is usually done automatically. The warm air near the ceiling of a room is drawn out and supplied to the solar collector where it is heated further. It can pass over the hot-water preheat heat exchanger and be routed out of the building. The warm air drawn from the building is replaced by cool air from the outside that enters through dampered vents.

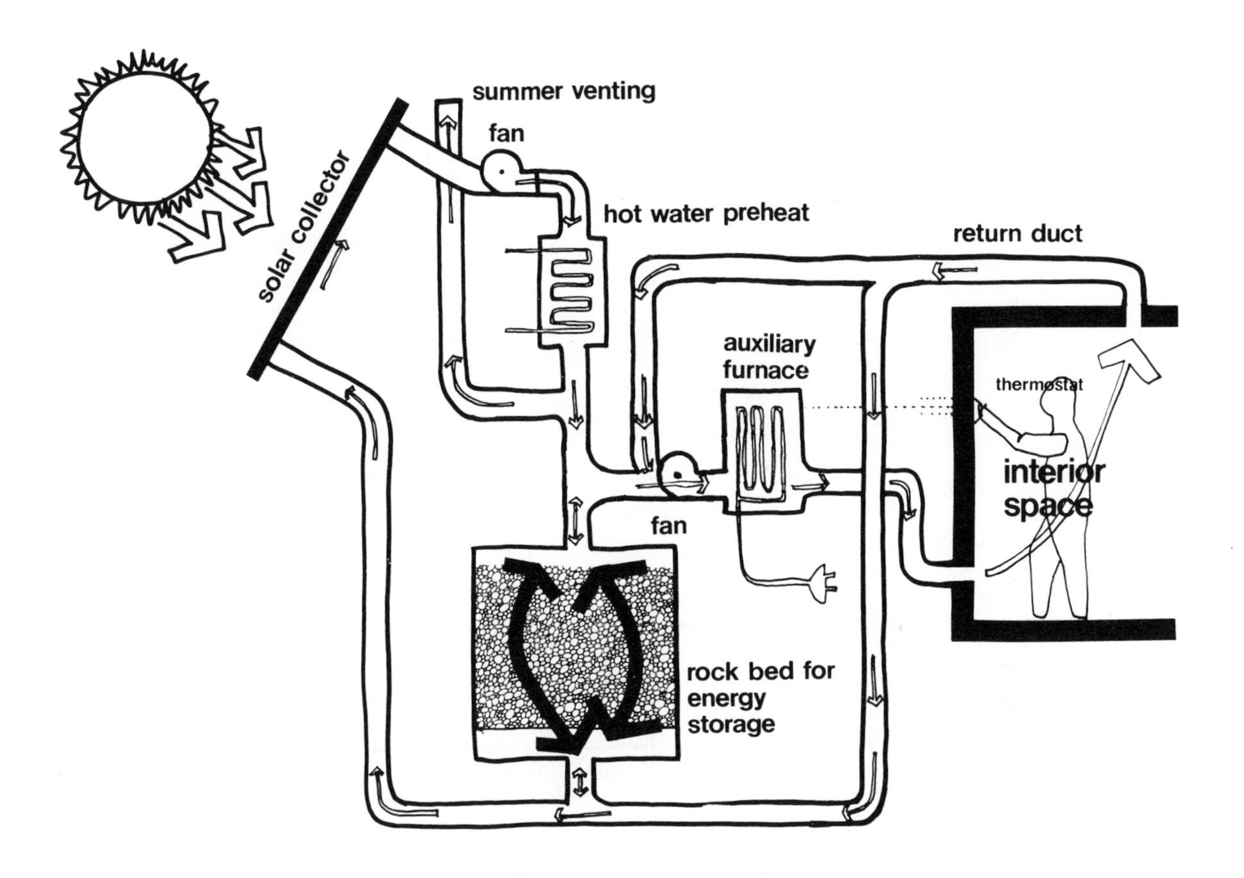

air-type collector system

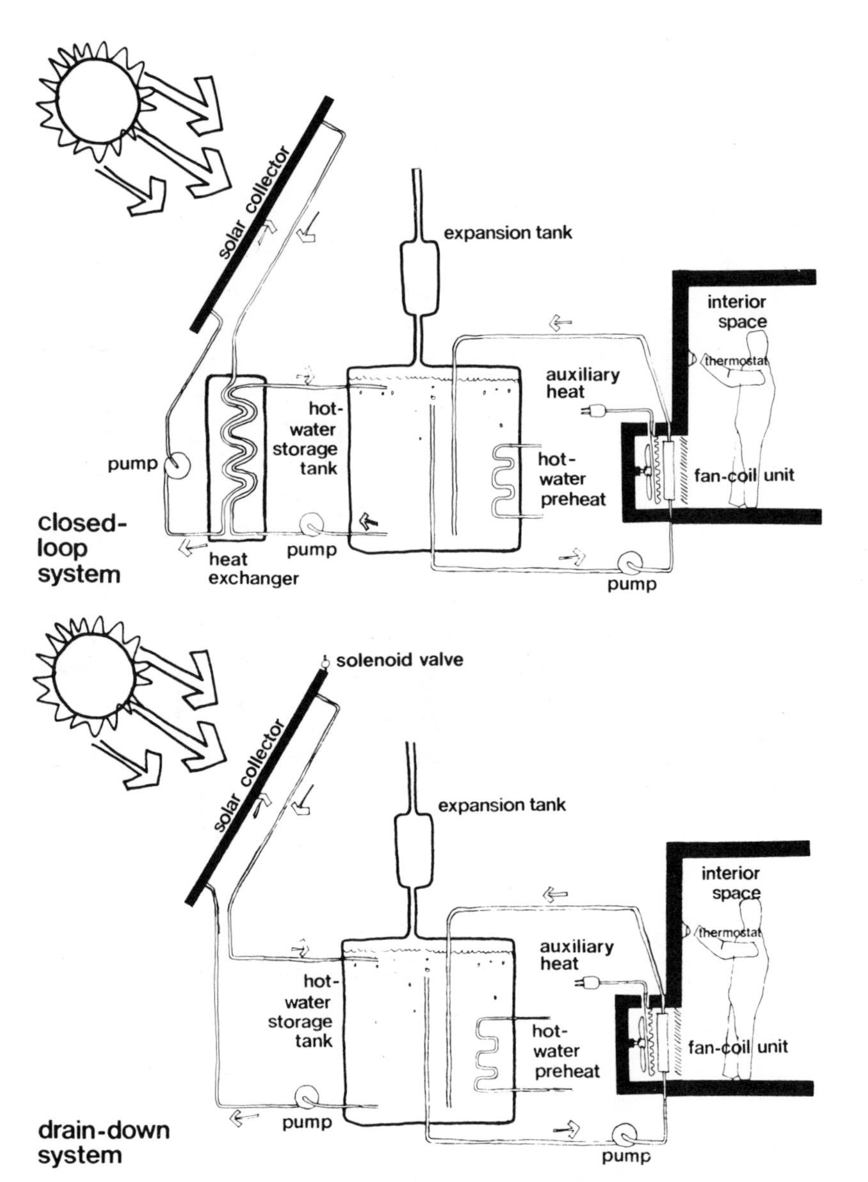

liquid type collector systems

Liquid-Type Collector System

A liquid-type flat-plate collector system can work in several different modes:

Collecting Heat

A liquid (antifreeze solution or water) is pumped to the base of the collector (see section on collectors), from which it is distributed uniformly behind the absorber plate. The temperature of the liquid increases as it takes on heat from the absorber.

Storing the Heat

Depending on the type of system, one of two routes is taken by the heated liquid. If the liquid is an antifreeze solution, it travels to a heat exchanger where its heat is delivered to the water in the storage tank. If the liquid is water, it is routed directly to the storage tank, where it mingles with the heated water already there. Both these methods raise the temperature of the entire storage tank, since convective currents in the tank will tend to bring the entire tank into thermal equilibrium.

Using Storage for Direct Room Heating

When necessary, the building may be heated using only the heat in the storage tank. Water is pumped out of the tank and circulated through a fan-coil unit or other heating device where it loses much of its heat to a room. The cooled water is then routed back to the storage tank.

Auxiliary Heat

When the temperature of the collector (because of limited sun conditions) or the temperature of the storage is too low for space heating, the system should automatically activate an auxiliary heat supply. Depending on the type of system, this may heat the water before it enters the fan-coil unit, or it may be a forced-air system.

It is necessary to have an "expansion" tank located near the heat-storage tank to protect the main tank from any damage that could be caused as a consequence of hot water having a higher specific volume than cold water.

Preheating Domestic Water

By using a heat exchanger, the stored heat can be used to preheat domestic water. In the summer this system can supply almost 100 percent of the domestic hot water. Another excellent use of this energy when it is not required for space heating is to heat swimming pools.

Photovoltaic Collector System

The photovoltaic collector system converts radiant energy directly into electricity. Photovoltaics have been developed mainly for space application where size, weight, and independence of all systems are more critical than their cost. For all other applications, they have the disadvantage of being extremely expensive. New techniques for growing the crystals needed for photovoltaic cells and recent advances in thin-film technology should decrease their cost, as should fully automated assembly systems.

In some photovoltaic systems the photocells are mounted on a surface with a water-type solar collector behind them. The water or fluid helps to keep down the temperatures of the cells, while functioning as a conventional heat-collection medium. The output of these cells increases as temperature rises to about 82°C (180 F) and then drops rapidly.

Another photovoltaic-collector concept under development by the Solar Energy Research Institute uses a luminescent plate for solar concentration. In this configuration, organic or inorganic luminescent materials dissolved in a flat, polished transparent plate absorb photons

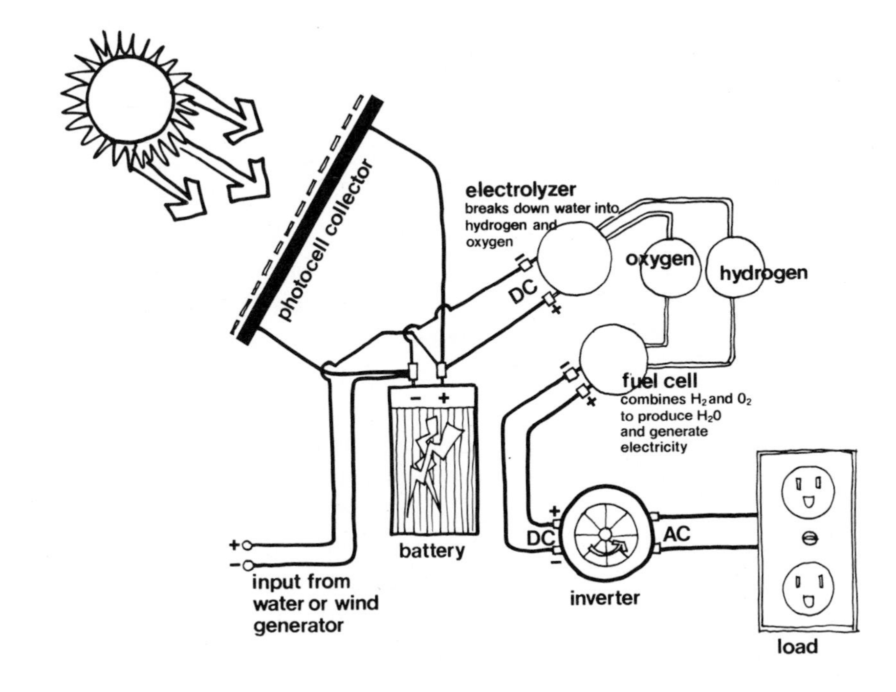

and re-emit them in a narrow spectral band. The emitted photons are trapped within the plate by total internal reflection until reaching an edge where they are converted into electricity by an appropriate photovoltaic (PV) cell.

The luminescent solar concentrator could reduce heat dissipation problems, improve diffuse-light-collection efficiency, optimize spectral response of PV cells (through selective reemission of photons), and eliminate the need for sun tracking to achieve effective concentration ratios. Ultimately, luminescent concentrator windows in a building could provide electric power without the expense of roof-mounted collectors.

Photovoltaic cells generate DC (direct current) electricity (the electricity in homes is AC, alternating current), which can be stored in batteries or fuel cells.

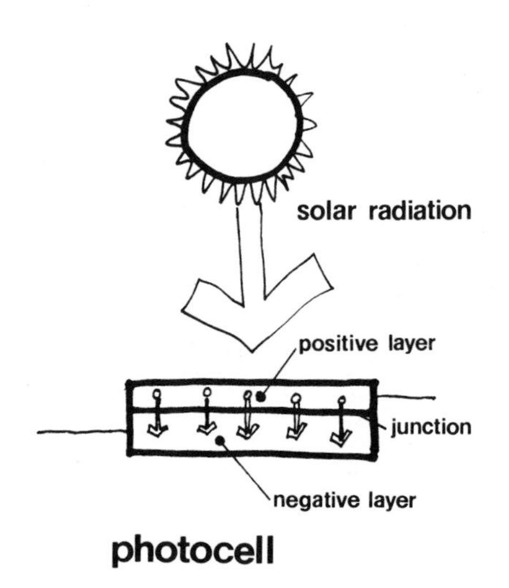

photocell

Wet-cell batteries store electricity as chemical potential energy by charging particles and plates in a solution; they can be recharged many times, but most have a life expectancy of about three years. Photovoltaic systems can operate at different voltage levels and may provide AC power through batteries using inverters.

Silicon for photovoltaic cells is "grown" in large, single crystals; then wafer-thin strips are carefully sliced from the crystal to form the basic unit of the cell. The silicon wafer is then coated with another material, such as boron, to produce a positive electrical layer that interacts with the negatively charged silicon layer. The interface between the two layers forms the p-n (positive-negative) junction.

When photons, the basic energy particles of light, strike the silicon cell, they are converted to electrons in the p-n junction. The positive layer accepts the negatively charged electrons, and the negative layer rejects them, producing direct-current (DC) electricity.

In fuel cells hydrogen and oxygen are combined to form water and electricity. In the opposite process, electrolysis, water is separated into hydrogen and oxygen by using an electric current. This results in energy storage in the hydrogen and oxygen components. Fuel cells produce direct current (DC), which can be converted to alternating current (AC) by an inverter.

An auxiliary generating system, such as a windmill or watermill, can be added to a photovoltaic or fuel-cell system since they both generate electricity.

Eutectic Salts

Latent heat is absorbed or released when a substance changes phase or state. This is the principle that is exploited when eutectic salts are used for energy storage. Compared to most other materials, they will store a large amount of heat in a relatively small volume. The melting and boiling temperatures of these salts are separated by a wide range. Once they have melted they can reach very high temperatures before vaporizing.

The salt starts out as a solid. As heat is added to it, its temperature rises until it reaches the temperature at which addi-

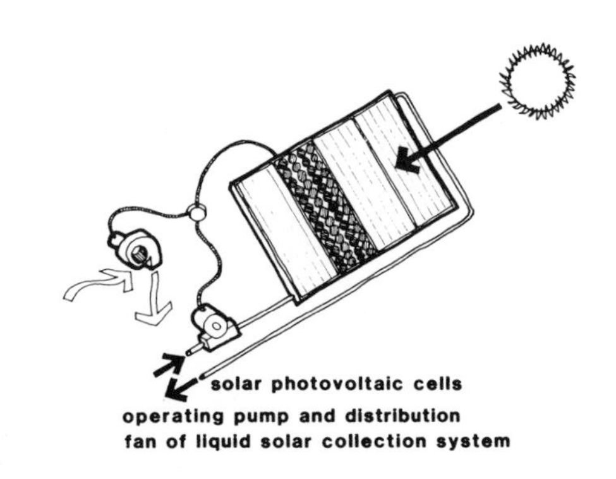

operating pump and distribution fan of liquid solar collection system

tional heat causes it to become a liquid. A large amount of heat is stored in this transition. When heat is removed from the liquid salt, it solidifies, giving off the latent heat that was stored in it.

Heat from a solar collector can be stored as latent (phase change) heat in eutectic salts. When heat is required for space heating or other purposes and it is not available directly from the collector, it is retrieved from storage causing the salt to reform as a solid. Theoretically, the salt can cycle from a solid to a liquid and back an infinite number of times, taking on or giving off latent heat at each phase change.

There are many problems encountered with eutectic salts that have kept them from gaining widespread acceptance as a heat-storage material. Many are incompatible with metal containers and cause corrosion and deterioration of the metal. Others are toxic and must be handled with extreme care. The most severe problem with eutectic salts is their short-lived reversibility. Many will stop cycling from solid to liquid and back after many uses, and some will form solids that become permanently deposited on the bottom of the storage tank. Some eutectic salts react violently when exposed to air or moisture. Many advances are currently being made in their development to eliminate the problems that prevent them from being universally used as a heat-storage material.

Solar-Powered Cooling

In most areas of the United States solar energy can also be used to cool buildings. It can be used directly to run an absorption-refrigeration system or a Rankine-cycle cooling system. In areas where there is a significant difference between wet-bulb and dry-bulb temperatures, the sun can be used in passive systems to ventilate, cool, and temper the air. This capability will be discussed in subsequent chapters.

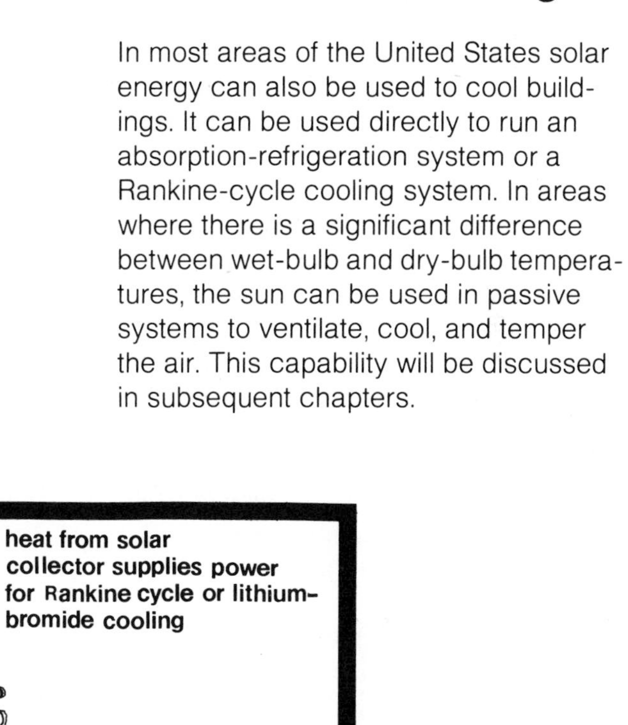

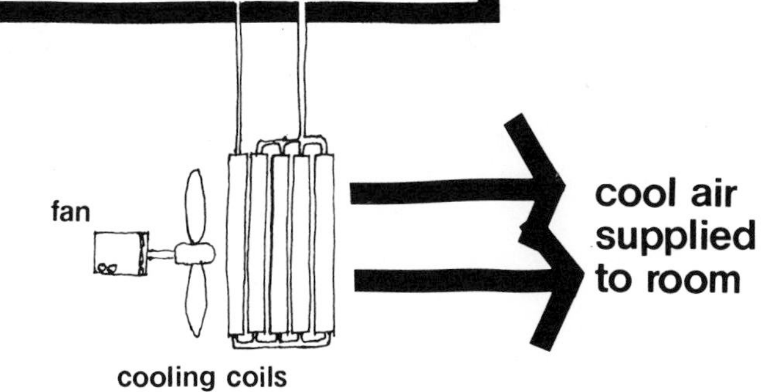

Heat pumps requiring a small amount of electricity can be used to heat (solar assisted) and cool buildings. Under certain conditions the amount of electricity they require is less than the amount required to run conventional air-conditioning units. For heating they require about half the amount of electricity that resistance heating for the same output requires. The coefficient of performance (COP) of a heat pump operating in the heating mode is the ratio of the useful heat delivered to the building over the heat equivalent of the energy input to the heat pump. Even without solar assistance, some heat-pump systems attain a COP of two or more, using earth, water, or air as low-temperature heat sources.

Absorption systems that use heat from a solar collector to run their generators require much equipment and piping and involve a large initial investment. Systems using the Rankine cycle for cooling use solar energy to power a turbine that operates a compression-refrigeration unit in the system. Both of these systems can be used to remove heat from the air. The cooled air is then distributed throughout a building.

Solar Pond

Another device for the collection of solar radiation is a solar pond. Basically, this is a shallow body of water that is itself an energy absorber and can be fitted with a cover plate if desired. The temperature of the water is increased by the direct absorption of photons or by conduction from the bottom of the pond. The amount of energy absorbed by the bottom is increased if it is lined with black plastic or some other black material. As the temperature of the water increases, the water expands, decreasing its density. The hottest water is the least dense and normally rises to the surface. Once in this location the heat is readily lost to the outside air, by conduction and convection, making this an inefficient collector. In most ponds the cover plate, fabricated from clear plastic, is used to keep it free of dirt, leaves, and other debris and is not a barrier to heat.

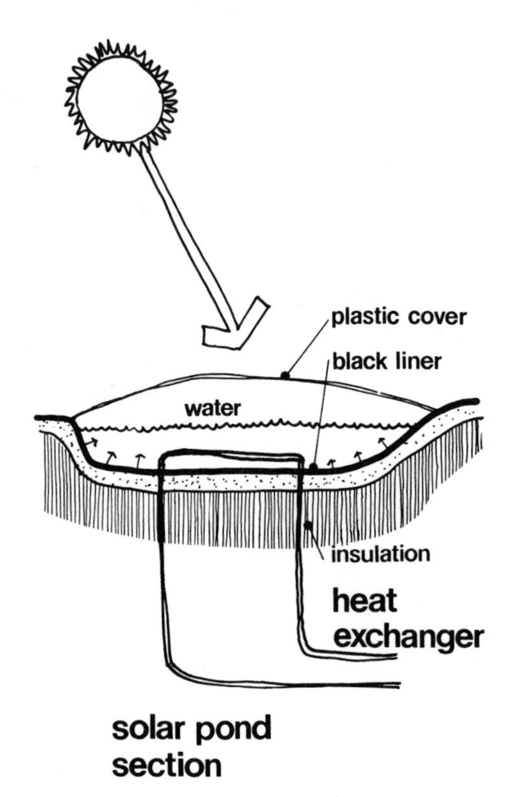

solar pond section

Another method, relying on chemical properties, is used to reduce heat loss. This produces the class of ponds known as saline-gradient solar ponds. Their successful operation depends on two properties: The first is that the amount of salts that can be dissolved in water increases rapidly with rising temperature; the hotter the water gets, the greater the amount of salt that can be dissolved in it. The second property is that as the concentration of salt increases, the weight per unit volume, density, increases.

For water the increase in concentration of certain salts held in solution helps to overcome the reduction in density caused by expansion. It is possible to obtain a stable situation in which the hottest fluid is also the densest and is located at the bottom of the pond, producing gradients in both temperature and salt concentration, with decreasing magnitudes of each toward the surface. Salt will tend to diffuse from the highly concentrated zone at the bottom to the zone of lower concentration at the top. Periodically, salt must be removed from the surface and reintroduced at the bottom if the gradients of both salinity and temperature are to be maintained.

The advantage of these ponds is that because the saline gradient induces the hottest water toward the bottom, convective energy losses are reduced, increasing the amount of heat that can be extracted from the pond.

For small ponds, insulation is needed to decrease the amount of heat lost to the ground. In many designs the greatest losses would occur through this region.

Heat exchangers, located at the bottom of the pond, are used to extract the heat, which can be used to heat interior spaces or domestic water, as well as to power absorption-cooling equipment. In experimental ponds in Israel, equilibrium temperatures have been found that approach 100° C (212 F).

Certain problems are encountered with solar ponds that decrease their efficiency and, hence, their desirability. In the winter, snow and ice may cover the pond, blocking incident radiation and reducing water temperature. Even if covered, ponds need to be cleaned periodically to decrease their opacity. The cleaning process disturbs the thermal and saline gradients, which take time to reestablish themselves.

Thermosiphon Collector System

Probably the simplest solar-water heaters are ones that operate on the thermosiphon principle. These consist of an insulated storage tank, a water-type flat-plate solar collector, and various pieces of insulated pipe. The bottom of the storage tank is situated at least 30 centimeters (1 foot) above the top of the collector and must be high enough above it that cold water will flow by gravity from

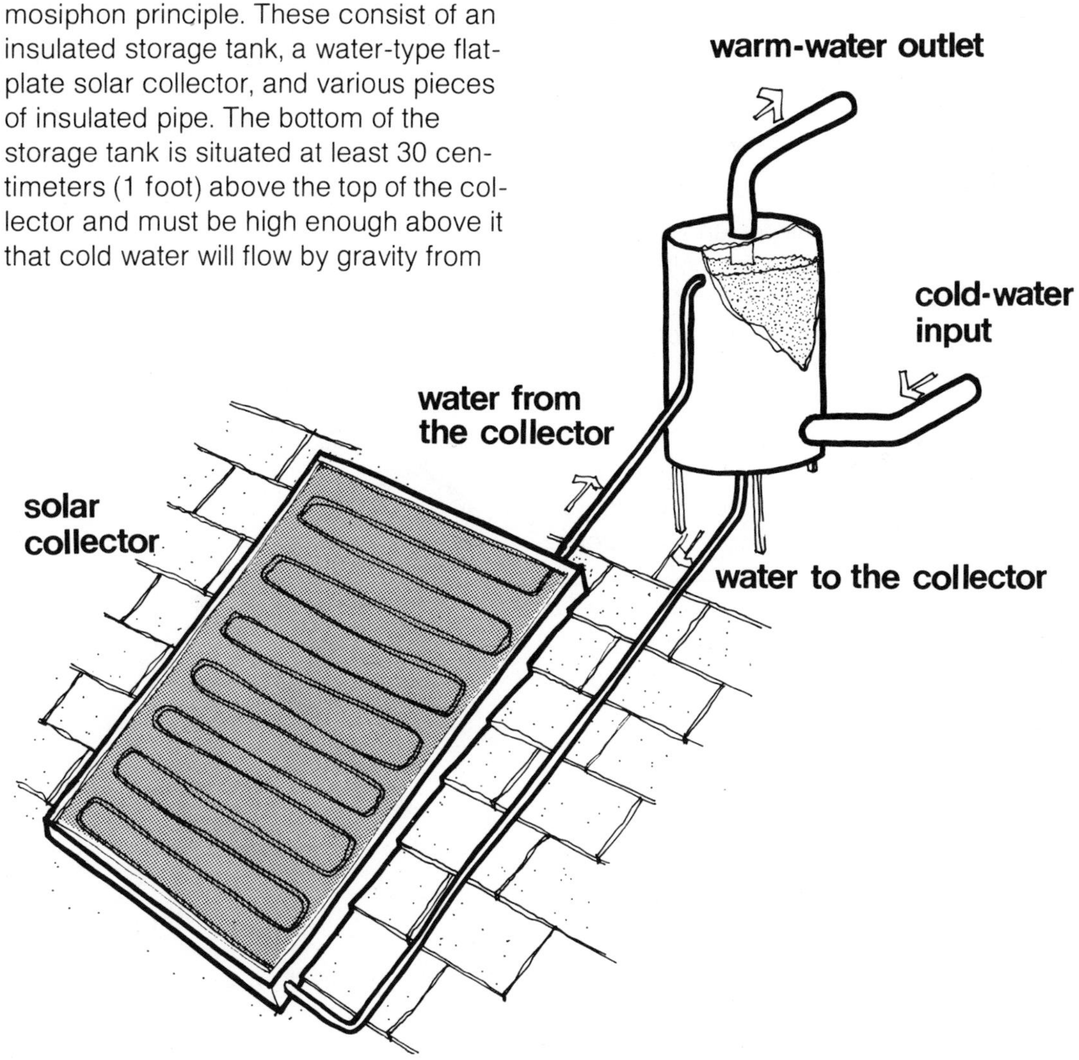

warm-water outlet
cold-water input
water from the collector
solar collector
water to the collector

storage to the collector base. Water is heated in the collector, and because hot water has a higher specific volume than cold water, it will rise to the top of the collector and back to the upper portion of the storage tank. As the hot water flows up the collector and back to the top of the tank, cold water is drawn from the bottom, setting up a continuous circulation of water. Valving is needed to prevent reverse convection at night.

This action will continue as long as there is solar radiation on the collector. The temperature of the water in the upper part of the tank will vary from approximately 74° C (165 F) on hot summer days to about 46° C (115 F) on cold winter days, unless it is too cold for the system to be operating. This type of system is well suited to preheating water before it enters a conventional hot-water heater.

The storage tank, feed and return pipes, outlet pipe, and the back of the collector must be well insulated. In most warm climates one pane of glass covering the collector is sufficient, but for colder climates two sheets will be necessary. For the best flow, the cold-water feed pipe to the collector should be continuously sloped from the storage tank to the base of the collector, avoiding horizontal sections.

Solar Domestic Water Heating

One of the fastest-growing uses for solar energy is the heating of domestic hot water. Most collectors for home heating can accommodate the addition of a hot-water preheat cycle to raise the temperature of incoming city water to a higher temperature before it goes to the hot-water heater. This preheat leaves the hot-water heater with a smaller difference to make up between the temperature of the incoming water and of the outgoing "hot" water, conserving "make-up" heat.

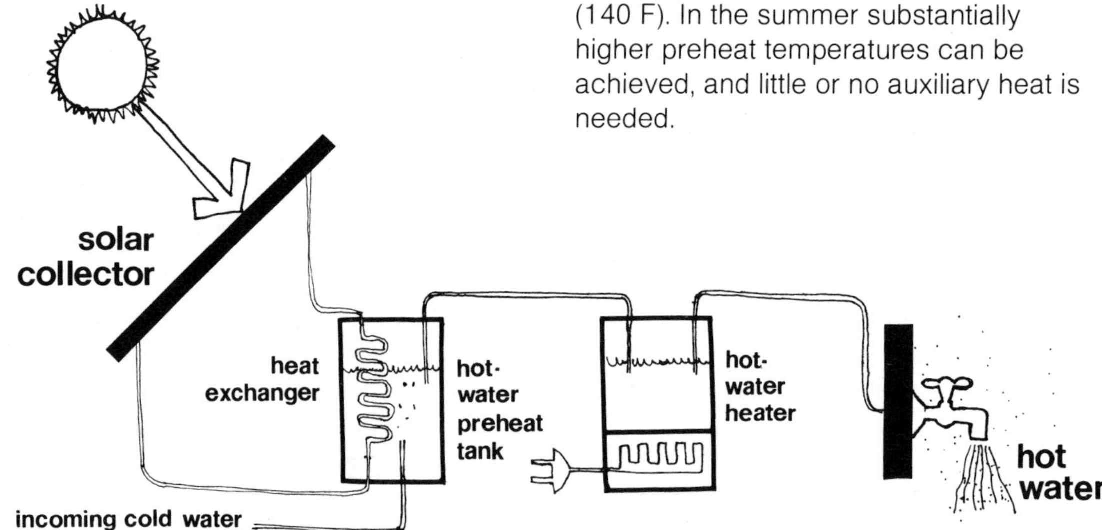

If incoming city water is at 4° C (39 F), and the hot-water-use temperature desired is 60° C (140 F), then the hot-water heater must raise the water temperature 56° C (101 F). Since each pound of water requires one Btu to raise its temperature one degree Fahrenheit, 101 Btu per pound are needed to raise the water temperature to 60° C (140 F). The addition of a solar-powered hot-water preheat unit to the system could raise the temperature of the water to 38° C (100 F) before it enters the hot-water heater. With the incoming water to the hot-water heater at 38° C (100 F) and the outgoing still 60° C (140 F), the temperature makeup is only 22° C (40 F) and will require only 40 Btu per pound of water to raise the temperature to 60° C (140 F). In the summer substantially higher preheat temperatures can be achieved, and little or no auxiliary heat is needed.

In areas with moderate temperatures all year long, such as southern California and Florida, solar heating of homes is rarely necessary, but solar heat can be economically used for heating of domestic water. The largest number of manufacturers and users of solar hot-water heaters are found in these geographic areas. The size of a solar collector used for water heating only is much smaller than one used for both space heating and domestic water heating. Solar water heaters can also be used to heat swimming pools and whirlpool baths. The collection system for hot-water heating is similar to the liquid-type flat-plate collector system, and many operate on the thermosiphon principle.

The size of any storage tank and collector system depends on its initial cost, the rate of hot-water use, the temperature to be maintained, and period of time during which there occurs minimum sunshine. Some collectors incorporate their own storage systems, recirculating and reheating the water in a closed loop. The collector can range from simple black-plastic bags on the roof to moderately complicated collectors with automatic temperature controls and heat exchangers to maximize the energy use. In climates where freezing occurs, self-draining collectors with insulated tanks are necessary to avoid damage to the system.

Solar Water Distillation

Solar-powered distillation (purification and demineralization of water) and desalination (removal of salt from water) projects are possible using the sun's energy to evaporate and purify water supplies. One simple system has a glass cover over the water supply. The cover is sloped toward a collection point or pipe. Usually the surface below the water is colored black to absorb the maximum possible incoming radiation. As the space enclosed by the glass heats up, the water temperature increases and its surface evaporates, leaving behind the minerals and salts that were dissolved. The water will condense on the lower side of the glass cover, which, because of its exposure to outside air, is cooler than the heated chamber. As the water droplets form, they will merge together and then flow to the lower end of the cover and be collected as distilled water, which may then be bottled or otherwise distributed. The minerals and salts deposited on the bottom of the still must be cleaned away periodically.

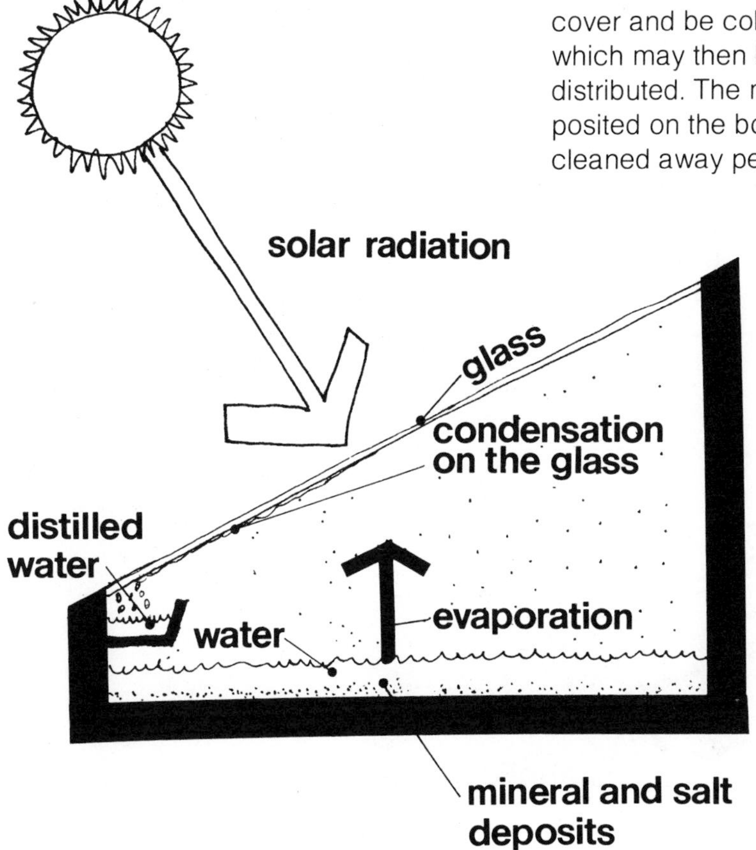

Large-Scale Solar Collection

Solar collection can take place on a large scale to generate electricity or provide steam to heat a district. To generate electricity high temperatures are required. These can be derived using collectors with reflectors and concentrators.

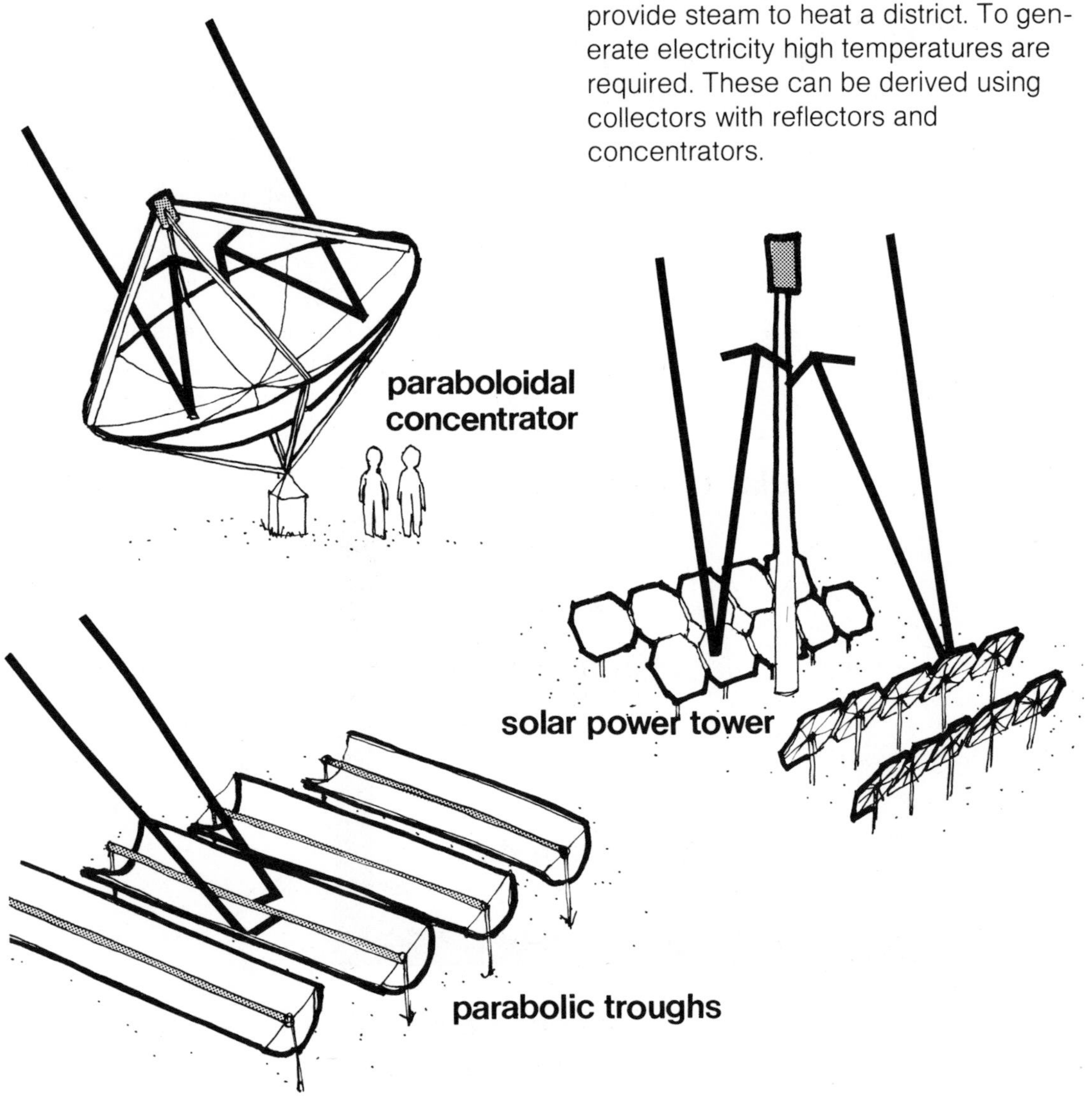

paraboloidal concentrator

solar power tower

parabolic troughs

Listed below are three proposed schemes for using solar energy on a large scale:

Use an array of parabolic reflectors to concentrate solar energy on small focal areas to produce high temperatures for steam production and turbine operation.

Focus the sun's rays on piping with an array of linear/parabolic trough reflectors. Water run through the piping will be turned to steam by the high temperatures. The steam under pressure can be transported over long distances and be used in large power generation plants.

The "power tower" arrangement uses a focal-point tower 75 to 100 meters (246 to 328 feet) high situated among several thousand flat mirrors that track and reflect the sun to the focal point of the tower. Steam is then generated in the tower.

These systems would convey steam to generators, which would then provide electricity.

Another proposed, yet currently cost-prohibitive, system for large-scale generation of electricity from solar radiation would use photovoltaic cells. Large-scale operation would work similarly to smaller home systems, but batteries would prove impractical in terms of cost and materials. The energy collected could be best utilized as mechanical potential energy, such as pumping water up to a higher level when energy use is low and then allowing it to flow to generate electricity for high-demand periods. Also possible, yet expensive, is energy storage by the electrolysis of water, producing hydrogen and oxygen and their subsequent recombination in a fuel cell.

To make solar-heat collection systems cost-effective in a building, energy conservation must be made a top priority. New designs and forms will result in a climate-responsive ethic, which may soon override our more familiar historic architectural motivations.

Typical urban and suburban houses are major energy wasters, and most cannot economically facilitate the integration of solar energy. Some manufacturers market solar systems on the basis that one can add these units to the exterior of a home without any additional changes to the building. This is uneconomic unless energy conservation is made a primary effort. Without the use of energy-conservation measures, any form of energy, whether solar or fossil fuel, is poorly and wastefully utilized.

4 Wind/Air

Wind and air movement are the results of the sun's warming of different layers of the atmosphere and parts of the earth's surface. As the sun strikes the earth, areas near the equator are heated more than areas near the poles. This causes equatorial air to rise to upper levels of the troposphere (the atmospheric zone closest to the earth's surface). Cold polar air moves to the equator to replace the air that has risen and sets up circulating planetary wind patterns. The rotation of the earth has an effect on these winds, in what are known as Coriolis forces (after the French civil engineer who identified them in 1843): Cooler air moving along the earth's surface toward the equator is diverted toward the west, while the warmer air at the upper levels of the troposphere tends toward the east. These effects initiate large counterclockwise circulation of the air around low pressure areas in the northern hemisphere and clockwise circulation in the southern hemisphere. Topological features, day-night heating and cooling cycles, and man-made structures further complicate airflow patterns and generate microclimatic winds.

Localized wind movement and direction depend on the location of high and low pressure areas (equalization of pressure areas), the density of air (barometer readings), the amount of cloud cover (temperature differentials caused by shade or sun), topography (shape and contour of the land), land/water formations, and the position and tilt of the earth in respect to the sun. All of these factors affect wind in such a manner as to generate a much more complicated pattern than is described by the simplified cell diagram. It shows, however, that basic air-movement patterns are caused by the sun.

Man has utilized wind energy for many thousands of years but has never taken advantage of its full potential. Wind has only recently been reemployed to propel large ships, although its use for the propulsion of sailing vessels dates back to the Egyptians, circa 3000 BC. The development of the steam engine and other mechanical systems gradually supplanted the use of wind energy to propel large sailing vessels.

Wind can propel many types of vehicles; sails are even available for bicycles. The movement of the air exerts a pressure on any surface with which it comes in contact. With a constant wind speed the force acting perpendicular to a surface increases at the same rate as the area of the surface increases. The surface can be either stationary, such as the side of a building, which must transfer the force of the wind to the ground; or movable and be set in motion by the force of the wind. Wind will exert a pressure of approximately 2394 newtons/square meter (50 pounds/square foot) on a surface, at a wind speed of 240 kilometers/hour (150 miles/hour).

Wind energy may be harnessed for the movement of a sail or a blade. Many old windmills (for milling grain) in Europe used sails stretched over wooden frames to generate mechanical movement. The movement of the rotating sails was then transformed to useful work by wooden gears and shafts. Windmills are generally thought of as producing direct mechanical energy while wind generators and wind turbines produce electrical energy. The majority of the wind machines being developed today are for the generation of electricity, yet the function of most units currently operating is to pump water.

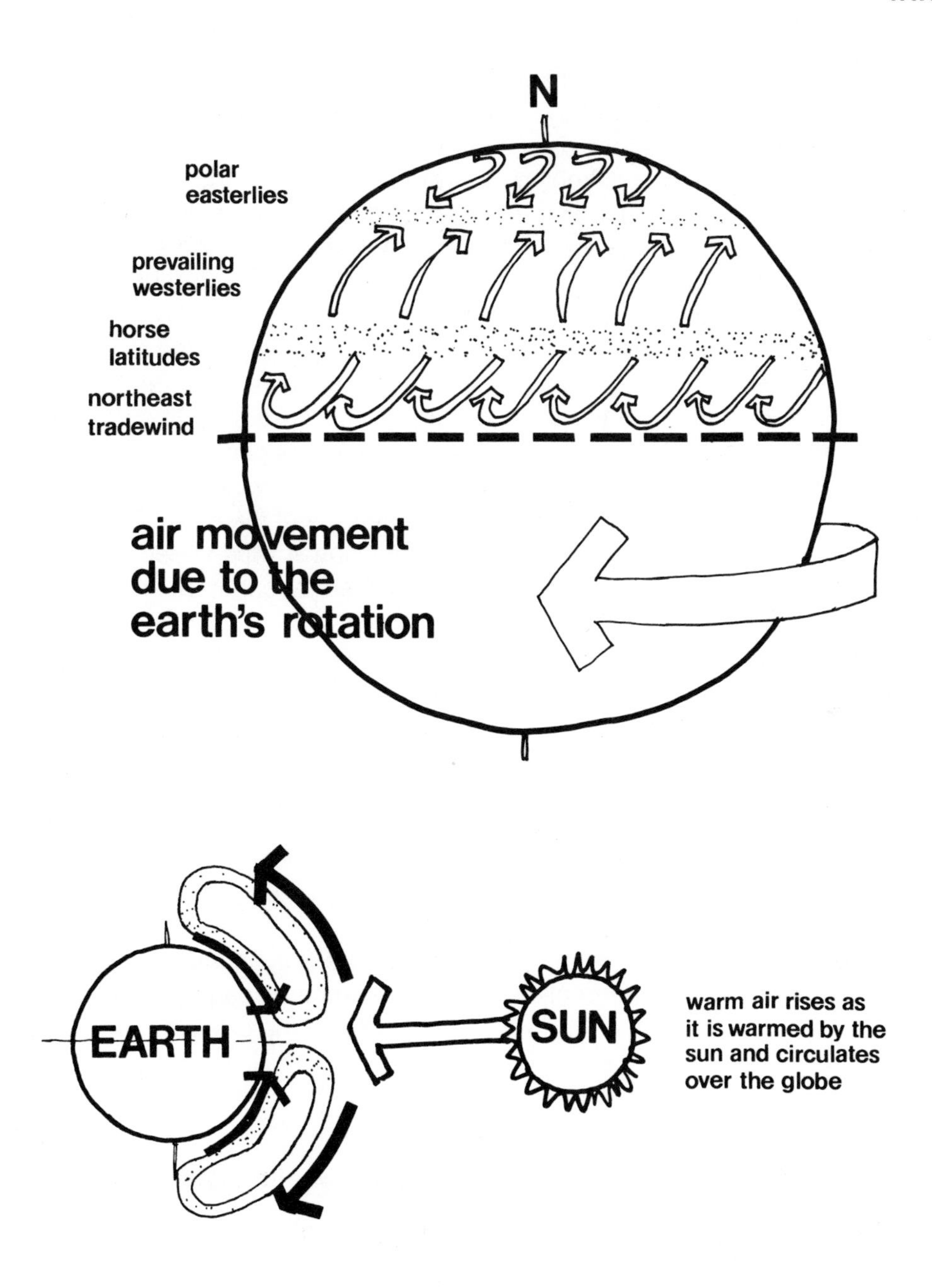

The theoretical maximum amount of kinetic energy that can be extracted by a wind machine from the windstream is only 59.26 percent. This factor is known as the Betz coefficient, first derived in 1927 by the German engineer A. Betz. His calculations assumed 100 percent efficiency within the wind-machine system itself. Since no machine can be perfect in construction and operation, and since energy is lost in electrical transmission lines, only about 40 percent of the wind's kinetic energy can be practically utilized in buildings.

The amount of available wind energy varies depending on the location, elevation, and orientation of a wind machine. Climatic data have been collected on the force of wind and its direction for many areas of the United States. For some locations the wind speed and direction are plotted on a polar-axis graph (graph with north, south, east, and west indicated) in terms of wind strength and frequency. The resulting plot is called a wind rose map, in reference to the shape most often generated; percentages indicate the portion of time (annually in this example) during which wind speeds reach the figures shown. Also helpful in determining the approximate amount of power that can be extracted from the wind are calculations for monthly average wind speeds and annual average velocity-duration curves.

The quality of wind also affects the amount of power that can be extracted from it, as illustrated. There are significant amounts of shear and compression in a horizontal windstream flowing over the surface of the earth. This shear results in lower wind speeds close to the earth's surface than at altitudes at which free flow occurs. Laminar flow is the movement of air in parallel layers over an even landscape or calm body of water. An interruption in topography causes turbulence in the wind pattern. A tree, a house, or a hill can cause multiple redirections of wind in a manner that would cause instabilities in a poorly located wind machine.

Therefore, it is usually suggested that wind machines be placed as high as possible off the ground to eliminate surface effects. A customary minimum distance is 10 meters (33 feet) from ground level to the bottom of a turbine or airfoil. In nearly all geographic locations, winds become laminar above this height, and their velocity increases sharply. (The power of the wind increases as the cube of the velocity.) For large-scale wind generators, heights may reach 65 meters (213 feet) or more.

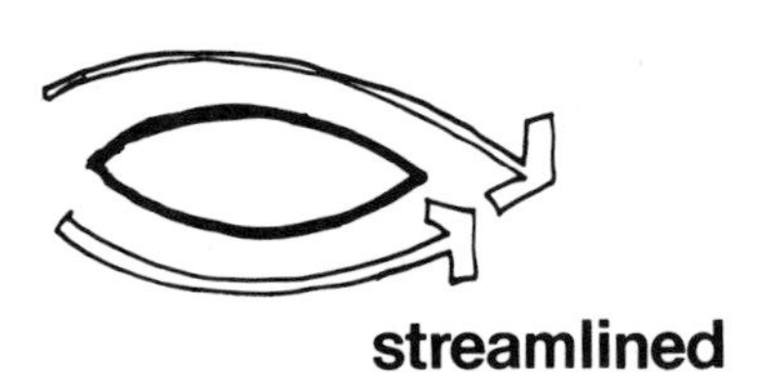

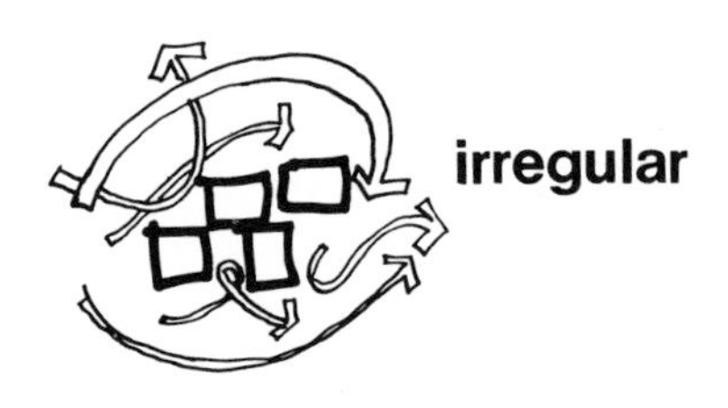

Great care must be taken in choosing a suitable site for a wind machine. It is possible to increase the average power output by siting a machine to take advantage of local topographical features. Flow is accelerated as it passes over a rounded hill or through a narrow valley. Because of local anomalies, it is worthwhile to make a detailed wind survey before choosing a site for a wind machine.

Wind machines convert the horizontal force of wind into rotary or oscillatory mechanical motion. Limitations on the mechanical efficiency as mentioned above restrict the amount of energy that can be removed from the wind. One important principle in the design of wind machines is the tip-speed ratio, the relationship between blade tip speed and actual wind velocity perpendicular to the plane of rotation. The higher the tip-speed ratio is, the more efficient the machine will be. Many old-style windmills produce tip-speed ratios of 1 to 2 and develop more torque (turning power) than is necessary for electrical

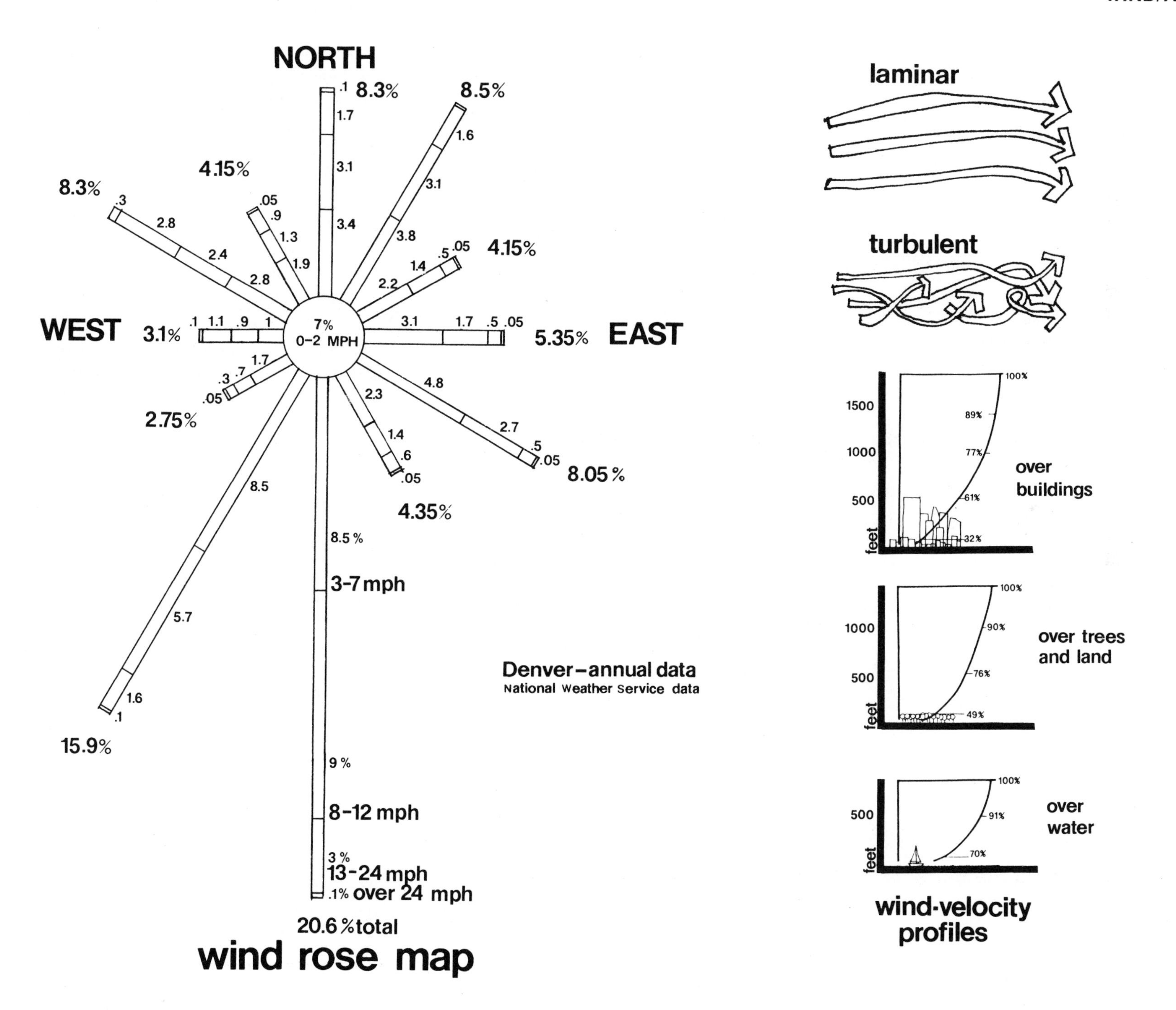

wind rose map

wind-velocity profiles

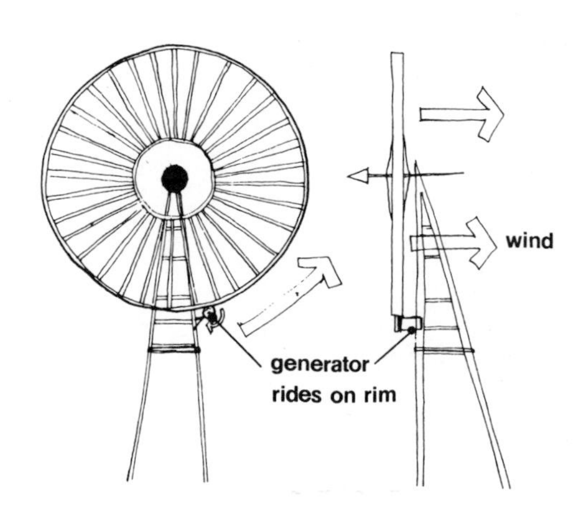

bicycle wheel turbines

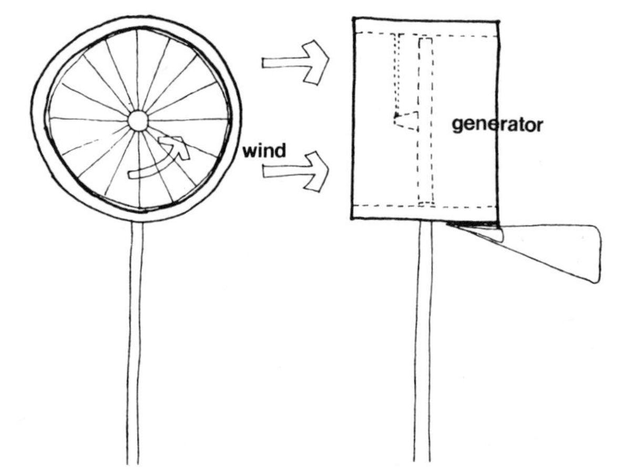

Gillette wind turbine

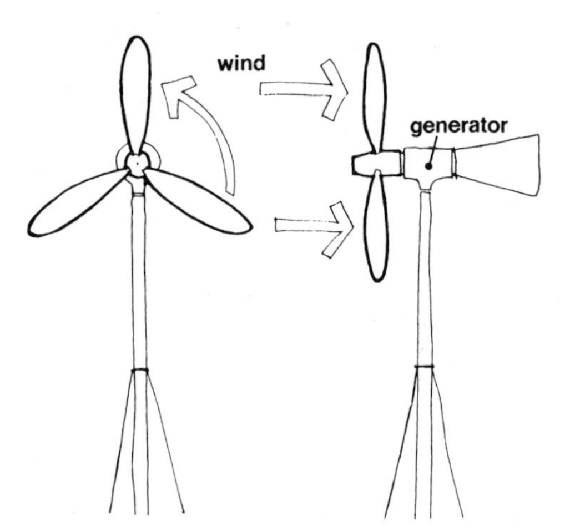

3-blade wind machine

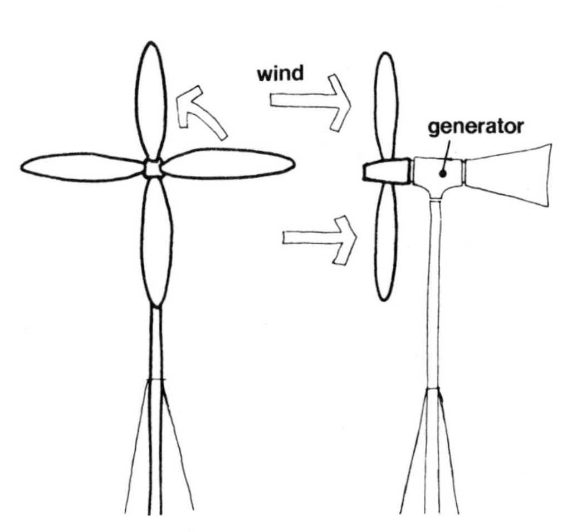

4-blade wind machine

generation. The most recently developed wind turbines and generators usually feature tip-speed ratios of 6 to 1 and 8 to 1, facilitating more efficient electrical generation than is possible with lower tip-speed ratios.

Wind machines produce useful power between certain minimum and maximum rotational speeds. Most machines need at least 12 to 16 kilometers/hour (7.5 to 10 miles/hour) wind to generate electricity. Older windmills can use wind speeds as slow as 3 to 4 kilometers/hour (1.87 to 2.5 miles/hour) to pump water. The top utilization speed of wind machines averages approximately 65 kilometers/hour (41 miles/hour). No additional useful power can be extracted, without adjustments, from winds above that speed. Some wind machines will let the turbine, blades, or rotors spin on a free-clutch system or centrifugal governor in high-velocity winds, whereas others have blades that can be mechanically modified to maintain a constant rotational speed in variable winds.

Much research has recently been done on airfoils, sailwings, hoops, rotors, single propellors, paired contrarotating propellors, and turbines to determine their suitability as part of a wind-machine system. Each of these mechanisms may be categorized as either a horizontal- or a vertical-axis device.

Horizontal-axis machines are typified by old windmills, in which the plane of rotation is, ideally, always perpendicular to the prevailing wind direction. These machines must be equipped with a mechanism to turn them when the wind direction changes. This can only be accomplished via the tracking process in which axis and blades all pivot into the wind. Because of tracking, some mechanical energy is lost. Horizontal-axis devices have nonetheless enjoyed popularity for centuries, since they require only a low wind speed for fairly successful operation.

Horizontal-axis wind machines vary greatly in their performance and design. Wind turbines consist of many blades that convert wind power into rotary motion. They can have an electrical generator connected to the central shaft or one positioned to ride the outside circumference of the wheel, as shown in the illustration. The speed of the central shaft is generally too slow to power a generator without gearing upward at some cost in efficiency. In contrast, generators operating on the outer circumference of the wheel can take advantage of the high velocity of the blade tips. Shown is a variety of horizontal-axis wind machines.

The rotor of a vertical-axis wind machine revolves in a horizontal plane. Because of this orientation the machine does not have to pivot to face into the wind. This feature is a distinct advantage on wind machines being used in areas where sudden changes in the wind direction occur, because high stability is exhibited and little "adjustment energy" is lost to any wind-direction or velocity change. Therefore, wind gusts would cause power surges rather than disruptions.

Vertical-axis machines have been used intermittently for centuries, but they have been less applicable to small installations than to large-scale milling operations and water pumping.

Many vertical-axis wind machines produce high torque at a low speed. This is useful for operating an irrigation pump, but it is too slow to power an electrical generator directly. Some wind machines, of both the vertical- and horizontal-axis varieties, use gearing mechanisms to convert high-torque, slow-speed rotary motion into low-torque, high-speed motion sufficient to power an electrical generator. Advancements have recently been made in the development of electrical generators that operate at low speeds.

The two most established types of vertical-axis wind machines are the Darrieus hoop rotor and the Savonius rotor (also known as the S-rotor).

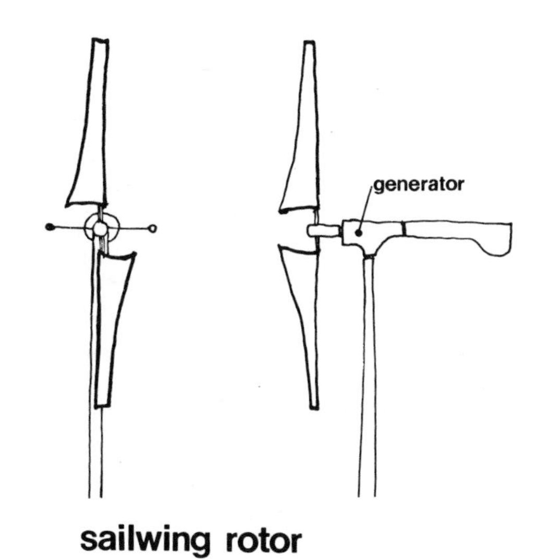

sailwing rotor

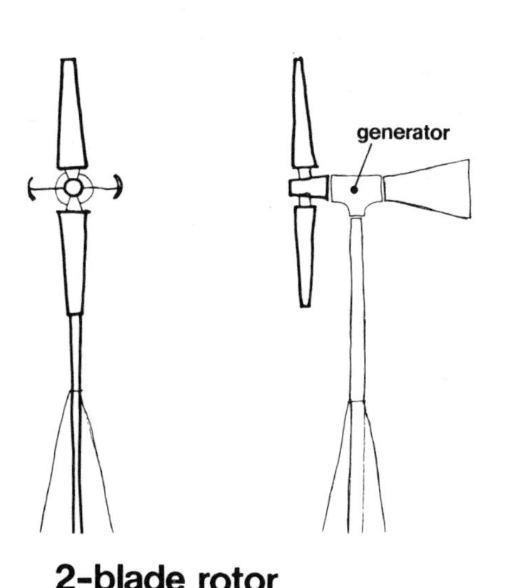

2-blade rotor

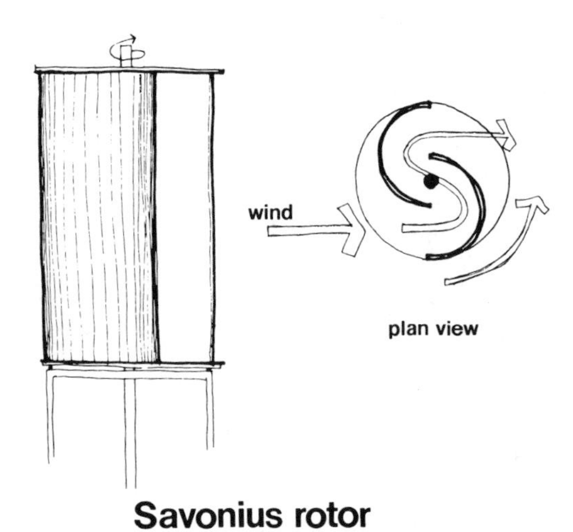

Savonius rotor

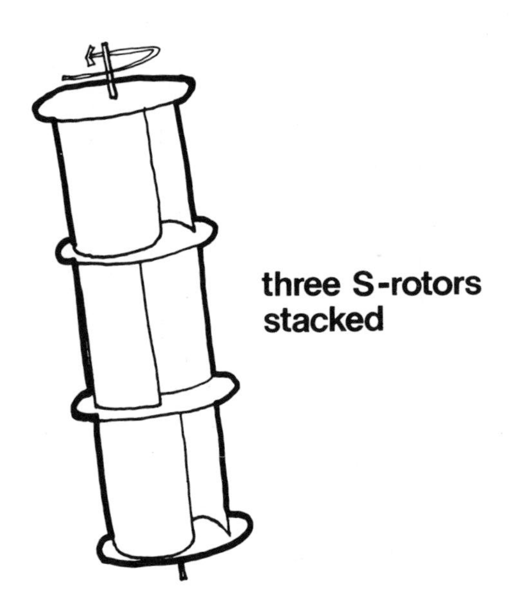

three S-rotors stacked

The Savonius rotor was patented in 1929 by Finnish engineer S. J. Savonius. This is predominantly a drag device, although it does provide some lift. The vanes of the rotor catch the wind, striking the back of the vane and exiting behind the rotor shaft. A single rotor can have more than two vanes, in which case the torque may become more constant and the axial balance more refined. Another configuration accomplishing this result is the stacking of several two-vane S-rotors at different angles along a central shaft (see illustration). The S-rotor operates at a maximum efficiency of 31 percent with a tip-speed ratio of 0.8 to 1.8. The S-rotor can be economically built from oil drums cut in half vertically and then welded between end caps to form scoops; these in turn are mounted on a central shaft, and the shaft is connected to a diaphragm pump or electrical generator.

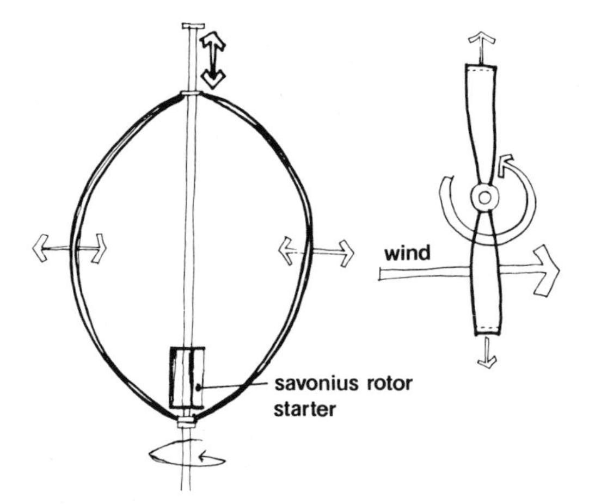

Darrieus hoop rotor

The Darrieus hoop rotor was patented in 1931 by the Frenchman G.J.M. Darrieus. Research has been conducted recently on the rotor by NASA in the United States and by the National Research Council of Canada. Both teams have found high-potential efficiency in this configuration, within a certain wind-speed range. The blades of the rotor are flexible and bend in response to the wind. Darrieus rotors are lift devices and are characterized by curved blades with airfoil cross sections. They have relatively low starting torques but relatively high tip-to-wind speeds and high power output for a given rotor size, weight, and cost. Usually the rotor needs some type of starter system to initiate its rotation. Frequently, a small S-rotor is mounted at the base of the central shaft to start the Darrieus rotor spinning. Such an addition increases the weight and cost of a system, and so trade-offs between maximum power output, starting efficiency, and cost must be considered in developing an optimum design for a given application. The efficiency of the Darrieus rotor is approximately 35 percent with a tip-speed ratio of 6 to 8, depending on the type of rotor. The main advantage of this type of system is its low cost.

Energy Storage

One of the main problems facing the use of wind for small-scale electrical generation is the present limited ability to store the energy. Batteries are the most efficient chemical-storage system now available, but banks of batteries are costly to install and to maintain. The power generated by a wind machine is not constant enough to guarantee a reliable quantity of electricity or a stable AC frequency. Therefore, most wind generators produce DC (direct current) voltage. Electricity in this form is compatible with storage in batteries, which themselves yield only DC voltage. Resistance appliances such as heaters, toasters, and incandescent light bulbs can be operated on DC current, but most large appliances contain induction motors that operate only on AC. It is possible to use an inverter to change DC to AC, although these devices are quite expensive. Some authorities foresee a possible resumption of the general use of DC motors to power large appliances.

Two of the most promising developments for wind-energy storage appear to be the flywheel storage system and the electrolysis/hydrogen-fuel storage system.

A flywheel can store a much higher quantity of energy per mass weight than a battery. The principle of the flywheel

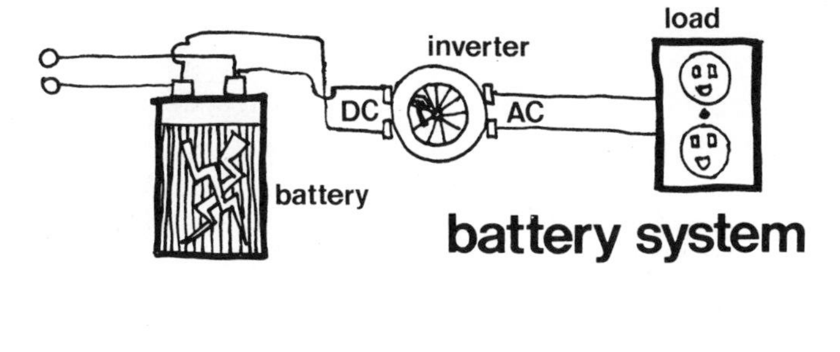

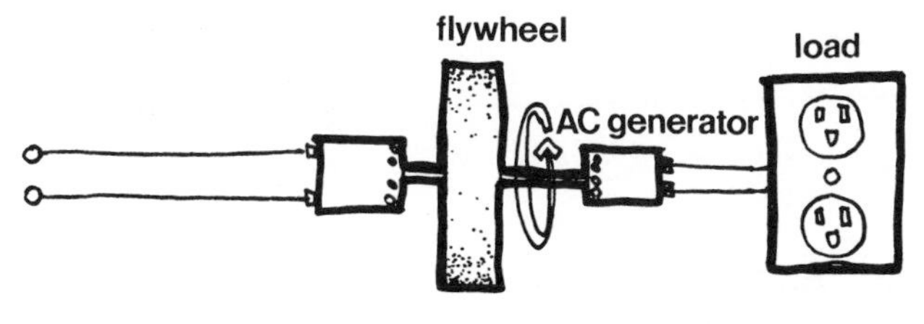

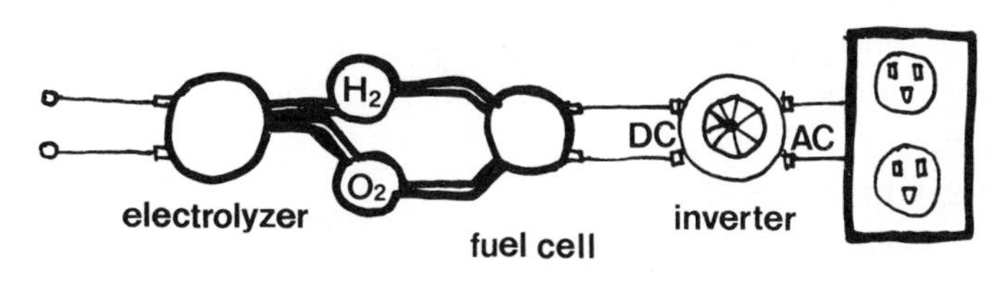

system is that a spinning wheel can store and accumulate energy as momentum and release it for later use. Advances in material technology and low-friction bearings have made the flywheel principle feasible for energy storage. Flywheels can be mounted in a vacuum to decrease air friction. As the mass of a flywheel increases and the speed at the rim increases, there is an increase in the amount of stored energy.

The amount of energy that can be stored in a flywheel is a function of the material from which the flywheel is made, its size and shape, and its speed of rotation. Some heavy materials develop greater internal stresses than lighter materials at a given speed of rotation. Lightweight flywheels, however, must be spun faster to store the same amount of energy as the heavy ones. A multi-ring or radial-fiber composite material flywheel could store thirty to forty times as much energy per pound as a lead-acid battery. The flywheel would be attached to a motor/generator unit to recover the energy.

A motor powered by electricity from a wind machine can spin a flywheel and continue to add energy as long as the wind blows. When the wind speed is sufficient to generate continuous electricity, the motor can turn off and the flywheel will continue to spin. When electrical needs can no longer be met directly by the wind generator, the flywheel could then spin its own generator. It is estimated that a flywheel sealed in a partial vacuum and riding on extremely low friction bearings (possibly magnetic suspension) could store energy for months, although this has not yet been demonstrated.

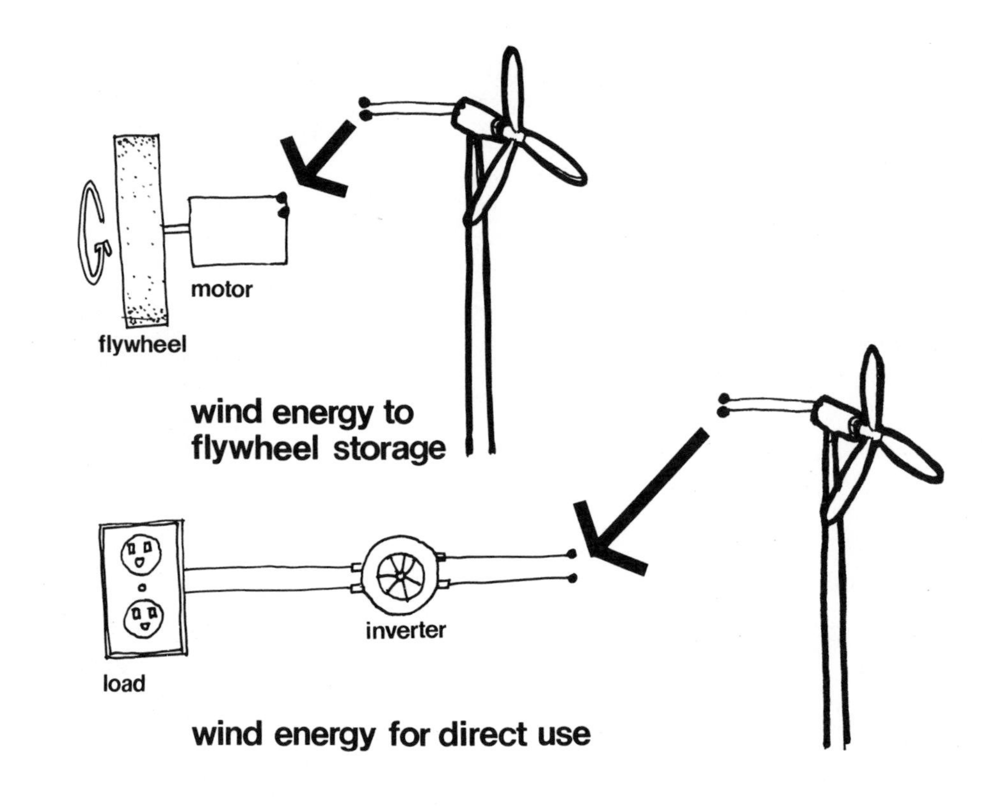

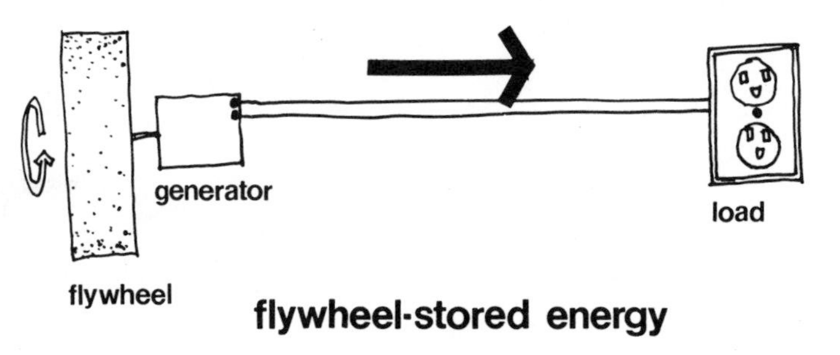

The electrolysis of water to produce hydrogen and oxygen is another alternative for energy storage, in which electricity generated by the wind machine breaks down the water into its two components, hydrogen and oxygen. These two gases are then stored in tanks for future use. Energy is released as electricity in a hydrogen fuel cell, where the hydrogen and oxygen are recombined to give off water vapor as they produce electricity.

Large-Scale Power Generation

The use of wind energy is by no means limited to small-scale applications. It is theoretically possible to build wind generators that can deliver several megawatts of power. There are certain constraints on the size and power output of wind machines because of limitations on the strength of materials, which in turn limit the allowable stresses for blades, bearings, towers, and other components of the system. Cost is also a limiting factor. In certain applications it may be cheaper to build two small machines of equivalent total output rather than one large one.

The most effective application of wind energy would be to use it directly to pump water for irrigation or to generate electrical power. Wind systems without energy storage could be used in conjunction with a fossil-fuel-powered generating plant to save fuel when the wind is blowing. For most applications, however, energy storage would be required so that uninterrupted power could be delivered. Storage could be accomplished by the use of pumped water systems, compressed fluids, stored electrolytic hydrogen, flywheels, and other forms mentioned previously.

Three types of wind machines would be suitable for large-scale applications: those having open horizontal-axis rotors, open vertical-axis rotors, or vortex generators. (Small-scale vortex generators may also be practical in the future.)

Wind machines with horizontal-axis rotors for large-scale applications are similar in form to the ones already discussed for small-scale applications. In a 27-kilometer-per-hour (17 mph) wind, a machine with an 18-meter-(59-foot) diameter rotor has a rated output of 100 kilowatts (kw). One with a 50-meter-(164-foot) diameter rotor is rated at 1.0 megawatt (mw), while one with a 136-meter (446 foot)-diameter rotor is rated at 10 megawatts, but would entail the height of a 45-story building in addition to its mounting height. The safety of life and property in the vicinity of a wind machine of such scale is a primary design consideration. In case of a rotor failure, buffer zones would have to be established in which occupancy is limited. Ultralarge machines could be placed in remote areas, or in the ocean, where structural rigidity would not be as critical as on land. Smaller machines could be used near populated areas and possibly arranged in sizable arrays.

Vertical-axis wind machines of the Darrieus type could also be used for large-scale applications. In a 27-kilometer-per-hour (17 mph) wind, a machine with a 21-meter-(69-foot) diameter rotor has a rated output of 100 kilowatts. One with a 58-meter-(190-foot) diameter rotor is rated at 1.0 megawatts, while one with a 158-meter-(518 foot) diameter rotor is rated at 10 megawatts. It is also necessary to take safety precautions and properly locate this type of machine, although sudden wind reorientation is far less a challenge than with horizontal-axis turbines.

Another concept in wind machines is the vortex generator. These machines spin the wind to increase the power output of a turbine located in or near the vortex. Two varieties are presently under study. They are the unconfined vortex type and the confined vortex type. One employs winglike structures to deflect the wind and create an unconfined vortex around the turbine. It is estimated that an unconfined vortex generator with the same diameter rotor can be designed to provide up to six times the power output of a conventional system.

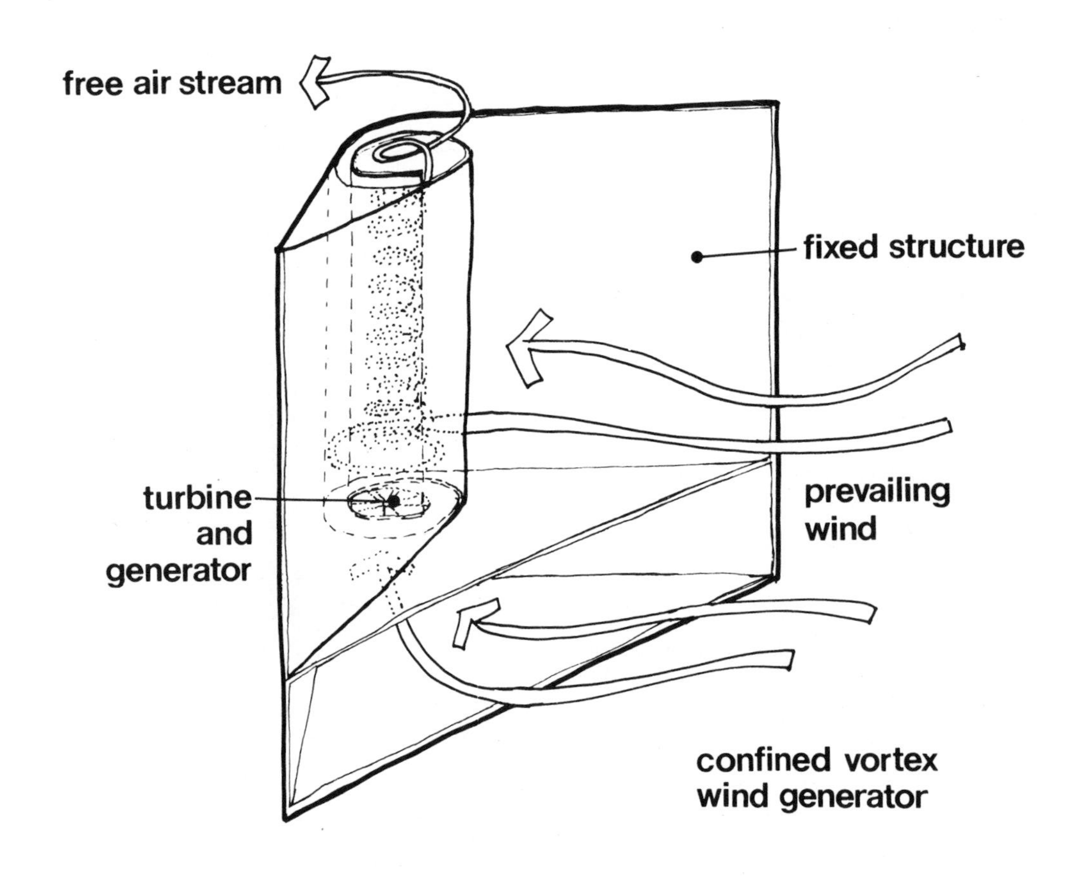

confined vortex wind generator

The Grumman Aerospace Corporation is developing a confined vortex system in which the pressure drop across a ducted turbine and the wind velocity through it are augmented by using additional ambient wind to produce a confined tornadolike vortex in a tower located at the exit of the duct. For a typical confined vortex system, the diameter of the tower might be three times the diameter of the turbine, and the height of the tower might be three times its diameter, or nine times the diameter of the turbine. Dr. James Yen of the Grumman Aerospace Corporation estimates that large-scale wind-energy systems, which have ducted turbines interconnected to vortex generators, may be designed to have power outputs that are 100 to 1000 times those of conventional systems, of the same rotor diameter, operating at the same ambient free-flow wind speed.

With these systems, smaller high-speed turbines can be used to obtain the same output power as the large-bladed conventional systems, thus avoiding the large weight and inherent stresses on the blades of the conventional systems.

These concepts can yield significant improvements in the design of wind machines. They indicate that high-power output can be derived from relatively small installations and that the structure needed to deflect the wind can also shroud the turbine, decreasing potential dangers.

The cost of generating electricity with wind-vortex systems can be markedly reduced by incorporating the vortex tower in a multipurpose building.

There are a number of additional ways to store the energy produced by large-scale wind machines; most require interconnecting the machines with other systems.

In pumped-water storage systems, the wind machines are used in conjunction with a hydroelectric system. Water from below the dam is pumped into storage behind it, using wind power. This can be accomplished with direct mechanical pumping or with electric-powered pumps. The water behind the dam flows through turbines, generating electric power, and is retained in a holding basin. It is then pumped behind the dam again, using wind machines, and the cycle starts over. With this type of system a steady supply of electric power can be supplied.

While wind machines are among the simplest energy-conversion devices, the wind is an unpredictable force. The selection of an optimum site for a wind machine and the choice of optimum designs for particular applications are often difficult and complex problems. Much research and development is currently under way on wind machines; undoubtedly, innovative designs will continue to emerge. Wind machines, varying with geographic location, appear to have a high potential for making significant contributions to the goal of meeting our future energy needs.

Wind energy is derived from the sun's energy, and its use has no discernible ill effects on the environment. Harnessing the wind's energy produces no thermal pollution, no air pollution, no radioactivity, and no other by-products that are dangerous to populations.

5 Water/Precipitation

There are three major forms of water energy on earth: gravitational, tidal, and thermal gradient. Ocean water moves because of the constant change within the hydrologic cycle and the effects of the earth's rotation, as well as lunar influence (in larger bodies of water). Evaporation from bodies of water brings about cloud formations, which redistribute the water to higher land elevations in the form of rain and other types of precipitation. This water then flows into streams, creeks, then rivers, and back into the ocean. The sun provides the energy for evaporation and condensation in clouds and also propels the clouds and their inherent payload of water to new locations.

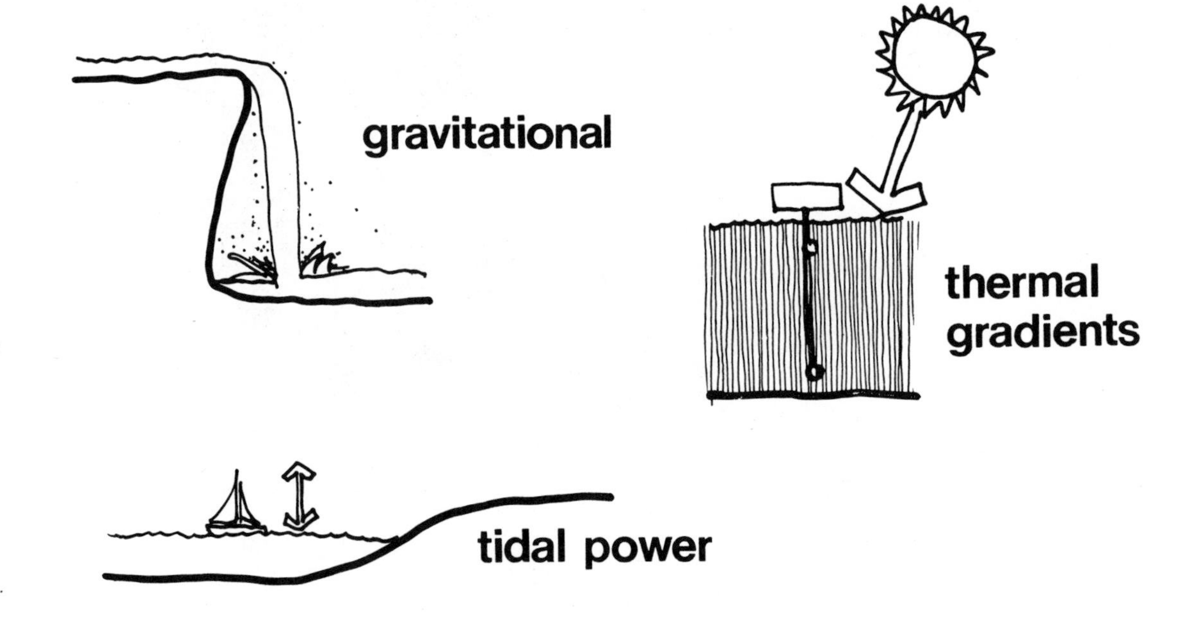

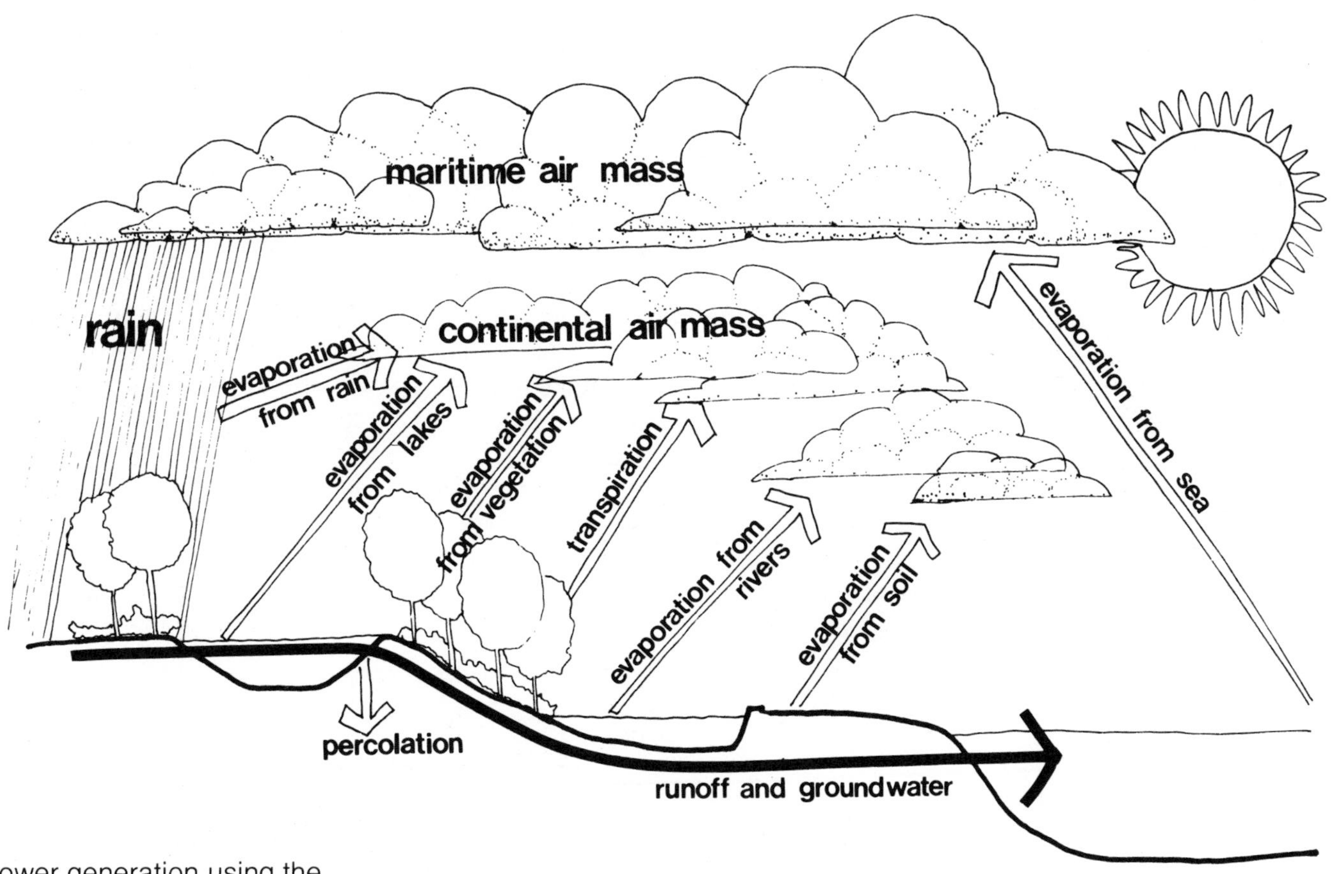

Electrical power generation using the gravitational potential of water is the most developed method of extracting energy from the hydrologic cycle. The earliest known use of water power was to drive pumps for irrigation and power mills for grain processing. Waterwheels were later used to power textile plants in Europe and the United States. The wheel supplied power to a central shaft, and by means of belts and pulleys the energy was transferred to individual machines.

hydrologic cycle

Adapted from Into the Hidden Environment with the permission of Roxby Press Ltd.

The two most common types of waterwheels are the undershot wheel and overshot wheel. The undershot wheel can be installed adjacent to or over an existing stream. Paddles or buckets on the outer rim of the wheel are propelled by flowing water, developing a turning or torque power of the wheel.

The overshot wheel needs a cliff, dam, or sluice to raise the level of the water behind it to the top rim of the wheel. In general, a sluice, a small channel, is set up wherever terrain prohibits the more natural forms of channel. As the water leaves the sluice, it drops into the scoops on the wheel. The weight of the water forces the scoop down, turning the wheel.

Both the overshot and undershot wheels work at slow speeds delivering high-torque power useful for mechanical purposes.

Another system using water motion is the turbine, which can develop much higher rotational speeds than waterwheels and is used for generation of electricity. With the turbine, water is forced through a pipe or tube. As it exits the tube under pressure, it strikes the blades of the turbine, causing it to turn. Most hydroelectric dams use large-scale turbines to generate electricity. To develop the water pressure necessary to operate a turbine, a dam is needed to develop a vertical column of water, or pressure "head."

When a river or stream is dammed, the ecological impacts of the dam must be taken into consideration. The dam interrupts the flow of water and prevents fish and other water creatures from reaching areas downstream and upstream. The water behind the dam will evaporate at a much higher rate because of its changed surface-area-to-volume ratio. These extra pressures can result in shifts in geological formations. New vegetation may start in areas adjacent to the banks where high-water-level conditions did not exist before. In areas where the amount of water varies because of precipitation changes, reservoirs can become mud ponds during periods of little precipitation and the accompanying low water levels.

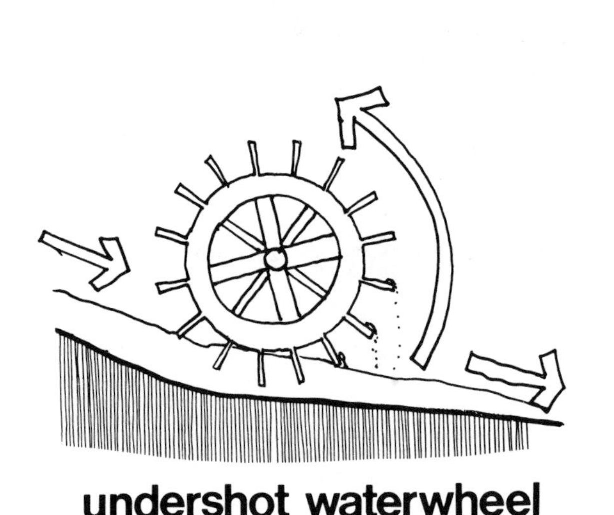

undershot waterwheel

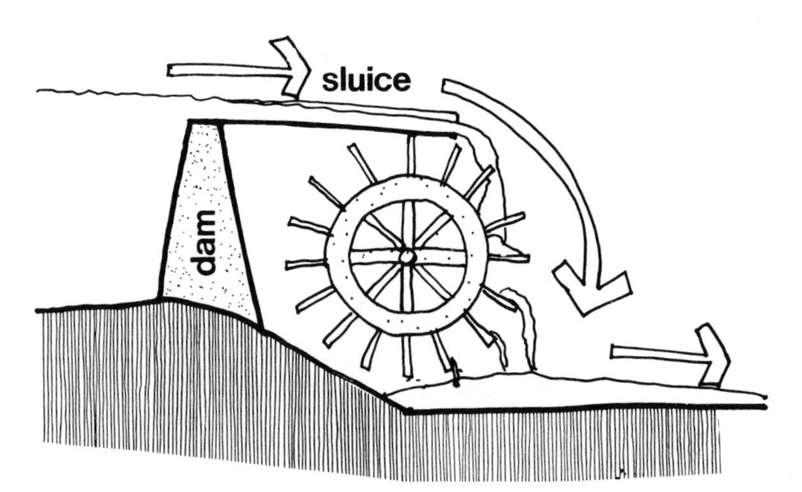

overshot waterwheel

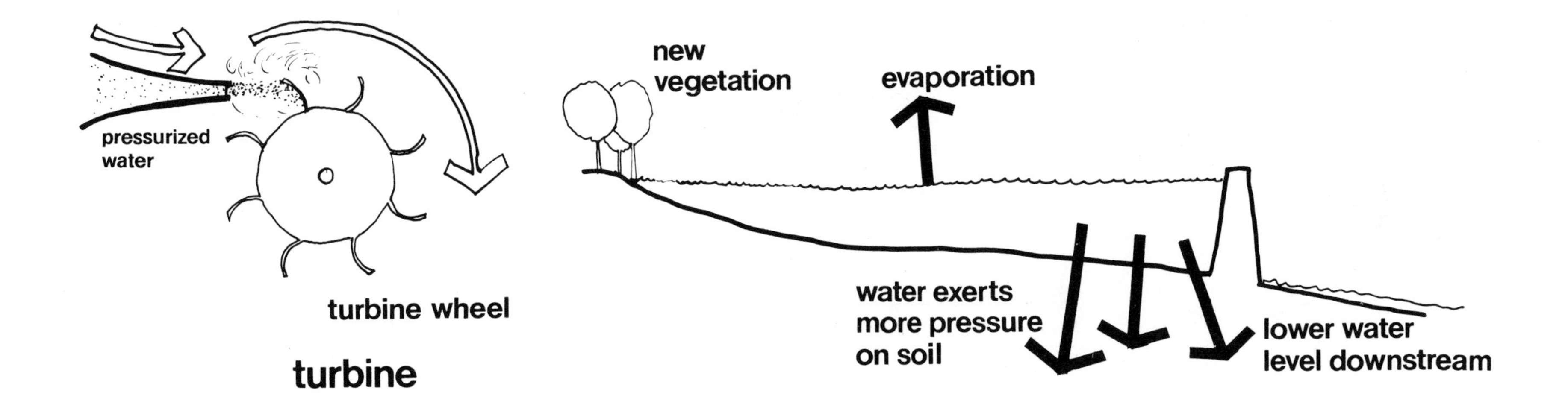

The earth's tides are not directly related to solar energy but are another force of nature. As the moon travels around the earth, the gravitational force between the two acts on the oceans. The effect of the moon orbiting the earth and the earth revolving on its axis produces high and low tides on the earth's surface. The motion of this water represents a tremendous amount of power.

Generators mounted between or within dams could be constructed to economically convert this power into useful energy—electrical power, for example. Such a system would trap alternately high and low tides as illustrated; the height of the water head and the actual horizontal motion would combine to offer extremely large-scale cyclic-power potential.

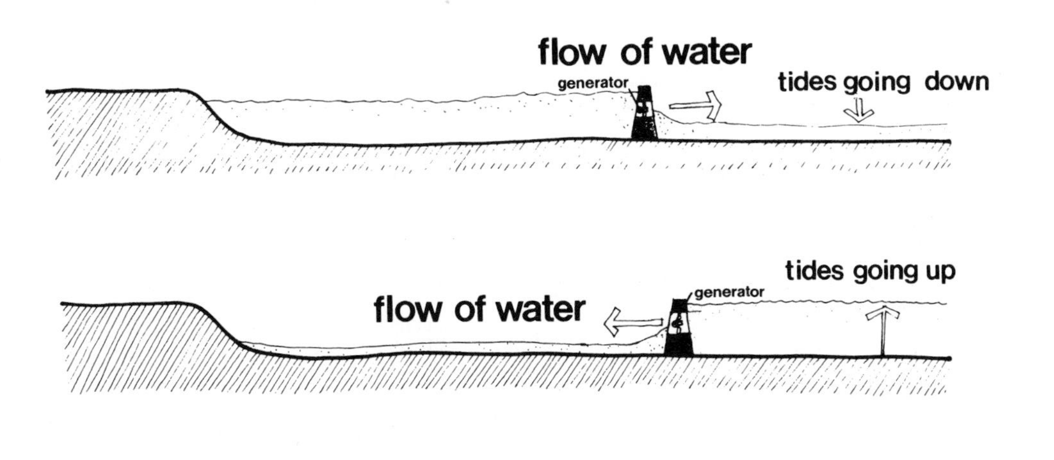

tidal power

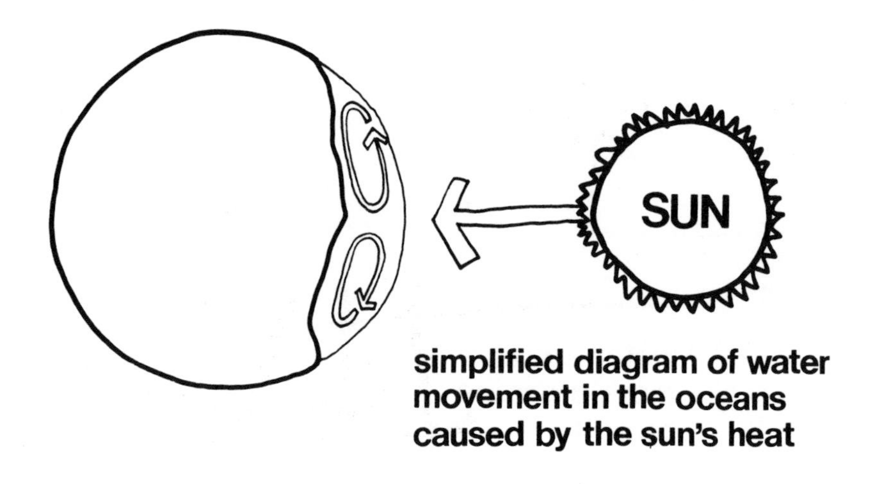

simplified diagram of water movement in the oceans caused by the sun's heat

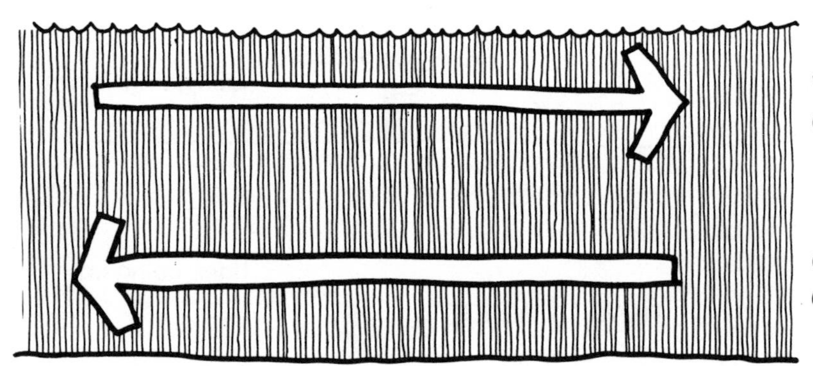

Thermal gradients are differences in temperature at varying depths; in natural bodies of water other than geothermal ponds, these gradients are a result of the heating of the upper water levels by the sun. As the sun radiates to the water, currents are induced (similar to wind currents) in which warm water rises to the surface and then flows to cooler water areas where it drops in temperature and, flowing to the bottom, it completes this cycle. As with wind, this is an oversimplified rendition of changes that, interrupted by turbulence, actually occur. Currents are affected, for example, by variations in the ocean floor terrain, geological irregularities producing geothermal warming, and by air currents above the ocean.

The low levels of the ocean are generally colder than areas near the surface. These differences in temperature can provide power for Rankine cycle generators as illustrated.

Since the earth itself is a solar collector, and most of the earth's surface is water, the oceans are a most effective means of collecting and storing the thermal energy of the sun. The temperatures in the ocean vary as to the amount of localized cloud cover and the measurement of depth below the surface. Tropical and ocean temperatures may range between 4° and 27° C (39 and 81 F), suggesting, in some areas, the existence of strong convective currents.

The best collection areas lie in temperate waters between the Tropic of Cancer and the Tropic of Capricorn. In these regions also exists the highest ratio of water area to land area: nine to one. Water-surface temperature here averages approximately 27° C (81 F).

Large-scale electrical generating plants could utilize the temperature differentials with power generators located just below the ocean's surface. The condenser and boiler units could be several hundred feet below the surface, while the generator and turbine could be at a minimal depth (deep enough for clearance below ships and still accessible to divers). Stationing the generator below the surface minimizes danger from storms and their climatic disturbances. Water-intake points should be located where maximum temperature differentials occur. The cold water entering the condenser might be found as much as 610 meters (2000 feet) below the surface, with the warm water inlet being located considerably above it. Naturally occurring differences in temperature and pressure over such a distance would cause refrigerant fluid to power a turbine. Several refrigerants are presently being considered: propane, Freon, and ammonia. Approximately 10 percent of the energy that the plant would produce is needed for the pumps that maintain water-pressure levels. Estimates of the construction costs of such systems

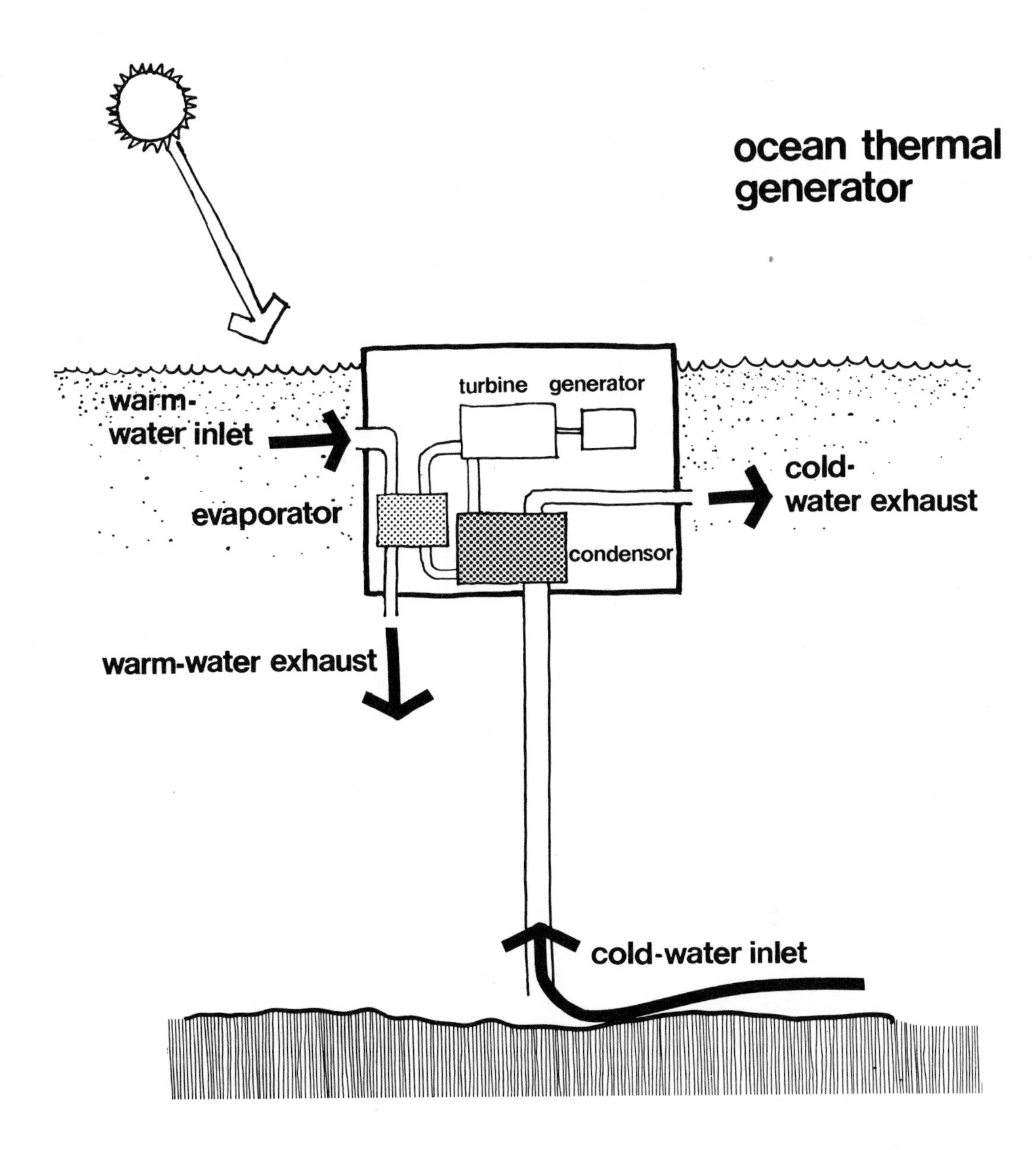

range from half to the same cost of a fossil-fuel plant of identical power output. A further advantage over conventional fuel plants is the abundance and renewability of ocean-thermal-gradient structures throughout temperate regions.

Some ideas for large-scale electrical power generation involve harnessing the energy contained in deep water waves and shoreline breakers. Waves are initiated by the sun (temperature differential), wind (direct force), and the gravitational attraction of the moon (tides). They contain vast amounts of power and have the potential of being used to generate electricity.

An object on the surface of the water bobs up and down with the passage of each wave, yet always seems to remain in the same spot. The wave is not a moving mass of water but a force that lifts the water's surface as it passes along. A large float, a closed cylinder filled with air, etc., on the water's surface will rise and fall with the periodic wave motion. Work can be extracted from this motion and used to operate a pump, which, through an appropriate mechanical arrangement, can be used to generate electricity.

The power from waves breaking on the shore can also be used to generate electricity. The upward breaking power is first converted to mechanical power and then to electrical power. One scheme would be to use the breaker power to compress air, which in turn would be used to power a turbine generator.

6 Earth

The earth is constantly generating heat that is transferred to its surface. This heat comes from the earth's molten center, where temperatures in excess of 1000° C (1832 F) are attained as a result of the natural decay of radioactive core materials and frictional forces resulting from solar and lunar tides. The earth radiates 54 calories per square meter per hour (0.0199 Btu/square foot/hour). As the distance from the earth's core decreases, the internal temperature level increases. For each kilometer below the earth's surface, the temperature will rise 30° C (54 F). The average temperature at the earth's surface is 10° C (50 F), but drilling to a depth of approximately 3 kilometers (1.8 miles) will yield a 90° C (194 F) temperature, which is economi-

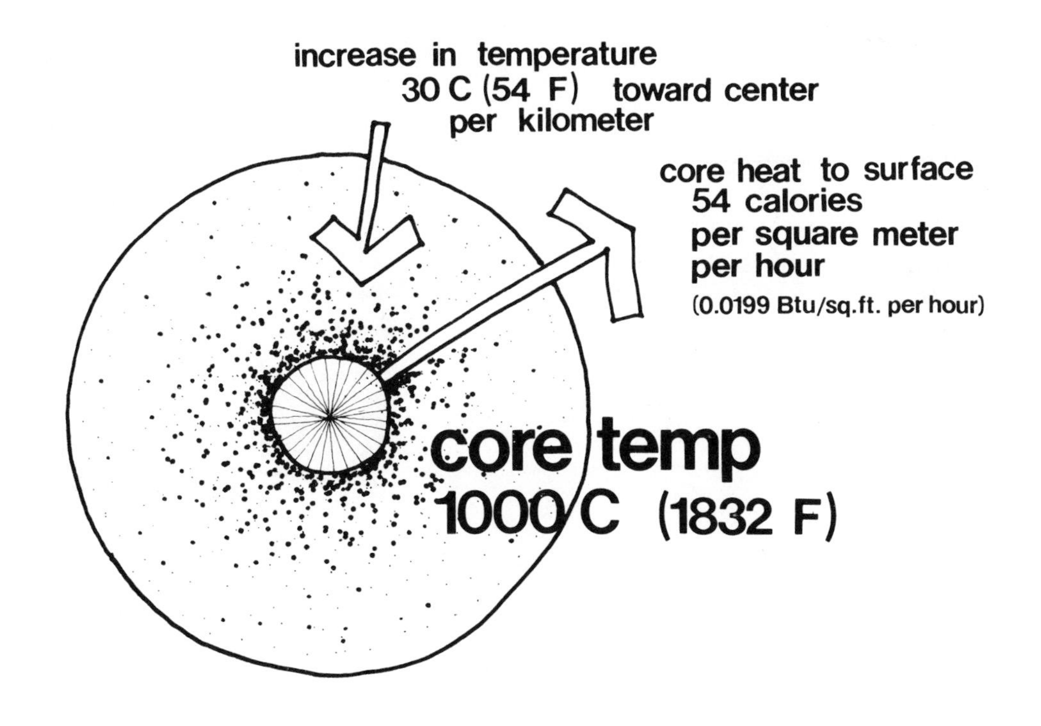

cally useful for space heating and many industrial uses. A depth of approximately 7 kilometers (4.3 miles) will yield a temperature of 180° C (356 F), which can be useful in the generation of electricity with the assistance of a low-boiling-point fluid to power a generator. The depth at which ideal steam-electric-generation temperatures (240° C or 464 F) are normally found is 9 kilometers (5.6 miles); this is near the present limits of well-drilling technology and suggests use for only large-scale power generation plants.

Irregularities in geological formations can bring about and indicate changes in the depth and location of useful geothermal-temperature gradients. In some areas geothermal sources of heat penetrate the surface as geysers and hot-water springs. The Geysers, an area in northern California, is known for its escaping steam. The Pacific Gas and Electric Company buys the steam and presently uses it to generate approximately 200 megawatts of continuous electricity. Other geothermal sources worldwide are now being used for power generation, industrial power, space heating, and air-conditioning in a lithium-bromide refrigerating system.

Geothermal

There are three major types of geothermal power: dry steam, hot water/wet steam, and hot, dry rocks. Each has different characteristics of depth, access, and economic feasibility. Many of these characteristics change with each potential site.

Dry steam is the cleanest and the most usable of the three, but it is very limited and occurs only within unusual geological formations. Temperature levels obtainable with dry steam are much higher than with other geothermal sources. Dry-steam temperatures at accessible levels are approximately ten times as high as the average corresponding depth elsewhere should yield. Shafts are drilled into areas containing dry steam; it is filtered and then channeled for elec-

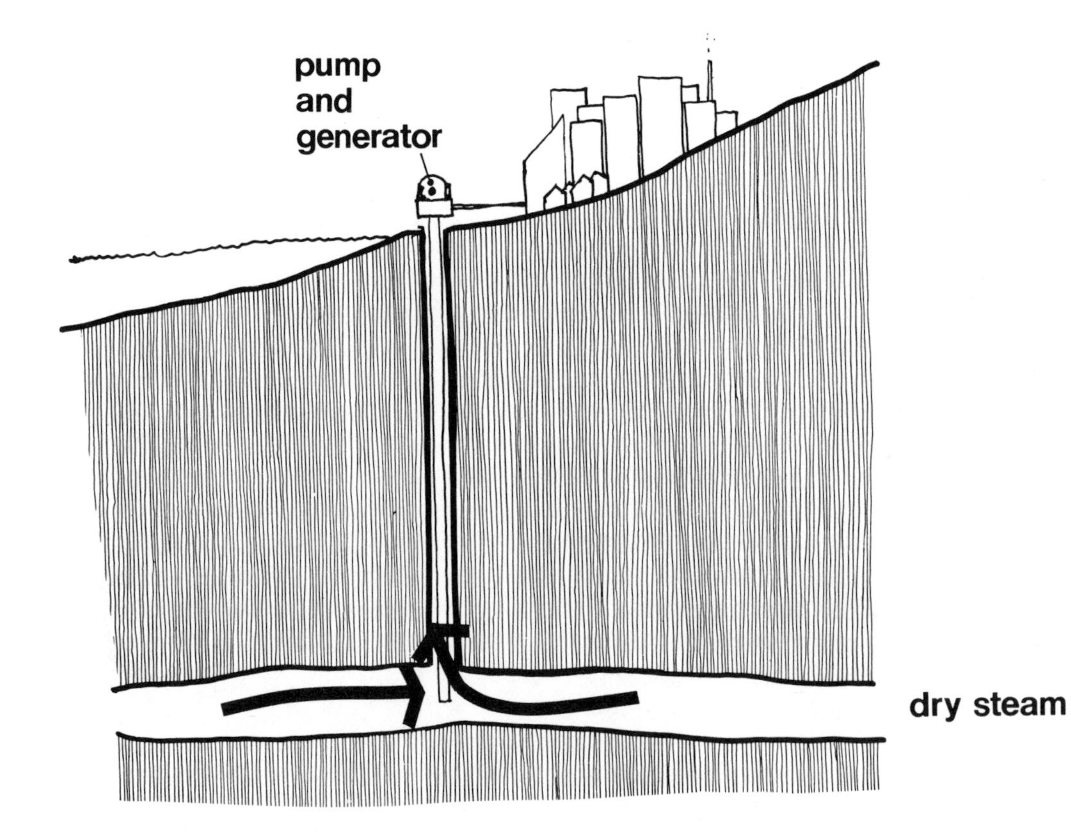

trical power generation. Water condensed from the steam can be a reserve of fresh water.

Wet steam/hot water is more abundant than dry steam. Hydronic heating systems can use the geothermal hot water directly or remove heat through a surface-mounted or "downhole" closed-loop heat exchanger. The temperatures are sometimes high enough for electrical generation, but they are usually more suited to space-heating requirements. Wet steam/hot water geothermal deposits are approximately three times hotter than similar drilling depths.

Much of the wet steam/hot water that is available under the geographic United States contains high percentages of mineral and salt deposits. These materials can cause a number of problems in the typical piping of collection and distribution systems. Hence, the closed-circuit system is preferable where water composition would corrode piping. Also, there exists a number of potential hazards regarding air pollution and stream contamination when such water is exposed to the environment. Unless water (following heat removal) is pumped back into the underground geothermal reservoir, large craters can result when the unsupported earth overlying the reservoir collapses. Several such incidents have occurred in New Zealand where geothermal water resources are widely employed.

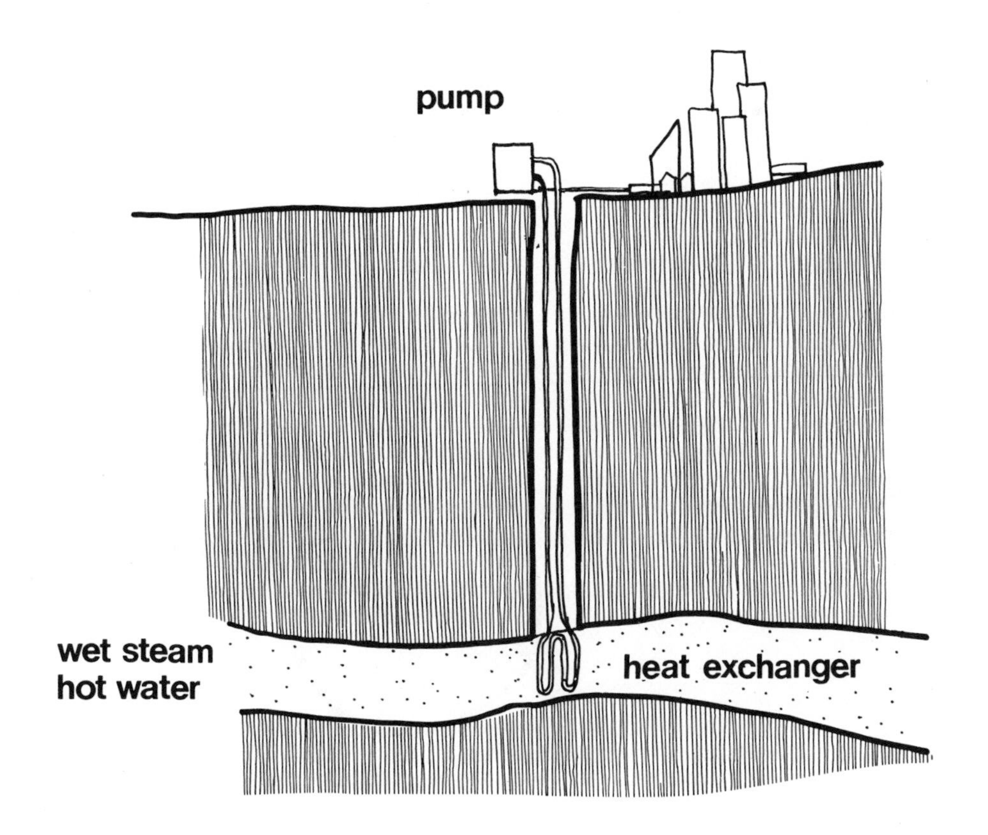

Both hot-water and dry-steam geothermal reservoirs are limited in reserve quantity; estimates range from fifty to 300 years of useful life for either of these geothermal sources.

Dry rock contains the primary form of the natural heat energy of the earth, and as an energy source it is practically unlimited. This heat can be removed by a number of methods. One of the simplest of these consists of drilling to a depth necessary for the temperature required and then pumping in neutralized water and returning heated water to the surface for heating domestic water or space heating.

Unless large areas in hot rock strata are fractured, the minimal exposed surface area will be insufficient for economical

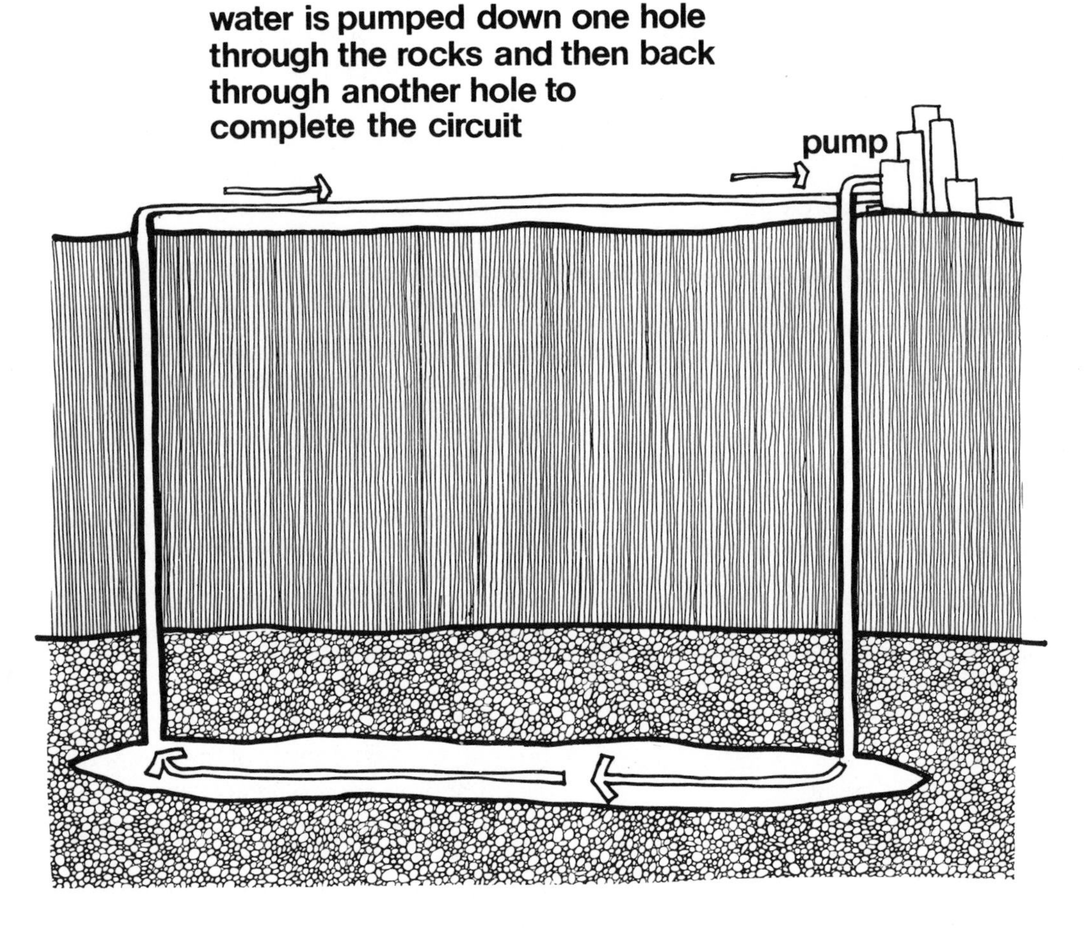

hot dry rocks

heat exchange. A method of correcting this, hydrofracturing, calls for pumping the water at very high pressure into the well to crack and fracture large areas of hot rock. An auxiliary hole is then drilled to another point along the fracture to complete a pipeless "circuit." Water is pumped down the first hole, is warmed by the hot, dry rocks, and is piped out through the second hole. The technology for such drilling has already been developed by various oil companies.

As obstacles in engineering and economics are overcome, geothermal heat could prove to be a major energy source for the future. Techniques are being developed for large-scale projects to supply space heating for housing and industry.

gravity is the attraction between two masses

Another source of energy provided by the earth is gravity—that is, the attraction between any two masses. The force of attraction between the two is proportional to the product of their masses and is inversely proportional to the square of the distance between their centers of mass. As the mass of either body increases or decreases, or their separation distance varies, changes take place in their mutual force of attraction or gravity. Smaller planets exert lesser gravitational forces than larger ones of the same density because they are less massive. Gravity is used as a force in many natural-energy systems, such as falling or moving water.

One use of off-peak electrical power involves the pumping of large volumes of water to an elevated storage reservoir for later release through turbine hydroelectric generators at peak-demand times. Such a process may save the capital cost of oversized conventional and standby generators, especially when implemented on a large scale.

Of course, geothermal and gravitational energy are only two of the types of energy provided by the earth. Others, including the fossil fuels (coal, oil, natural gas), and nuclear energy, have already been mentioned.

Natural gas is ninety-five percent methane, which is found in oil wells and coal mines. It is also produced as a by-product of the decay of organic materials in the absence of oxygen. The presence of methane has been detected in swamps, septic tanks, landfills, and the digestive systems of animals.

In a landfill at Palos Verdes, near Los Angeles, thirteen million tons of refuse have accrued since 1957. This refuse is now generating methane, and one contractor hopes to be able to recover 28.32 cubic meters (1000 cubic feet) of gas per minute from it. If this project is a success, there are numerous other landfills around the country to which the same techniques could be applied.

Methane can also be derived from a digester into which organic materials are

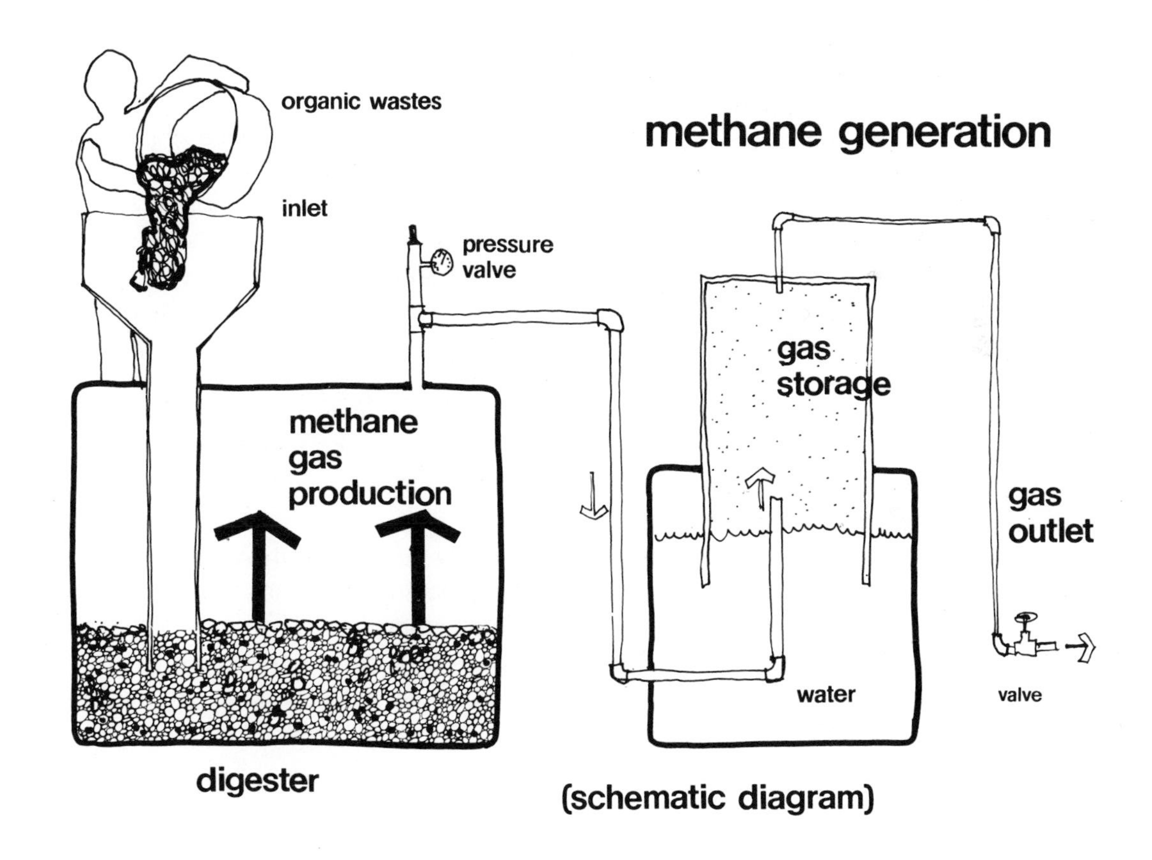

fed. This type of digestive process takes place in several stages. First, organic matter is mixed with water to form a slurry that is loaded into the digester tank. Aerobic bacteria—those requiring oxygen—begin to break down the material into water and carbon dioxide. When the oxygen inside the tank is used up, the aerobic bacteria die, and the anaerobic bacteria go to work. Without the benefit of oxygen, these bacteria are not able to break down the organic matter as completely as the aerobic bacteria. Methane is one of the products of this anaerobic fermentation and is collected for use. The uses of methane are as varied as the uses of natural gas, but methane produced in a digester is a renewable fuel and has applications for transportation, synthetic materials, and heating.

Biomass represents a potentially vast energy resource. Biomass is essentially plant material derived from a variety of sources, such as grains, sugar crops, food and animal wastes, crop residues, wood and forest product residues, and municipal solid wastes. (Wood and wood residues are the largest biomass resource in the United States.) The estimated potential of biomass energy in the United States is up to sixteen quads by the year 2000.

Bioconversion technologies include physical, chemical, and physical-chemical processes. Energy products derived from these conversion methods can include fuel gases, electricity, and a variety of alcohols and chemical feedstocks. Transportation and industrial, chemical, and utility-market sectors represent possible areas that could benefit from increased production of biomass-based fuels.

Methane can be a substitute for natural gas. Methanol and ethanol are considered the most promising alcohols for use as transportation fuels. When gasoline is blended with ethanol, which has more octane than gasoline, less octane need be added, and unleaded gasoline can be produced at a lower octane level. This can result in substantial energy and economic savings in the refining process.

Ideas for expanding the biomass resource include biomass plantations. Energy landform and aquatic biomass farming processes (aquaculture) could substantially increase our human food base as well as provide useful energy.

7 Historical Use of Natural Energies

Throughout history, at least until the Dark Ages in Europe, man has allowed his architecture to evolve in harmony with natural climatic conditions. The sun and the natural forces that are available to him have been utilized in various ways. Changes in his technology and culture have fostered changes in his attitudes concerning his utilization of nature's energy. The characteristics of available materials have also had an impact on the architectural forms predominantly used by various civilizations. Some of the numerous ways in which these natural forces have been used are presented in this chapter.

The energy of the sun in combination with the earth provides the food energy used by man and all other living creatures. Plants are the first major link in all food chains, as well as the receptors of the solar energy eventually stored as fuel. Many plant fibers are also used as the construction materials (wood, straw, etc.) that are the basic units of shelter. Other materials for shelter (steel, concrete, brick, stone, mud, etc.) are derived directly from the earth or are produced from natural materials that have been refined in some way.

Primitive man in temperate zones used the earth and its topography as he found it to provide himself with shelter from the environment and protection from wild animals and enemies. He used caves and the undersides of cliffs for his habitat and showed little desire to fabricate his own dwellings. His attitude toward his shelter produced the least possible negative impact upon the environment. These natural earth dwellings provided microclimates that were less severe than the outside climate. The thermal inertia and insulation of the earth produced a cool home in the summer and a warm home in the winter.

As man's needs developed, he started constructing his shelter using the materials he found around his encampment. He learned that entries and windows facing the sun and solid walls intercepting the harsh winter winds were the most comfortable configurations. This orientation was prevalent until the advent of industrial man, who, through his advanced technology, decided he could provide a comfortable interior environment by expending fossil fuels no matter how he oriented doors, windows, etc. Topography and the availability of resources played a dominant role in dictating how early buildings and communities would be situated. Landforms were used to separate classes within societies, as well as to provide protection from climatic forces and hostile neighboring tribes.

Egyptian designers understood how to use natural climatic forces to provide comfort in their buildings. They used the sun to warm space and to provide interior illumination. Thick walls, warmed by the sun, were used to create uniform nighttime temperatures; in turn, the stored heat was used to induce natural air movement, which ventilated buildings. The walls were effective even after the sun had set because of the heat they contained.

The Egyptians were the first known civilization to produce a sun-dried brick. Walls 20 to 30 centimeters (8 to 12 inches) thick, constructed out of these bricks, could develop a thermal time lag of six to ten hours.

Interior courtyards and pools of water were used in structures built by the Aegean civilization to provide natural lighting to interior and exterior spaces, as well as to control temperatures in these areas. The design of entryways and court areas encouraged the natural flow of air through interior spaces. Runoff water was collected in the pools located in interior courts. The water provided evaporative cooling and humidification for the building.

The Greeks were worshippers of the sun, and this devotion is amplified in their architecture. Their public meeting places, agoras, were squares situated so that they would receive the sun's natural warmth. They were surrounded on three sides by buildings, with the fourth side open to the south sun. Private residences were arranged so that each had a prominent southern exposure. Hippocrates, in his *Air Light Treatise*, declared that cities should be sited so that buildings and streets would be penetrated by the sun and wind for hygienic purposes and odor removal. (Many streets also served as open sewers.) He suggested also that cities should be located where pure water was available and that marshlands should be avoided. The Greeks realized and exploited the fact that the sun, used to cast shadows, would bring out the detail and form of buildings.

The Romans were located at much the same latitude as the Greeks and were equally fascinated with the sun. Vitru-

primitive man used the natural energies and materials around him

vius, a Roman builder and architect-philosopher, wrote the *Ten Books of Architecture* in which he presented many energy-conserving ideas relating to siting, orientation, and climatic response.

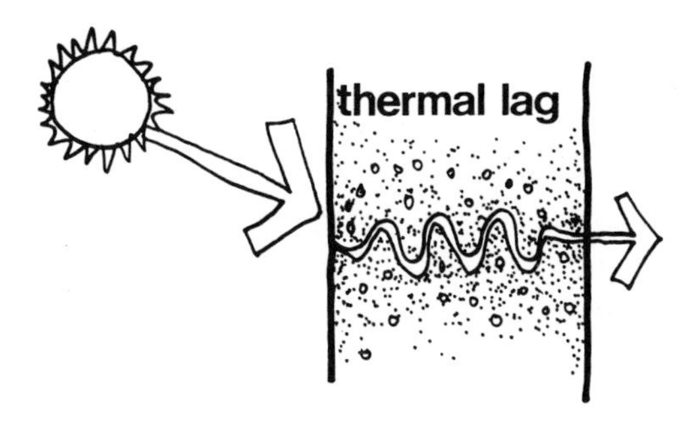

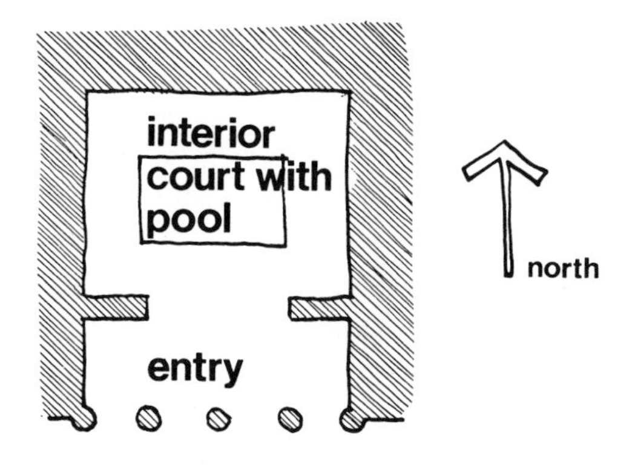

house surrounding court

> "In the north, houses should be entirely roofed over and sheltered as much as possible, not in the open, though having a warm exposure. But on the other hand, where the force of the sun is great in the southern countries that suffer from heat, houses must be built more in the open and with northern or northeastern exposure. Thus we amend by art what nature, if left to herself, would mar. In other situations too, we must make modifications to correspond to the position of the heaven and its effect on climate."

Vitruvius incorporated many passive energy systems into his designs. He was greatly concerned that his planning and layout of cities and buildings would allow the maximum use of natural-energy forces. He included many concepts in his designs that had been presented by earlier writers; most of these earlier works have since been lost. On the use of sun and natural light he wrote

> "There will also be natural propriety in using an eastern light for bedrooms and libraries, a western light in the winter for baths and winter apartments, and a northern light for picture galleries and other places in which a steady light is needed; for that quarter of the sky grows neither light nor dark with a course of the sun, but remains steady and unshifting all day long."

The need for climate-responsive architecture is reflected in his statement that, "One style of house seems appropriate to be built in Egypt, another in Spain, a different kind in Pontus, one still different in Rome, and so on with lands and countries of other characteristics."

The observations of Vitruvius have remained valid, though often ignored, through history. To be in harmony with natural surroundings, architectural forms must respond to the influences of regional climates. In all areas dwellings should respond to the natural forces of the sun and weather. They should harmonize with the environment and employ as many local building materials as possible.

In cold, humid northern climates, dwellings must be constructed so that heat loss is minimized and heat gain maximized in winter. In sparsely populated Arctic regions, the igloo is the most common type of building. An igloo is built in the shape of a dome, which gives it a number of advantages. The dome shape encloses the maximum amount of internal volume within the least external surface area—an important consideration in controlling heat loss.

To minimize the heat lost to prevailing north winds and to realize a small amount of solar-heat gain, the only opening is on the south side. The opening is actually a tunnel that dips below grade and then back up into the interior space. Since cold air does not rise, it settles in the lowest spot in the tunnel and does not gain entry into the interior. An igloo is constructed of ice and snow—local materials—which form a vapor barrier to the outside. Animal skins are stretched across the walls and ceil-

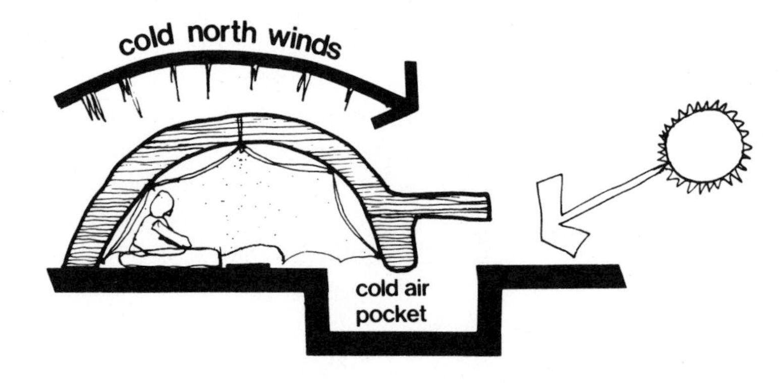

winter igloo

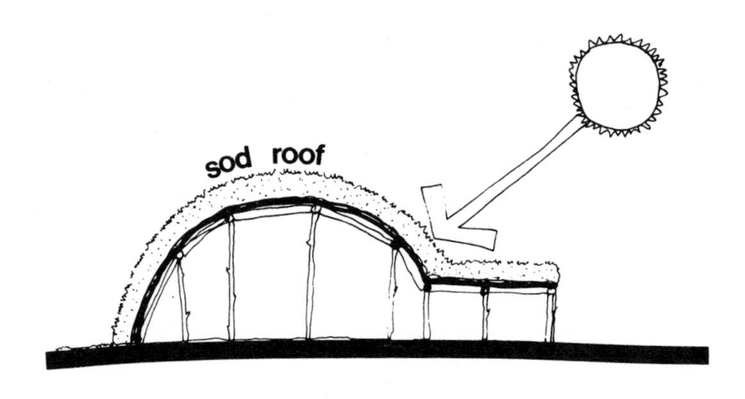

summer hut

ing areas on the inside; these create dead air spaces between the skin and the ice, providing insulation and restricting heat loss. Most of the heat in such a space is provided by the occupants, and the average indoor temperature is maintained at 4.4° C (40 F). The Eskimo's comfort range is considerably lower than that of the average American, and because of their warm clothing, they feel no discomfort at this low temperature. During extremely cold periods a fire can be built inside the igloo, with smoke vented through a wooden exhaust stack that penetrates the top of the igloo.

Summer huts are constructed in a manner similar to the igloo, but different materials are used. Sod is used on the roof to act as insulation and also to seal the surface to protect against moisture leaks. These huts are oriented with the back facing the north for protection from the cold night winds and the entry facing the sun for maximum solar-heat gain.

In wet, humid climates, cooling is the primary consideration. Buildings are left as open as possible to allow the maximum amount of air movement from the exterior to the interior. The area of the roof is minimized to decrease solar-heat gains through it. Usually roofs are constructed of materials that are light colored so that heat is reflected from them. A grass or thatched roof will allow air movement through it but will not allow water leakage. Frequently, penetrations are provided in roofs to permit ventilation of upward rising warm air. In regions with high precipitation, buildings are raised above the ground to prevent damage resulting from water runoff. All the materials used in the construction of these shelters have low thermal mass and high thermal transmission values to prevent heat buildup during the day.

In hot, dry climates, cooling of the air is necessary and humidification is advisable. These can be accomplished by using the maximum possible air movement and evaporative cooling. In parts of India and Africa, many buildings are close together, restricting the potential for cross-ventilation, so other means of inducing air movement are employed. Air scoops that catch the prevailing winds and direct them downward into the building are attached to roofs. The air travels down a large duct (approximately one square meter) to the basement or cellar area in which jugs filled with water are kept. The jugs are made from a porous clay to allow for maximum evaporative surface. As the air passes over the water, it is cooled and picks up moisture. The conditioned air is then directed upward through the building by warmed air leaving the stack above.

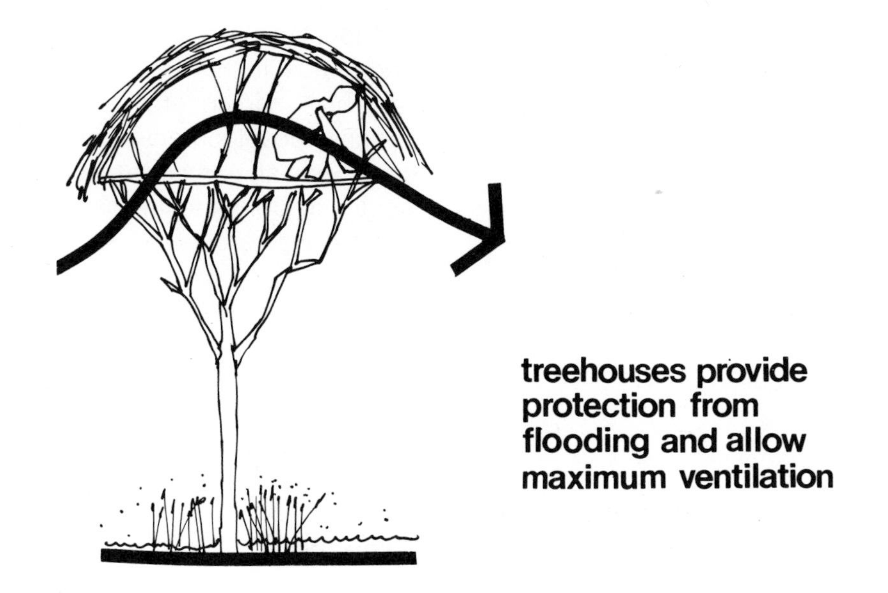

treehouses provide protection from flooding and allow maximum ventilation

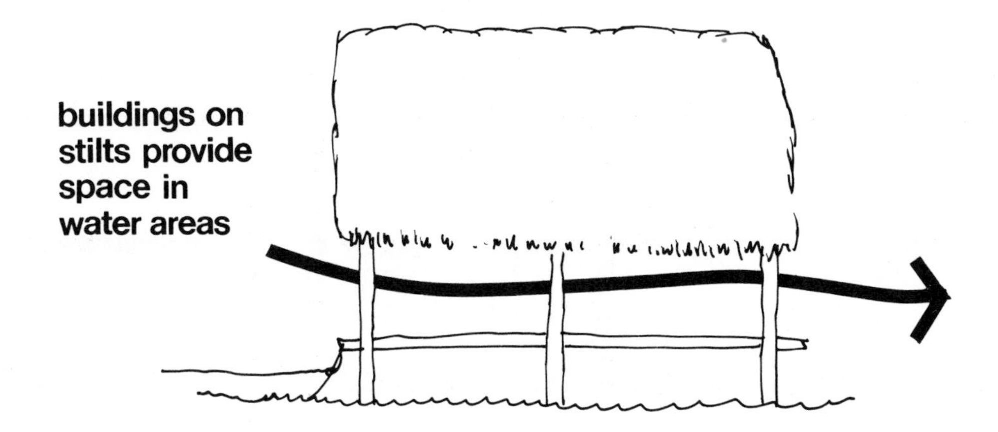

buildings on stilts provide space in water areas

Most houses are situated around an interior court which acts to promote air circulation and collects any precipitation. Since most buildings are very close together, very little heat from the sun is gained through exterior walls. Roof surfaces are light colored, reflecting heat upward.

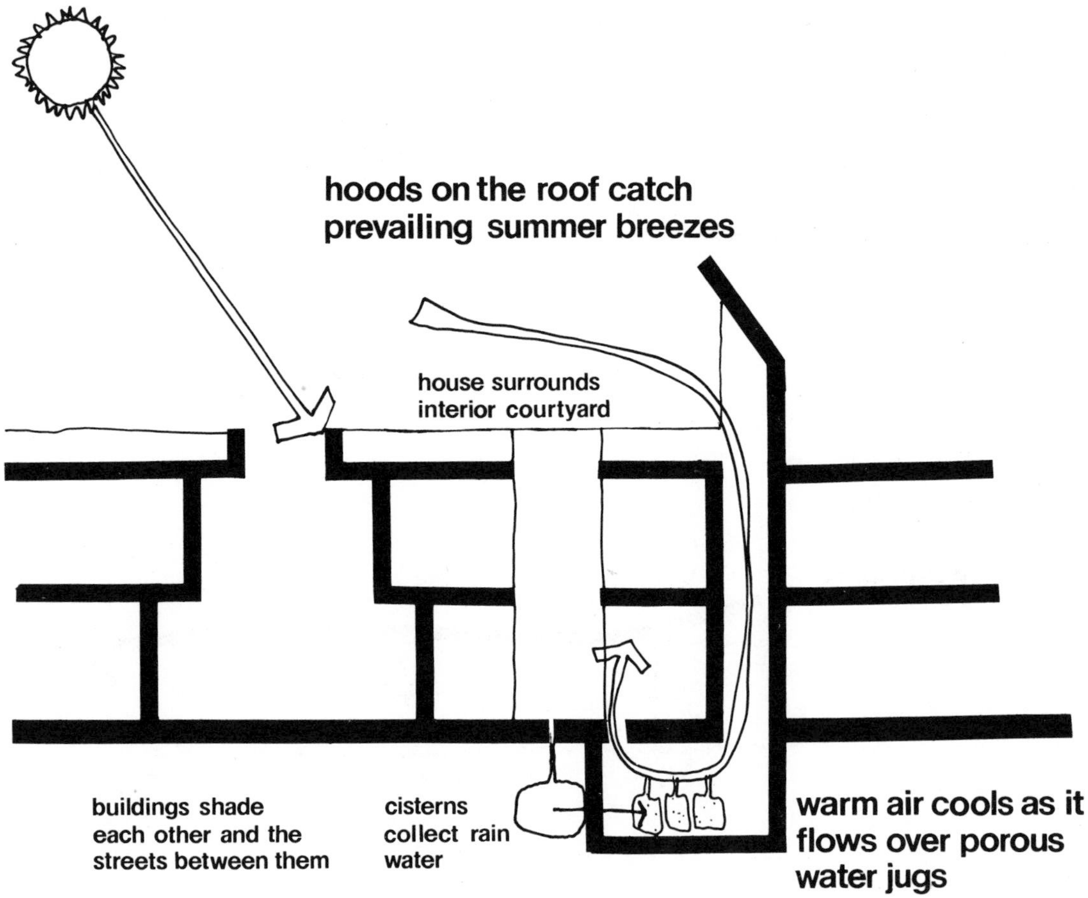

A hot, dry climate is also prevalent throughout the Southwest region of the United States. In this area the American Indians built many energy-efficient community developments. One of these has the Spanish name Mesa Verde and is located in Colorado. In this area the sandstone was eroded by wind and water, forming recesses below the edges of cliffs. These recesses were cavelike and provided shelter and a natural defense barrier for the earliest people in the area. As the culture of the region developed, dwellings were built in the backs of the caves, using the local materials of adobe and short lengths of timber. The overhanging cliffs not only provided natural defense from enemies but also offered protection from the intense summer sun. The design of the community provided for collection of solar radiation during the winter. Because the winter sun is low in the sky, buildings received sun during the day. The heat was stored in the massive adobe walls and then reradiated to the cave area at night. The cave and the buildings contained in it had large surface areas for heat gain. The development is sheltered from cold northern winds but is open to the prevailing summer breezes. The land above the cliffs was used for farming; hence the name Mesa Verde ("green table"). Because the people of this area had protection from harsh climatic elements and used natural energies to provide their comfort, they were able to enjoy a higher standard of living than other tribes living in this region.

South of Mesa Verde, in New Mexico, was the Acoma pueblo. The Indians who built this complex placed it on top of an inaccessible mesa, so that it could be easily defended. It had many design features that modern man should incorporate into his buildings. The entire complex was tiered, which allowed the maximum surface area to be exposed to the sun. Each level had a specific function that promoted the maximum utilization of natural energy.

The lowest level had very few openings and was used for storage. Since heat rises, this was also the coolest area. The roof over the storage space was used as a terrace and was accessible from the second level, which served as a sleeping area. The upper story was used for living, cooking, and eating. Because of the stacking of all these levels, natural ventilation was supplied to the upper level to exhaust heat and odors. The pueblo had numerous houses tiered in this way and connected side by side. With only two walls exposed to the outside, heat loss was reduced. The dividing walls between units were several feet higher than the roof. This provided privacy but, more important, provided shade for the terrace work areas. The walls were constructed of adobe and reinforced with timber, and consequently they were high in thermal inertia, which moderated inside temperature year round. As the sun struck the walls, the heat generated convective air currents along them, which added to natural ventilation. The roofs of the units were made of lightweight materials, offering low thermal inertia and high heat transmission; because of them, summer heat did not accumulate in the buildings.

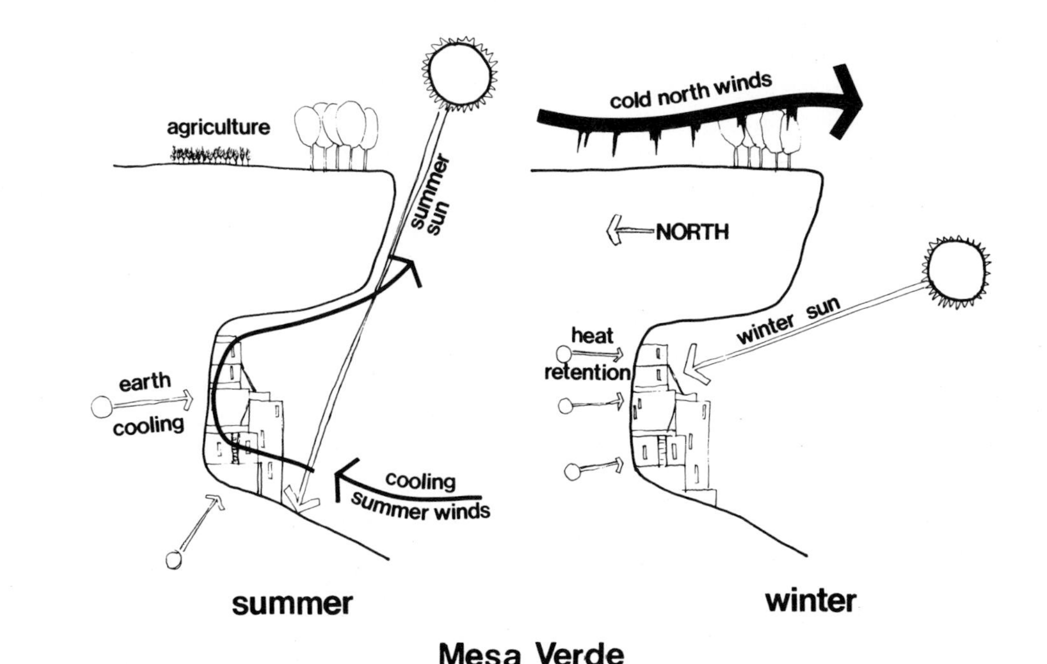

living/cooking
sleeping
storage
section

site plan

Acoma pueblo

Another Indian development in New Mexico was discovered by western man in 1921: Pueblo Bonito in Chaco Canyon. It is similar in design and construction to the Acoma pueblo, having three levels, but it was further unified by a curvilinear plan oriented to the south. High walls on the sides of the pueblo controlled the time of day at which the sun entered and exited interior areas. This pueblo was built in two stages; remarkably, the addition of the second phase further increased the overall energy efficiency of the complex. Similar materials were used in the Acoma and Bonito pueblos, and these maximized the utilization of energy in each.

pueblo Bonito

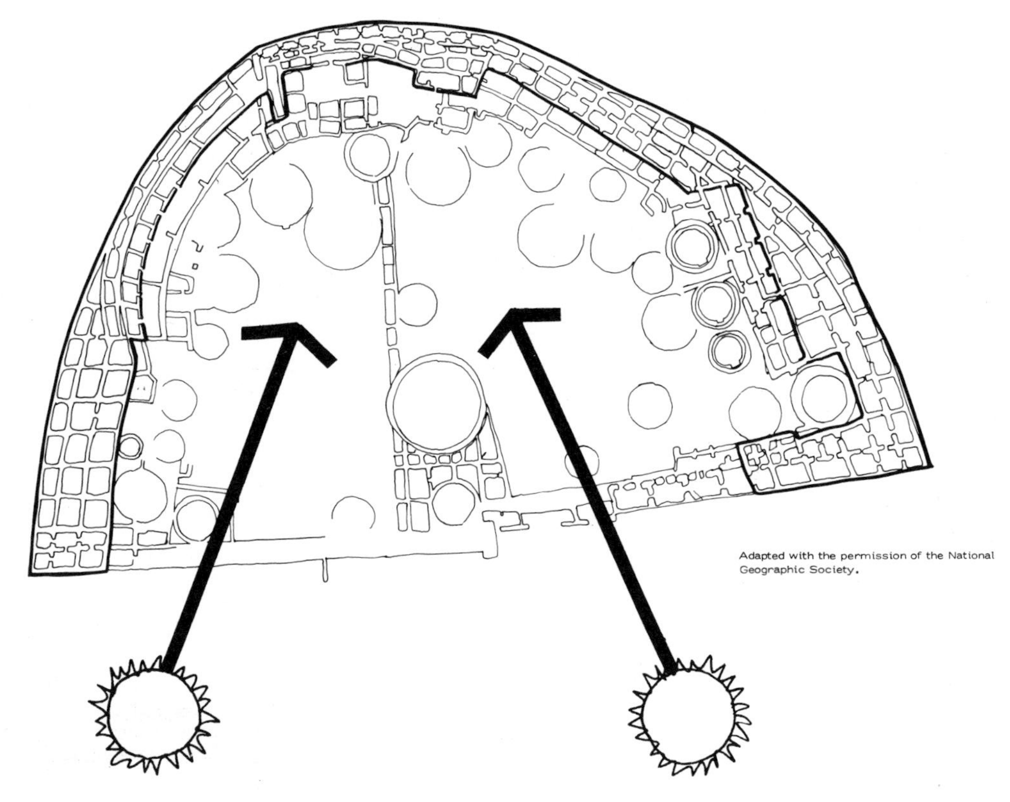

Adapted with the permission of the National Geographic Society.

site plan

The Indians were not the only Americans who built their structures to be responsive to climatic forces. Barns in many parts of the country were built with a long sloping roof facing into the direction of prevailing winter winds and terminating several feet above the ground. The space between the ground and the roof was filled with hay, straw, or other fibrous grasses to provide an insulating barrier. Snow covered the grass, which added to its insulating value. The south wall was painted red ("paint" was made by combining skim milk, iron oxide, and lime) to absorb the solar radiation striking it. With the protection from winter winds and insulation on the north, the absorption of solar energy on the south, plus the body heat of the animals on the inside, the barn functioned as a passive solar collector, providing warmth and protection to the farmyard occupants. This roof form also distinguished the numerous New England "saltbox" houses of the same era.

Early settlers in the Rocky Mountains used the materials they found there to build cabins that provided them with year-round comfort. Structures were set into the ground at least 61 to 91 centimeters (2 to 3 feet) below grade to take advantage of the stable ground temperatures. Large logs were used to construct the walls and roof, which provided insulation, thermal inertia, and heat time lag. In addition to the logs, the roof was covered with a layer of sod, forming an insulating barrier that minimized the heat loss and gain through it. In the winter, snow would accumulate to the height of the lower edge of the roof, producing a streamlined shape to minimize the heat lost to winter winds. The furry animal skins that covered the windows, plus the snow depth, greatly decreased heat loss through wall and window areas. In the summer, the skins were removed and the cooling breezes blew freely through the interior. The roof reflected most of the day's heat, while ground temperatures helped maintain comfort.

Many other examples can be cited in which people throughout history, faced with adverse climatic conditions, adapted their architecture to high energy efficiency. Many ideas found in primitive architecture can be incorporated into modern designs to promote harmony with nature instead of dominance over it. To waste our energy and that stored in our limited resources simply to overcome nature is a futile undertaking.

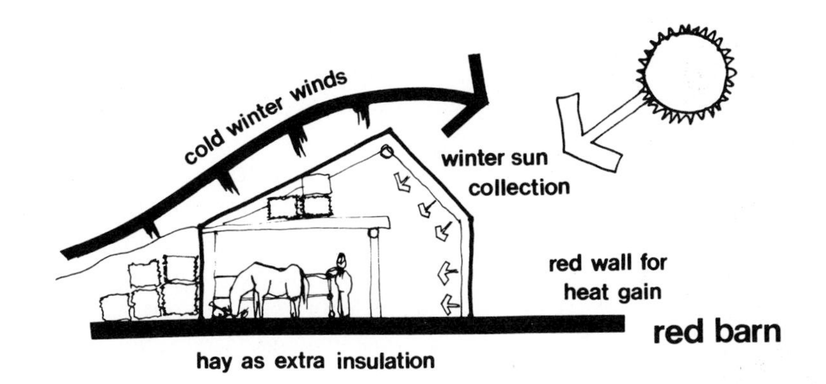

red barn

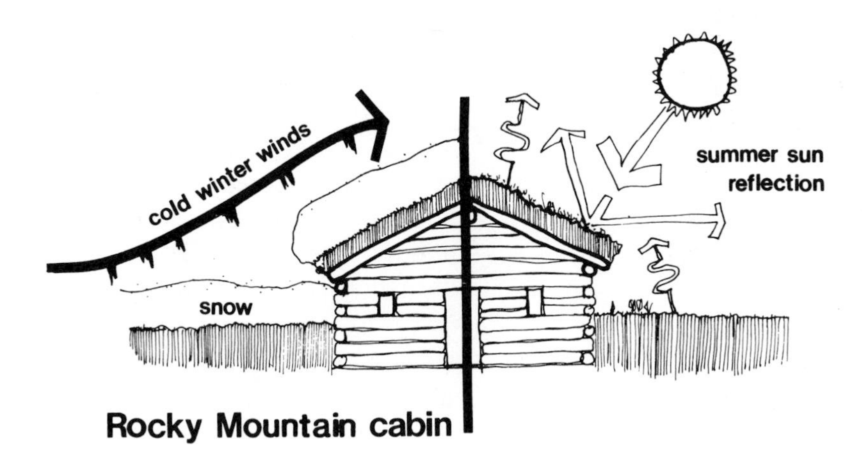

Rocky Mountain cabin

Most of the examples show how solar and other natural energies have been used in a passive way. Historic examples of active systems are also available.

One of the first recorded uses of solar energy was made by the Greek mathematician and engineer Archimedes. When the Roman fleet was attacking Syracuse harbor in 212 BC, he allegedly set the ships on fire by focusing the rays of the sun onto their sails using approximately 4000 gold and bronze shields. It is reported that the sails were black, which helped to make this amazing feat possible.

A twelfth-century writer reported that a similar solar display was mounted at Constantinople in AD 600 when a mirror-focusing system designed by Proclus was used to ignite ships in the fleet of Vitellius. Other pyrogenic military applications have been reported through history.

One of the first machines that could harness the sun's energy and convert it to mechanical work was invented by a Frenchman, Solomon de Caux, in 1615.

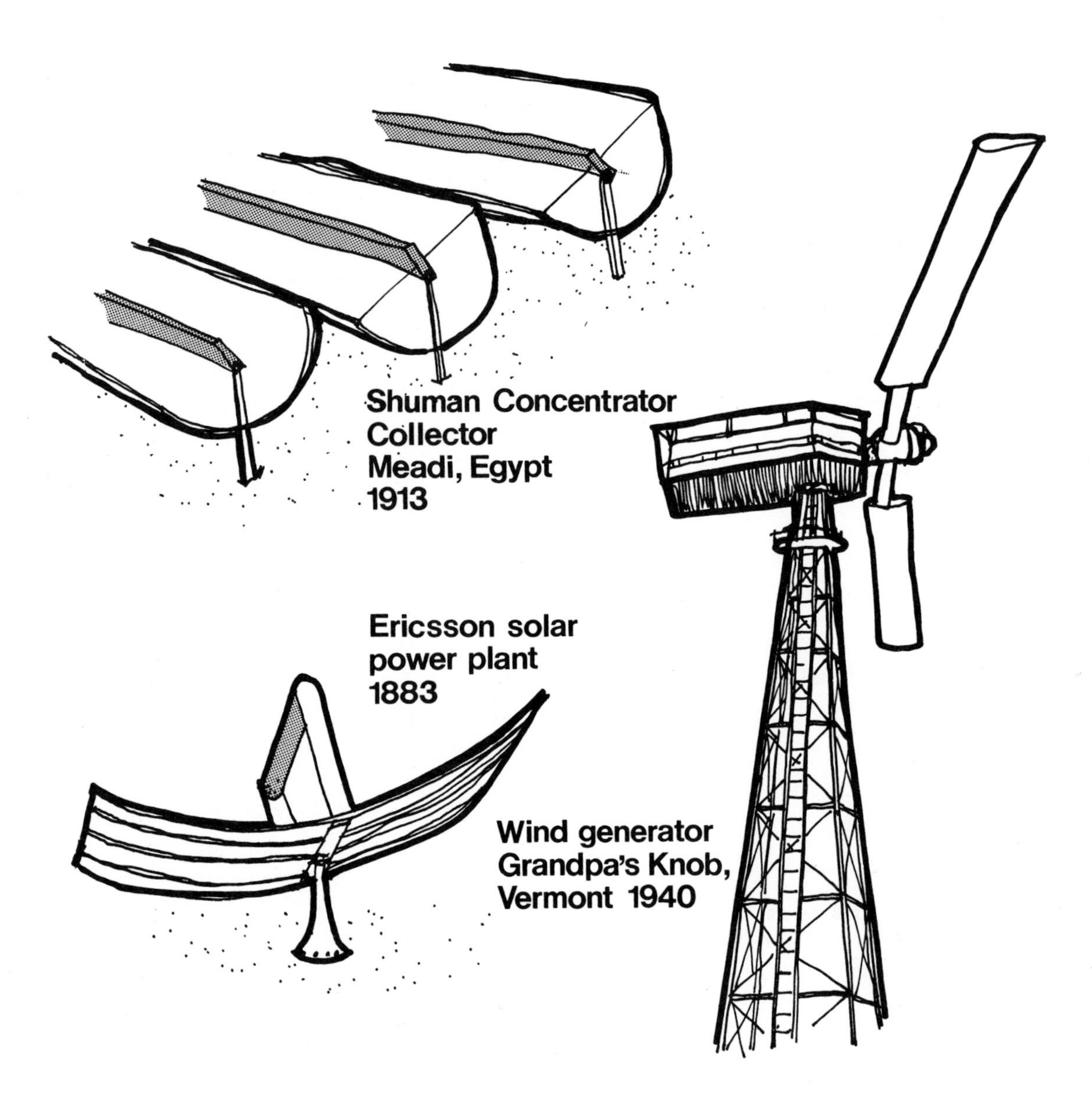

In his machine air was expanded by the sun's heat. This expansion caused an increase in internal pressure, which was then used to pump water.

Another Frenchman, Antoine Lavoisier (1743–1794), used a 128-centimeter- (50-inch) diameter lens to concentrate the sun's rays and develop temperatures of up to 1755° C (3190 F), which were sufficient to melt metals. His lens was filled with white wine because it had better optical properties than the glass that was available to him. This was the predecessor of today's solar furnaces.

In the 1870s another Frenchman, August Mouchot, developed one of the first solar-powered steam engines, which was used to drive water pumps; the steam was used to produce ice by an ammonia-water absorption process (1878).

In 1859 oil was discovered at Titusville, Pennsylvania. This heralded the beginning of an era which oil and gas would be readily available at low prices. Because fossil fuels were readily obtainable, many people lost interest in using the sun to power machinery, except in the most remote desert areas where transportation was difficult.

Isolated examples of solar design appeared in the 1870s, when John Ericsson, an American inventor, designed and built several solar-powered hot-air engines that adapted to conventional steam power for cloudy day operation. Even though these were the most effective such engines up to that time, he admitted that they could not compete with the low initial cost and high reliability of steam power generated by the burning of coal.

In 1901 A.G. Eneas built a large solar reflector that was used to power water pumps on an ostrich farm in Pasadena, California. The pumping system was rated at 3 watts (four horsepower) and could pump 5,300 liters (1,400 gallons) per minute against a static head of 3.65 meters (12 feet).

In 1908 Frank Shuman, another American, established one of the first solar flat-plate collector companies, Sun Power Company, which in 1911 built a demonstration plant of more than 929 square meters (10,000 square feet) in collector area. The system operated a steam-driven water pump. In 1913 the company supplied an installation with 1,207 square meters (13,000 square feet) of parabolic-trough concentrator area in Meadi, Egypt (south of Cairo), which powered a 75 watt (100 horsepower) steam engine.

In the early 1940s, a large-scale wind-driven electrical generator was installed at Grandpa's Knob, Vermont. The machine was conceived by Palmer C. Putnam, an engineer, who coordinated the design and construction of the 1,250 kilowatt (1,676 horsepower) machine. It had a useful life of only eighteen months, eventually losing a rotor blade that was never repaired because cost calculations indicated that it could not compete economically with generators powered by fossil fuels.

By the 1930s small wind-powered electrical generators were gaining popular acceptance with people living in remote areas of the plains and western United States. During the Roosevelt administration of the 1930s, however, the Rural Electrification Administration (REA) was established to provide federally subsidized power to these areas, and privately owned wind generators soon lost their appeal.

Only recently has universal interest in solar, wind, and other natural energy been revitalized. The high costs of power and heat supplied by public utilities and the shortages of fluid fossil fuels have forced many people to reevaluate their long-term energy needs and the means by which they will be supplied. The necessity for climate-responsive architecture and systems has re-emerged.

8 Contemporary Use of Energy

Some buildings in the United States lose as much as 158 to 189 watts of heat per hour per square meter (50 to 60 Btu per square foot) during the winter. On a yearly basis, buildings directly consume 15 percent of all the energy used by this nation. If the energy required by buildings could be reduced by 80 percent through energy conservation and by using alternative-energy sources, a decrease in the nation's total energy usage of 12 percent could be realized.

The American fascination for detached single-family homes has produced numerous problems, the majority of which have perpetuated excessive energy usage. Each member of a family must use an automobile to travel to and from his single-family suburban home. The population density in most suburban communities is not sufficient to develop economical mass-transit or other systems for moving people. The transportation of goods and materials to suburban locations promotes further energy usage.

A major portion of the energy utilized for space heating is consumed by single-family houses. Consumption by larger buildings is controlled by many conditions that can be modified on existing structures and incorporated into the design of future buildings. The areas where the average house loses heat are illustrated in the accompanying diagrams. Each of these areas can be treated and improved separately to diminish winter heat losses. Many of these and other characteristics can also be analyzed in order to decrease summer heat gains.

A great deal of heat is transferred through doors. It flows through cracks in the wood of the door, around its edges, through keyholes, transoms, mail slots, adjacent windows, and is conducted through the door itself.

Heat is lost through windows when there are cracks in the frames or the glass. Losses through the glass itself can be excessive. Heat is conveyed to the inside surface of the glass by convection and radiation, conducted through the glass, and then lost to the outside by convection and radiation.

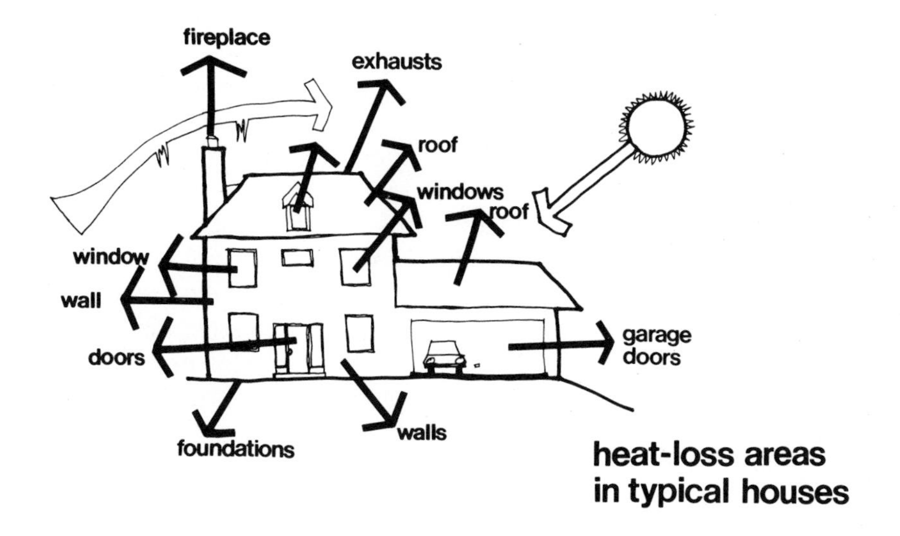

heat-loss areas in typical houses

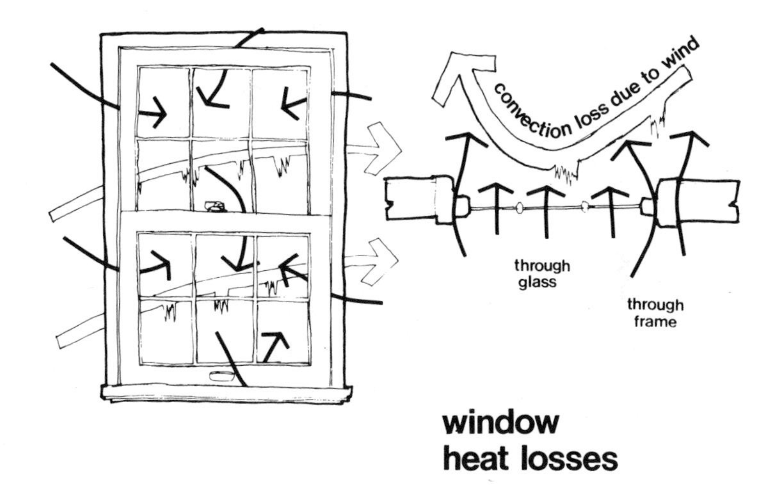

window heat losses

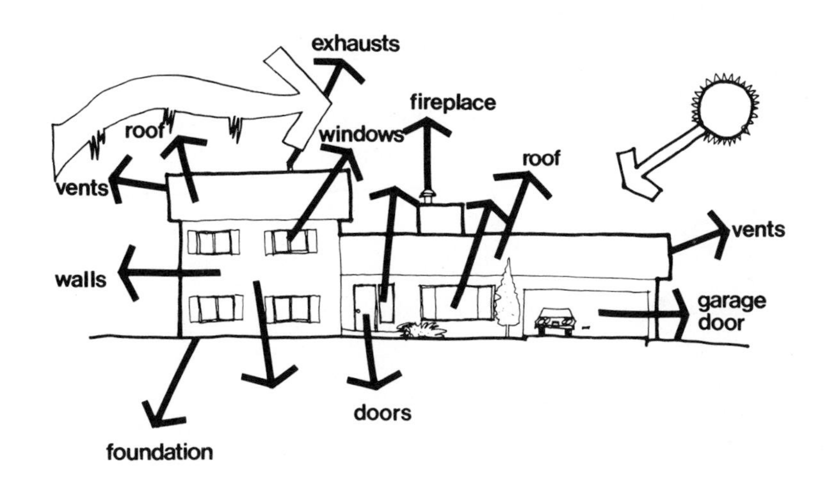

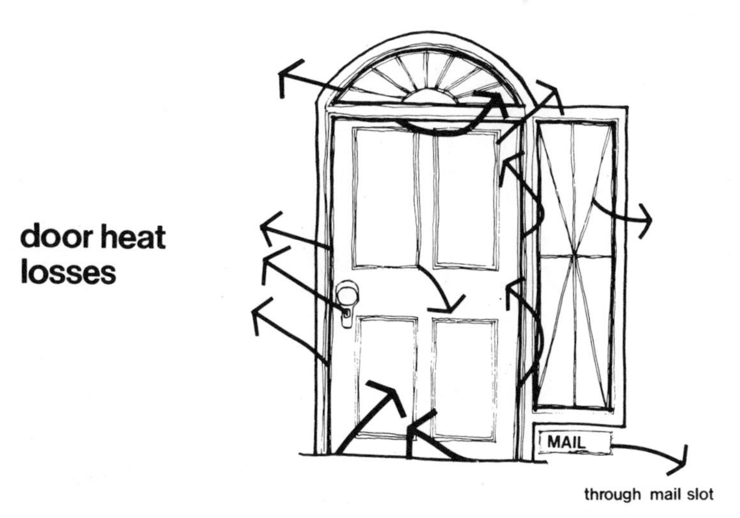

door heat losses

Fireplaces are traditionally identified with warmth and comfort. The negative aspects of continual interior-heat losses through the flue and chimney, which project through the outer wall or roof, and the depletion of interior space oxygen as a result of the direct-combustion process during cold periods are generally not considered. A large percentage of the heat produced by a conventional fireplace goes up the chimney. Fireplaces also contribute to the heat loss of a house by drawing warm air out of the room through the flue after a flow has been initiated. Dampers on most units leak, allowing heated air to flow out of the room, even when the fireplace is not in use. Whenever air is removed from a room, it is replaced by cold air that infiltrates from the outside. The net effect of most fireplaces is cooling of a house rather than heating it.

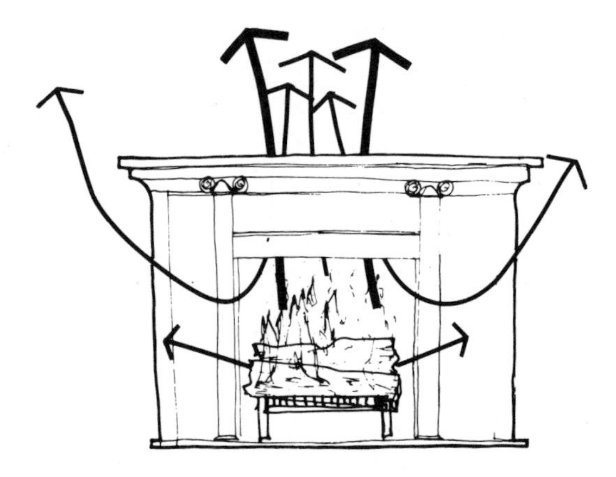

fireplace heat losses

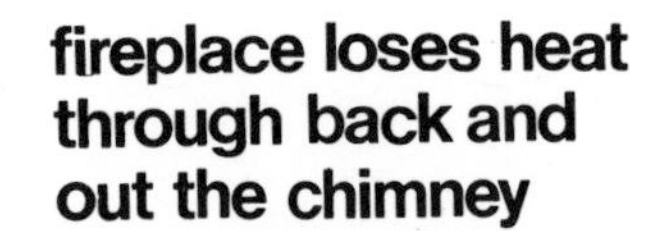

fireplace loses heat through back and out the chimney

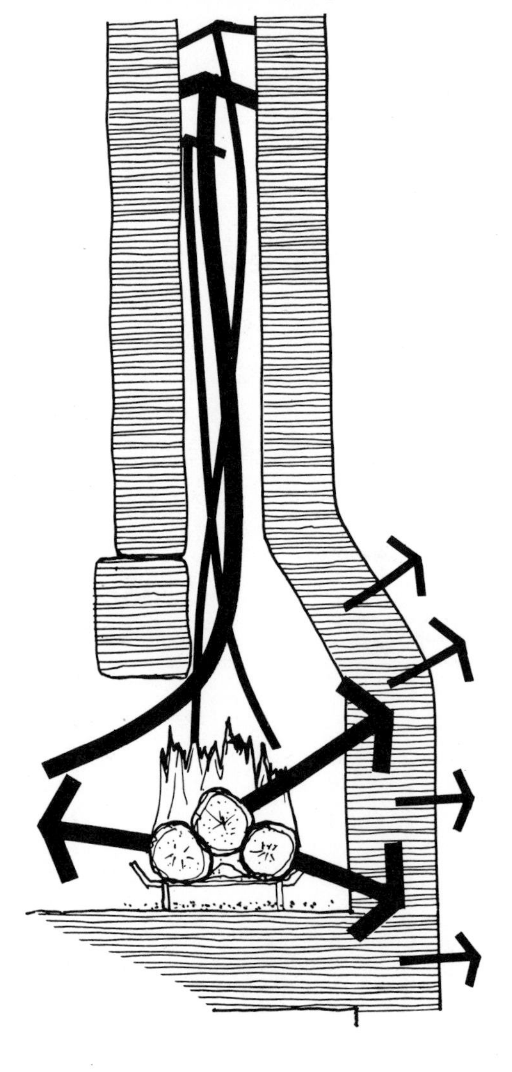

There are a number of ways that a conventional fireplace can be modified to overcome these energy-losing aspects and improve their efficiency. To understand why some of them are effective, it is necessary to analyze what happens to the heat liberated in a fireplace.

Much of the heat is dissipated up the chimney while some is radiated to the floor, back, and sides of the firebox, and the remainder is radiated to the house. Heat radiated to the sides and back will be conducted to the outside if adequate insulation is not provided. Insulation between the firebox and the outdoor projection of the chimney, or a chimney and firebox built within the insulated outer walls, will increase the quantity of heat radiated and conveyed to surrounding interior areas. If the walls and floor are masonry, they will store the heat and gradually release it to the house.

In this regard, the advantages of the interior-plan location of fireplaces are quite pronounced, because their lateral heat losses are confined within the building. Very tight or airlocked double dampers are needed to avoid continual losses through the flue. During the winter, they should be closed at all times, except when the fireplace is in use. During the summer, a properly designed flue may be used as a vertical stack to initiate passive ventilation. Direct outdoor air for combustion, brought in through a duct that is in use only when a fire is present, conserves energy and helps to maintain interior oxygen levels.

Many devices and prefabricated firebox units are available that employ convection to distribute the heat from a fireplace more efficiently. The prefabricated fireboxes have low-level cool-air-intake slots in the front or on the sides and warm-air outlets on the top. The air is circulated around the back and sides of the firebox, where it is warmed and then returned to the room or passed through a duct to adjoining areas. A number of manufacturers are marketing grates made from hollow metal pipes that are heated by radiation and convection. Air blown through the pipes absorbs heat before re-entering the room, thereby delivering heat that otherwise would have been lost.

Control over the supply of oxygen for clean but not too-rapid burning of the fuel should be considered in the choice of fireplace or wood-burning-stove details. Heat-tempered glass firescreens for fireplaces and Franklin stoves with draft control provide this function. Glass firescreens also afford safety by preventing sparks from being thrown into a room. Heat is radiated and conveyed to the glass, conducted through it, and then radiated and conveyed to the room. These firescreens help to prevent heat loss when the fireplace is not in use.

There are also devices available that attach to the stack or chimney and extract from the exhaust air heat that can then be routed back to the home. Some of these are heat exchangers of the air-to-air variety, while others contain heat pipes. The efficiency of these devices varies from one type to another; in fact, one confident manufacturer warns against cooling the exhaust gases too much, which might cause the chimney to stop drawing.

A well-designed fireplace, arranged as a central feature of a home or building, may be combined with an insulated chamber that would also retain either solar thermal or off-peak electric energy. This heat could be delivered to the building as necessary to maintain comfort.

Wood stoves have been regarded as a boon for passive-solar systems and a handy way to reduce or avoid winter heating costs. Unfortunately, wood stoves and fireplaces can grossly pollute the atmosphere, virtually blanketing a neighborhood with smoke and other particulate emissions, which contain known carcinogens. Creosote accumulation in chimneys and the general fire danger, especially in remote areas, are factors to consider when using wood stoves. Furthermore, interior oxygen depletion from tight, energy-conserving homes and the probable negative impact on forests of burning valuable hard and soft woods compound the problem; wood is a renewable but rather slow-growing energy form.

Recent advances in higher-efficiency wood-stove and furnace designs, including the use of catalytic converters, higher combustion temperatures, and heat-storage systems, should help reduce emissions, creosote problems, and the strain on forests and other wooded areas. Controlled intake air from outdoors for combustion in both fireplaces and wood stoves will prevent oxygen depletion from the indoor environment. This provision should be designed to fit the specific characteristics of the fireplace or stove, insulated and dampered as necessary to avoid indoor heat loss.

The installation of wood stoves for emergency use only is reasonably logical, but to depend on wood stoves as a major heat source is questionable.

Many factors affect a building's heating and cooling "loads," which can be thought of as the amount of heat that must be added in the winter and taken out in the summer to maintain the comfort of the occupants. Some of these factors are the size and shape of the building, the occupancy and tasks performed there, climatic conditions, the orientation of the building, the layout of the rooms, and the rate at which heat enters or leaves the building. The materials selected for construction and the thickness and type of insulation used control the rate at which heat will flow through the walls, ceilings, roof, and floor of a building. This heat-flow rate can be calculated in terms of Btu per square foot per hour per degree Fahrenheit temperature difference between the inside and outside surfaces (Btu/square foot/hour/degree F). To make this calculation, specific construction details must be known and certain properties of the construction materials must be determined.*

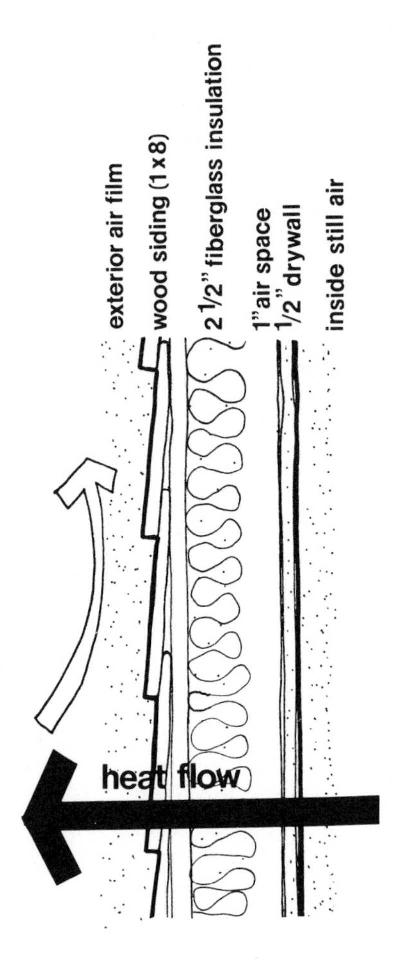

wall cross section

The following are sample calculations to determine the heat-flow rate through the walls, roof, floor, and openings of a typical suburban home. Some assumptions about the temperature differential are made, and the heat loss in terms of Btu per hour (Btu/hr) is determined. Suggestions are made for modifying the house to make it energy conserving, and the heat loss for the modified home is calculated.

The construction of the wall and the material inside the wall must first be determined. All the materials have heat-conductivity and heat-resistance values. One is the inverse of the other. Resistance factors are easiest to work with since they can be added directly to determine the resistance of a complete section. The larger the resistance factor (R factor) of a material, the better insulating qualities it will have. The films of air on the exterior and interior surfaces of a wall also act as insulators and are assigned an R factor.

Below are the R factors for the individual components of a wall with little insulation in a typical suburban house. Also given is the total R factor of the wall. The values are from the *ASHRAE Handbook*. This second printing of the 1977 edition does not include the Product Directory or Manufacturers' Catalog Data.

	R Factor
Exterior air film	0.17
1 × 8 (2.5 × 20 centimeter) wood siding	.78
1" (2.54 centimeters) air space	.90
2½" (6.35 centimeters) fiberglass insulation	7.80
½" (12.7 millimeters) drywall	.45
Inside still air	.68
Total of R Factors, R_t	10.78

The rate of heat transfer through the wall is expressed by the overall coefficient U, or U-value, which is expressed in units of Btu per hour per square foot per degree Fahrenheit (Btu/hour/square foot/degree F). The U-value is the reciprocal of the total R factor, $U = 1/R_t$. For this wall the U-value is $1/10.78 = .093$. The smaller the U-value, the better the insulating properties of the wall. For the wall section above, if 2.54 centimeters (one inch) of rigid polyurethane foam insulation (R = 5.88) were added in place of the air space, the total R factor would be 15.76, resulting in a U-value of 1/15.76 = .063. Changes could also be made to the siding, interior drywall, and other materials, which would decrease the U-value.

To calculate the amount of heat flow per hour through the wall, the U-value must be multiplied by the wall area and the temperature differential—that is, the difference between the outside and inside air temperatures. In Denver the design temperatures are -10 F outside and 70 F inside. This is a +44 degree C (80 degree F) differential. If the wall mentioned above, with a U-value of .093, had an area of 900 square feet (excluding doors and windows), the heat flow per hour could be calculated as:

Wall area × U-value × temp. diff.
= Heat flow
900 sq./ft. × .093 × 80 F
= 6,696 Btu/hr (1,961 watts)

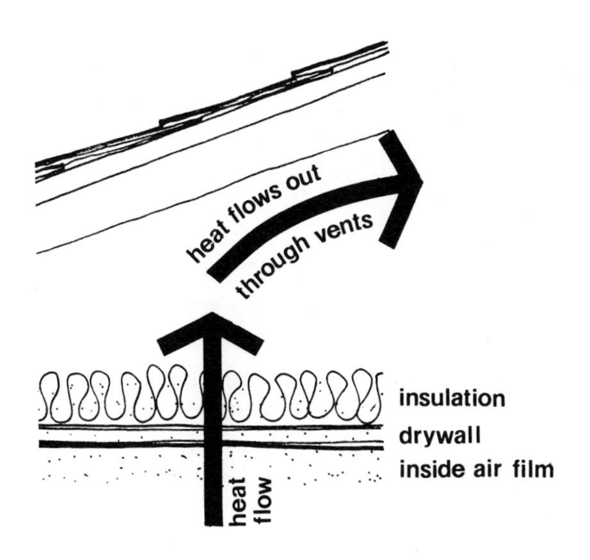

This type of calculation assumes that the temperature differential is constant throughout the wall height. It also assumes "steady state" heat conditions, as well as a homogeneous wall section. Strictly speaking, adjustments should be made to the heat flow for areas of the wall that are occupied by structural, electrical, and mechanical elements. In some cases, this could be in excess of 30 percent of the wall. The U-value in these areas would be greater than the one calculated, meaning more heat flow through the wall. These considerations also apply to the roof area and structural ground floors.

The heat flow through a wall with the extra insulation is decreased to 900 × .063 × 80 = 4,536 Btu/hr (1,328 watts), a decrease of 2,160 Btu/hr (633 watts).

The heat flow through the ceiling can be calculated in a similar manner. If the attic is vented to the outside, the calculations for a typical ceiling are presented below.

Ceiling	R Factors
Outside air	0.17
3" (7.6 centimeter) insulation	8.33
½" (12.7 millimeter) drywall	.45
Inside still air	.61
Total of R Factors	9.56

The U-value is 1/9.56 =
.105 Btu/hr/sq. ft./deg. F.
5,966 watts/cm²/° C

If the vents in the attic are sealed, the ceiling, attic, and roof contribute to the insulation value. With additional insulation and the attic vents sealed, the calculations become:

Ceiling, Attic, and Roof	R Factors
Outside air	0.17
Shingles (asbestos-cement)	.21
½" (12.7 millimeter) plywood	.62
Attic air space	1.40
6" (15 centimeter) fiberglass batts	18.72
Existing insulation	8.33
½" (12.7 millimeter) drywall	.45
Inside still air	.61
Total of R Factors	30.51

The U-value is 1/30.51=
.033 Btu/hr/sq. ft./deg. F.
1,875 watts/cm²/° C

Assuming an approximate projected roof area of 24 × 32 feet (7.3 × 9.8 meters) = 768 square feet (71.3 sq. meters), where 24 feet and 32 feet are the plan dimensions of a rectangular building, the maximum heat flow per hour is 768 × .105 × 80 = 6,451 Btu/hr (1,889 watts). If the alterations are made to the roof, the heat flow per hour is 768 × .033 × 80 = 2,027 Btu/hr (594 watts). This is a significant reduction. The projected roof area corresponds closely to the ceiling area of the house and is used instead of the actual area of the roof because the heat is assumed to be out of the house once it crosses the plane of the ceiling.

Depending on their construction and location in relationship to the grade, heat can be lost through floors. In this example it can be assumed that the house is atop a basement. The total floor area of the basement is 750 square feet (69.7 square meters) and the total basement wall area, exposed to the ground outside, is 800 square feet (74 square meters). For a groundwater temperature of 10° C (50 F) and a basement temperature of 21° C (70 F), the hourly basement floor loss is 2.0 Btu/square foot (6.3 × 10^{-4} watts/square centimeters) and the hourly basement wall loss is 4.0 Btu/square foot (1.3 × 10^{-3} watts/square centimeter).

The heat flow through the concrete is 750 × 2.0 + 800 × 4.0 = 4,700 Btu/hr 1,377 watts.

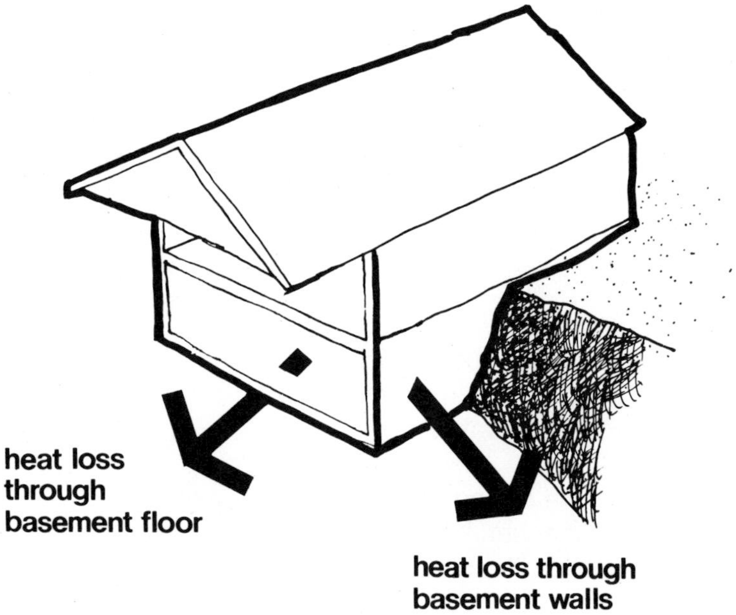

heat loss through basement floor

heat loss through basement walls

A great deal of heat can flow through the windows of a building. Assuming that the typical house has single-pane non-insulating glass windows, their U-value is 1.13. If the window area is 20 percent of the building floor area, then the heat flow through the glass is (768 × .20) × 1.13 × 80 = 13,885 Btu/hr. 4,067 watts. For a metal sash, in which the glass area is 80 percent of the window area, the adjustment factor is 1.0, with a resulting heat flow of 13,885 × 1.0 = 13,885 Btu/hr. 4067 watts.

If insulating glass (two panes with ½-inch [12.7 millimeters] air space between) is used, the U-value is .58. The heat flow through the glass can be calculated to be 153 × .58 × 80 = 7099 Btu/hr (2,079 watts). For a metal sash, double glass, and a glass area of 80 percent of window area, the adjustment factor is 1.20, and so the heat flow is 7099 × 1.20 = 8519 Btu/hr (2,495 watts).

If wood-frame insulating storm windows are used, in which the glass area is 60 percent of the window area, the adjustment factor is .80 and the U-value is reduced to .46, resulting in a heat flow of 153 × .46 × 80 = 5630 Btu/hr (1,649 watts).

The U-value is 1/30.51= .033 Btu/hr/sq. ft./1,875 watts/centimeters^2/° C

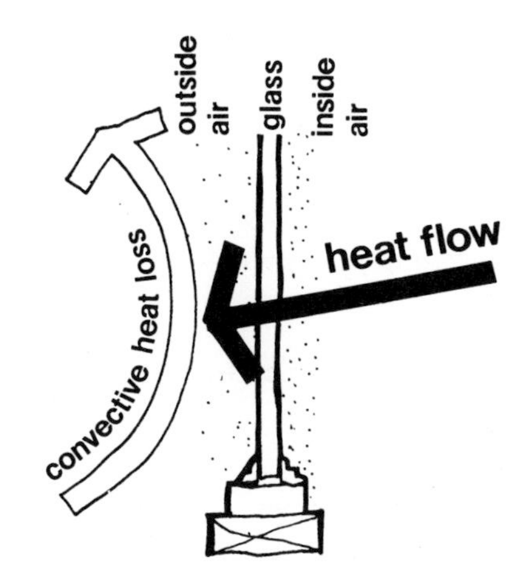

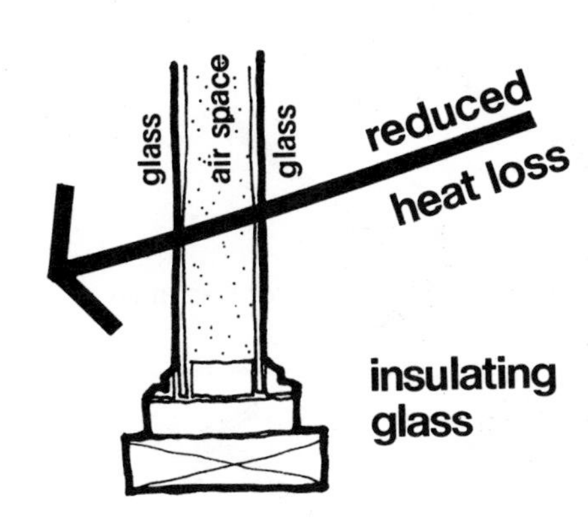

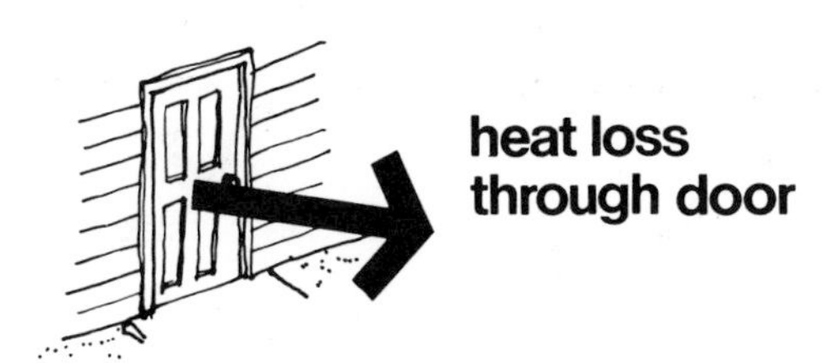

Heat is also lost through doors. In this example assume that a standard 1½-inch (3.8 centimeter) solid-core door and a metal storm door are used. The U-value for the doors is .33. There are two doorways, front and rear, each 3 feet × 7 feet. The heat flow is (7 × 3)2 × .33 × 80 = 1,109 Btu/hr (325 watts).

By replacing the conventional doors with insulated doors, the U-value can be reduced to .15 with a heat flow of 42 × .15 × 80 = 504 Btu/hr (148 watts).

Infiltration is the replacement of some or all of the air in a room or building by air from the outside. In the winter, cold winds blow outdoor air into interior spaces through cracks around windows and doors, as well as through other cracks in the building. The cold air forces out air that has been warmed by heating units. A great deal of heat is lost in this process. The procedure for calculating the amount of heat lost as a result of infiltration involves first determining the amount of air displaced per hour and then multiplying this by the amount of heat contained in a unit volume of air and by the design-temperature differential.

Different rooms of a building experience different numbers of air changes per hour, depending on the number of exterior walls, doors, and windows. The *ASHRAE Handbook* contains lists for this purpose.

In this example we'll assume that 30 percent of the building is composed of rooms with doors or windows on three sides, and the remainder has doors or windows on two sides. So, 30 percent of the building has 2 air changes per hour and 70 percent has 1.5 air changes per hour. To determine the amount of air displaced per hour, the room volume is multiplied by the number of air changes. The total floor area of the building is 768 square feet (71.3 square meters) and with 8-foot (2.4 meter) ceilings, its volume is 768 × 8 = 6,144 cubic feet (174 cubic meters). With 30 percent having 2 changes per hour and 70 percent having 1.5 changes per hour, the total amount of air displaced is (6144 × .30) × 2 + (6,144 × .70) × 1.5 = 10,137.6 cubic feet (287 cubic meters) of air.

The heat capacity of air is calculated by multiplying the specific heat of air (Btu/pound/degree F, or calories/grams/°C) by its density (lb./cu. ft., or grams/cubic meters). Representative values for these are 0.24 Btu/lb./deg. F and 0.0624 lb./cu. ft. (sea level air density is 0.075 lb./cu. ft.), with a product of 0.015 Btu/cu. ft./deg. F. The total heat loss caused by infiltration is 10,137.6 × 0.015 × 80 = 12,165 Btu/hr (3,563 watts). One of the largest heat losses of the entire building is caused by infiltration! This is typical of most buildings. It is necessary, in energy-conserving buildings, to decrease infiltration losses as much as possible. If weatherstripping is added to the windows and doors, the infiltration losses can be reduced by one third. This applies to all buildings, so for this example the heat losses are reduced to 8,110 Btu/hr (2,375 watts). Further reductions can be realized with double-door air-locked entries and by including a continuous vapor barrier in all walls and roofs.

A summary of all the heat losses is presented below along with the heat losses for the modified building.

By simple means of energy conservation, the original heat loss for the typical house was reduced by 43 percent. It is noteworthy that optimized interior planning, exterior reflecting devices, landscaping, and other measures would improve this figure considerably.

Similar methods for calculating heat flow apply to commercial and large residential structures. For many of these buildings the heat generated by lighting, internal equipment, and occupants may represent a large percentage of the total heat gain. In some cases these gains may be so large that heat has to be removed from the building, even during the winter. The occupancy of a building and the activities performed in it regulate the quantity of internal heat generated. For this reason these remarks are confined to characteristics of the building envelope and some mechanical systems that affect the heat gains and losses of large buildings and are directly related to enery conservation.

	Existing		Modified	
	Btu/hr	Watts	Btu/hr	Watts
Windows	13,885	4,067	5,630	1,649
Infiltration	12,165	3,563	8,110	2,375
Walls	6,696	1,961	4,536	1,328
Roof	6,451	1,889	2,027	594
Floors (basement)	4,700	1,377	4,700	1,377
Doors	1,109	325	504	148
Total	45,006	13,182	25,507	7,471

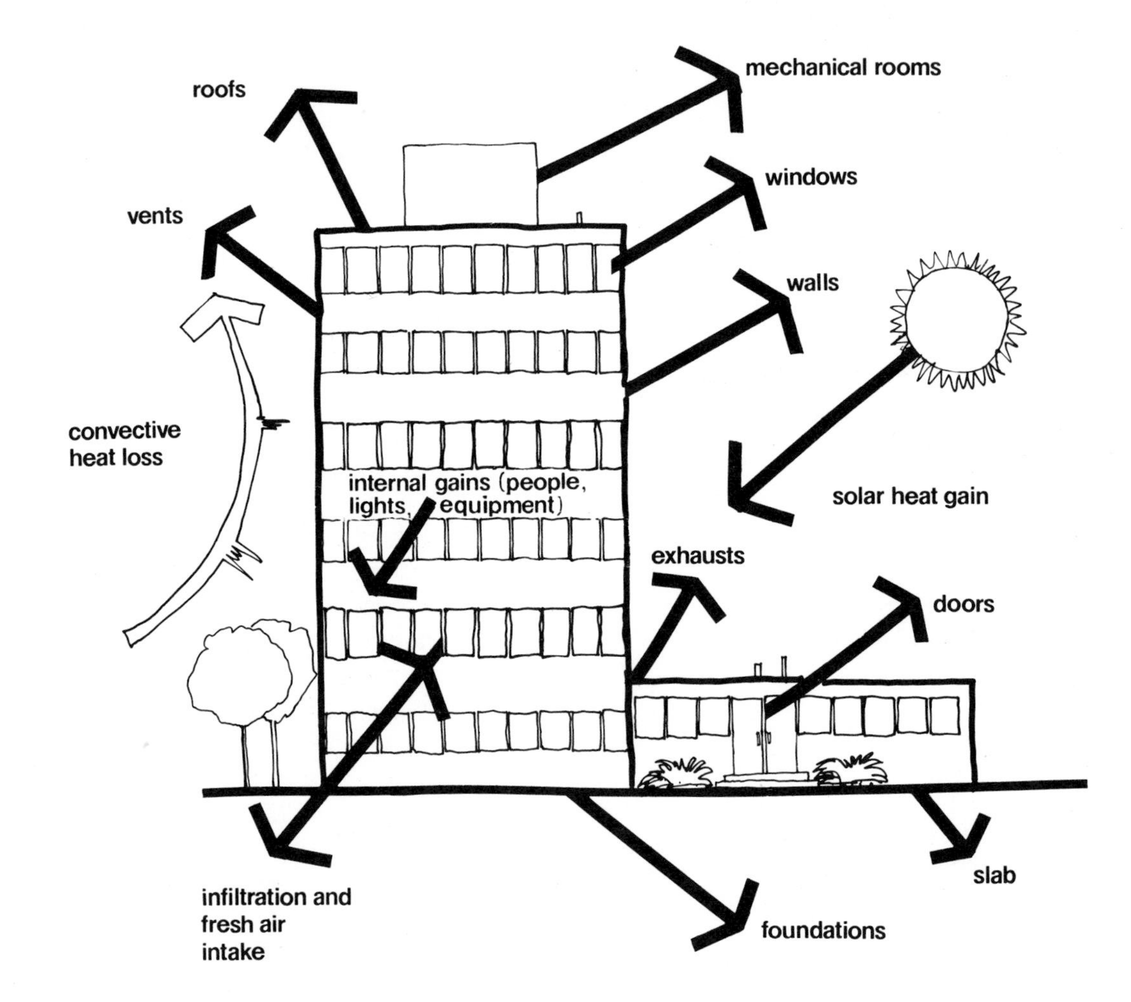

The areas in which typical large buildings lose and gain heat are illustrated in the accompanying diagrams. Each of these areas can be modified and improved to increase the energy efficiency of existing buildings, and they can be optimized during the design process of new buildings.

In the accompanying diagram the annual energy consumption of an average school in the northern United States is presented. This consumption is also representative of offices and other commercial buildings in this area, and it clearly indicates that the largest percentage of the energy is used for lighting. For this reason natural lighting should be a primary consideration in building envelope design, yet heat gains and losses as a consequence of excessive window area should not be overlooked.

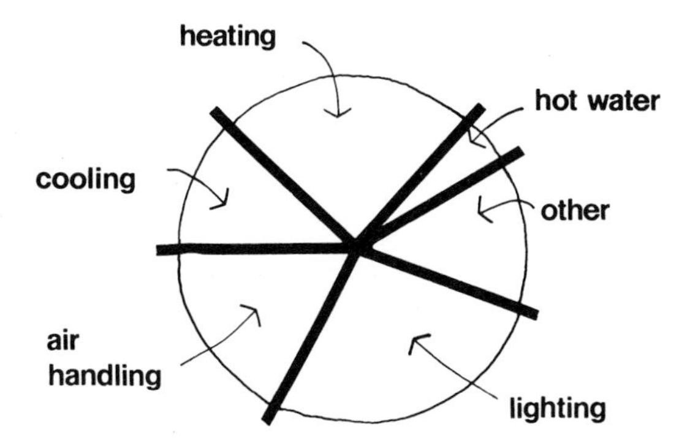

annual energy consumption of a school

Large window areas contribute to winter heat losses and summer heat gains, resulting in year-round expenditures of energy to maintain comfortable indoor environments. Glass is a notoriously poor insulator, with single panes of flat glass having winter coefficients of heat transmission, U-values, on the order of 1.13 (R = 0.88). This coefficient can be reduced by about 40 percent if insulating glass is substituted for a single thickness of conventional glass.

An even greater reduction can be accomplished by reducing the glass area. A decrease in glass area does not necessarily mean an equivalent reduction in light. Depending on wall, ceiling, and floor color, as well as the placement of windows with respect to interior surfaces, a 50 percent reduction in glass area may result in less than a 30 percent reduction in light.

The design and integrity of the building envelope contribute directly to the infiltration of outside air into a building. The cost of heating and cooling this air can run as high as 50 percent of the total heating and cooling bill.

In a well-sealed building, with weather-stripping, caulking, few cracks, and double "airlock" style doors, infiltration amounts to about one-half of the building's air volume change per hour. In a poorly sealed building the number of air changes attributable to infiltration may be five or six times this amount.

The insulating quality of walls, roof, and floors is among the factors contributing to the energy required for heating, cooling, and air handling. Adding insulation to these areas can be difficult and costly. However, by adding 5 centimeters (2 inches) of insulation to a brick cavity wall, the fuel required to offset heat losses and gains through the wall may be reduced by as much as 60 percent. Wall insulation, especially in multistory buildings, can significantly decrease heat losses and gains, with a corresponding savings in the fuel necessary to maintain comfort.

Orientation of the building with respect to the sun and microclimatic conditions has an effect on heating and cooling. Buildings with the majority of their windows oriented toward the south will receive beneficial winter sun but may receive too much summer sun. Properly designed shading devices can virtually eliminate summer radiative gains through such windows. On the other hand, north-facing windows provide for summer cooling but suffer tremendous amounts of winter heat loss. Further complicating this problem are buildings requiring cooling on the south to offset heat gains and heating on the north because of heat losses, with no mechanical system to transfer internal heat from one area to the other.

All too often energy is consumed to cool one area of a building while additional energy is used to heat another area of the same building. Proper zoning of the mechanical systems and heat-pump networks within the building can eliminate these wasteful practices.

Large classrooms, auditoriums, arenas, and other enclosed places of assembly can realize substantial heat gain even from seated occupants. It should be remembered that a sedentary human gives off approximately 400 Btu/hr (117 watts) to his/her surroundings. In a sports arena containing 20,000 spectators, hourly heat gain will easily exceed 2,343,000 watts (8 million Btu), exclusive of energy expended on vocal support (cheering) or abuse (booing)!

In hot climates cooling is of primary importance, and buildings with large surface-area-to-volume ratios are best suited to these areas. In climates where heating is the primary consideration, surface area to volume ratios should be reduced. Considerable energy savings are possible if ratios of this type are computed and adapted to local climates. There is no practical way, short of major reconstruction, to alter these ratios for existing structures. Computations of this type are beneficial mainly in the design of new buildings or in additions to old ones.

In conclusion, special attention must be directed toward identifying and quantifying internally generated heat. It may be the largest single energy input and is a factor to consider in the design of heating and cooling equipment. Building-envelope characteristics regulate heat flow in a number of ways, with the major concepts to consider being type, placement, and size of glass areas; quality, placement, and thickness of insulation; wall mass for thermal inertia and time lag; surface area to internal volume ratios; color and texture; and special shapes, wingwalls, overhangs, louvers, or shading devices. Building orientation for microclimatic conditions is essential. Prevailing winds steal heat and initiate pressure differentials that increase infiltration. Heat loss and gain caused by infiltration must not be underestimated, for it can be substantial in a "leaky" building. Ventilation may require exhausting large amounts of heat to the outside. For winter operation, heat exchanges should be considered. Mechanical systems make the building "work" and should be designed to selectively allocate internal energy before drawing upon a supplementary energy source.

9 Building Design for Energy Optimization

Life-styles as well as individual requirements regulate the amount of energy consumed by a society. Some life-styles are highly energy demanding while by comparison others are not. A procedure in the United States has been to ignore energy requirements and to plan communities that are extremely wasteful in their transportation and heating requirements. The amount of energy required to maintain and operate housing developments has been a secondary consideration in making decisions affecting land planning.

The environmental impact of these new developments has, until recently, received little attention. Business, industry, and small municipal governments have encouraged urban sprawl. Shopping centers have been built in outlying areas that helped encourage growth of new communities. Industries have supported sprawl because it promotes the placement of value on certain items, such as the automobile. Municipalities and outlying large urban areas benefit from growth because new residents mean a stronger tax base. Urban sprawl has also resulted in alarming increases in energy demand. All facets of delivering goods and services to people in divergent locations require energy.

Because the length of power-transmission lines must be increased to reach suburban locations, energy losses in them are increased. Commuting from the suburbs to jobs in the city requires tremendous amounts of energy and personal time. With each increase in urban land area more material and energy is used for buildings, utility supply lines, sewer lines, roads, and other distribution systems. Because people have tended to move away from the city core, additional freeways and roads are needed. More cars and trucks are used to move people and goods from one point to another. With services and buildings spread over larger land areas, the necessity for using automobiles increases and so does the consumption of energy and the degradation of the environment.

Many urban problems have been caused by the exodus of the urban dweller to the suburbs. The commuter derives his income in one area and spends the bulk of it in another, thus depriving the city of sales tax and job opportunities. A city loses property-tax money when someone moves out of it to the suburbs or when buildings are torn down and parking lots are constructed to accommodate commuters' cars. People move from the city to escape air pollution, then contribute to it daily when they commute to their urban jobs. In cities such as Denver, Phoenix, and Los Angeles, where low-density housing is predominant, there is a nearly exclusive reliance upon the automobile for commuting to work, shopping, and entertainment, with a resulting high ratio of automobiles to people. The air quality in these areas, meanwhile, has deteriorated significantly because of automobiles. The suburban commuter costs the inner city a great deal of money and contributes to its degradation by using city streets and services and not paying for them in his suburban taxes. He provides the city with its air pollution, parking problems, traffic-control problems, freeway traffic jams, yet he doesn't supply the tax money to solve them.

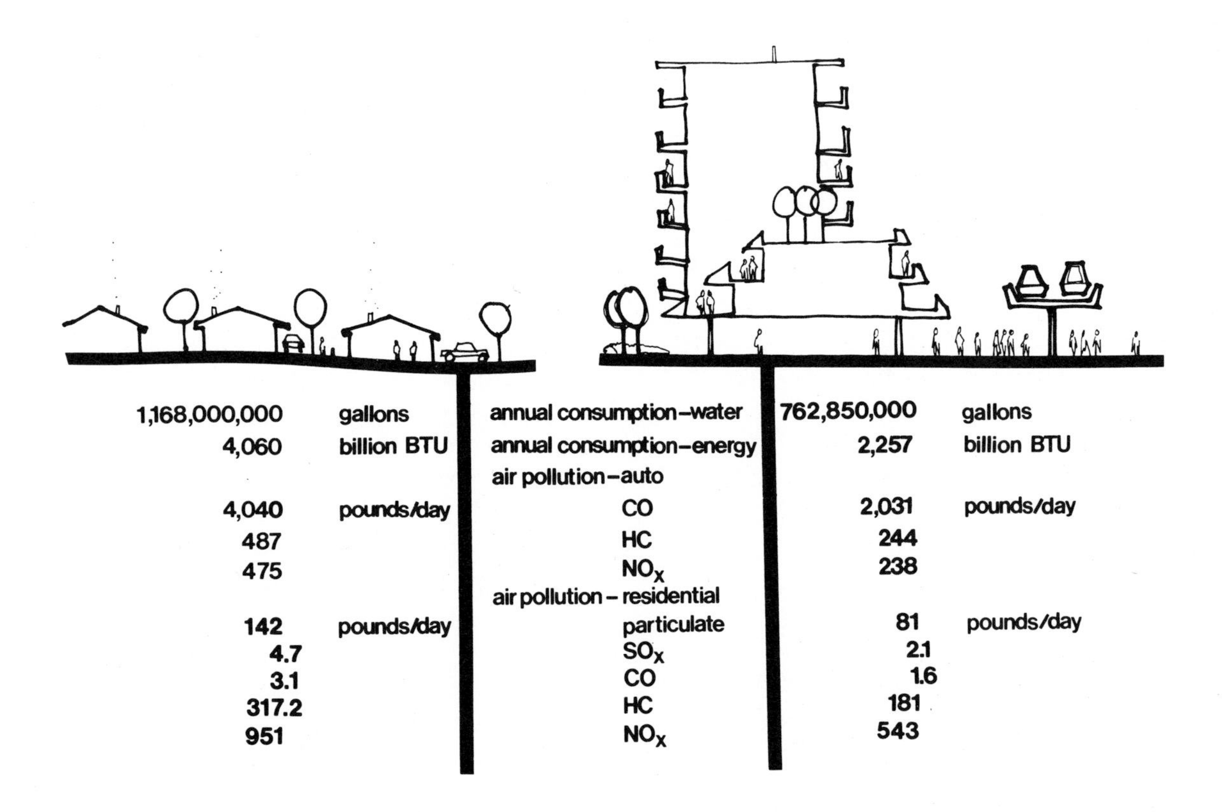

The closer people are to their work, the less they need their automobiles. This may be one major justification for mass transit, which shows increases both in cost-effectiveness and in logistic efficiency proportionate to population density. Few U.S. cities have yet discouraged the use of automobiles in the core city areas. In contrast, many European cities have encouraged pedestrian and bicycle traffic by closing off inner city areas to most motorized traffic. This latter trend has now reached the more progressive of American cities; often such downtown concourses enjoy the additional amenity of special bus transport in restricted lanes.

Energy and water usage are lower in densely populated areas than in suburbs. The accompanying illustration compares the usage of these resources and the amount of certain pollutants produced by a low-density community and a high-density community. Each has 10,000 occupants. The land area required for the high-density community is significantly less than the amount required for the low-density one.

Individual buildings and building complexes should be designed to implement maximum-energy usage in harmony with space use. One problem is the western custom of designating a single purpose to each room. Other cultures do not allocate nearly as much building interior space per occupant as do Americans. The number of rooms needed in a building decreases roughly as the multiple uses for each room increase. Many European families live in homes with less than 1,000 square feet (93 square meters), while most American families require 1,500 to 2,000 square feet (139 to 186 square meters) of area. Much of the additional space is poorly designed and utilized. By employing multiuse of space, the area required for buildings can be reduced.

Energy and resource consumption is minimized by urban buildings that are arranged in clusters and have multiuse rooms within individual living units and "common" spaces to accommodate activities for which the living units are not adequate. Buildings on the perimeter of the cluster buffer ones on the inside from extremes of climate, especially wind, and should be designed for the more severe conditions. Optimized energy conservation is facilitated by the cluster and multiuse concepts.

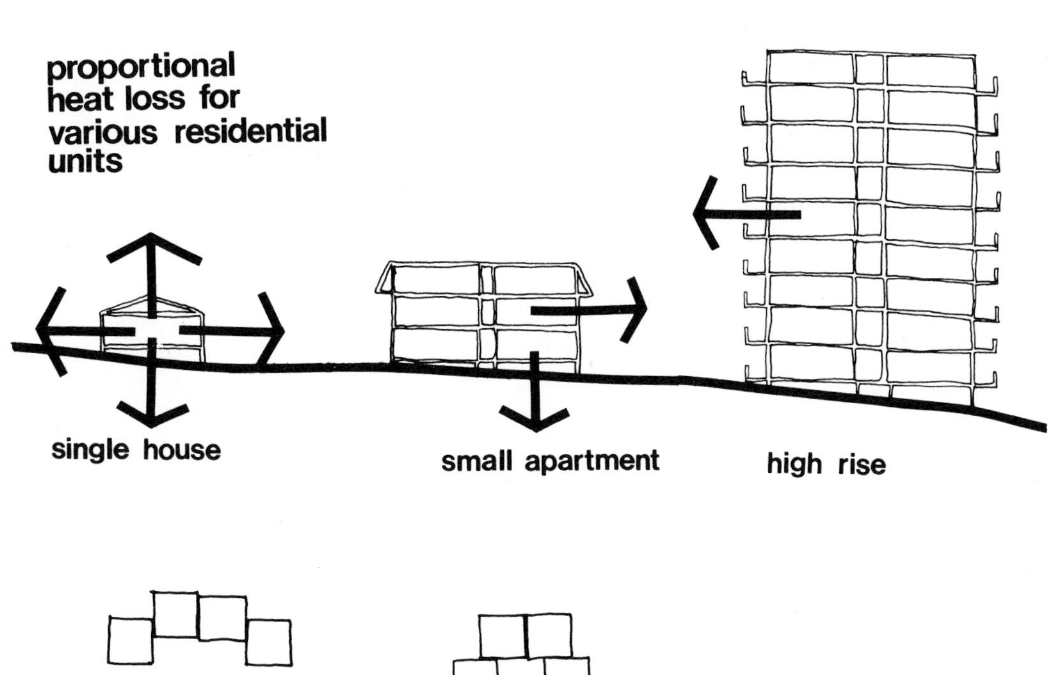

proportional heat loss for various residential units

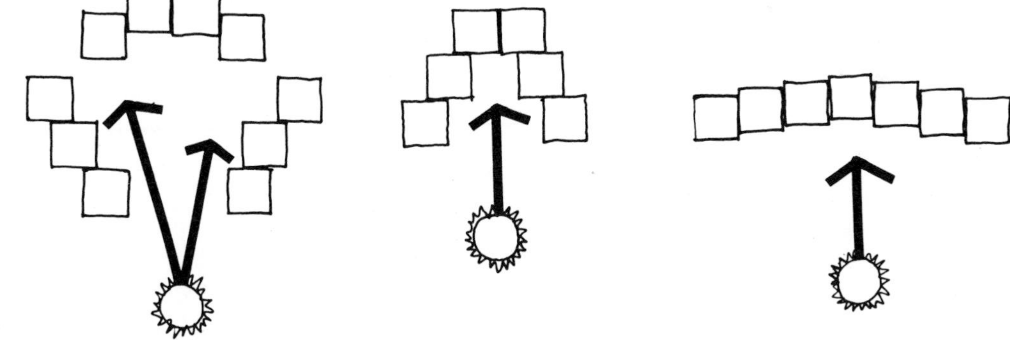

clusters and townhouses can be arranged to minimize heat loss and maximize winter solar collection

Because the total floor area needed for functional individual living units is decreased by multiuse rooms, energy demand for heating and cooling is decreased. Storage spaces for the support of activities performed in the multiuse room increases the effectiveness of the room and contributes to the building's decreased energy consumption. These spaces do not have to be maintained at comfortable temperatures so long as damage to their contents is not incurred. Depending on details of construction and the season, storage spaces can maintain temperatures as much as 11°C (20 F) above or below room temperatures and are another factor in minimizing energy consumption.

Common spaces are areas suited to recreational, social, and creative activities that require more space than is available in the individual living areas. People from a number of living areas share and support the common spaces. With this system the quality and diversity of the common spaces are greater than any individual could provide for himself. Of course, certain rules for use, conduct, maintenance, and repairs need to be established to the mutual agreement of all.

The cluster concept is easily adapted to the economical application of solar collectors for space heating, domestic water heating, or both. The type of collectors used would be a matter of economics and the needs the collectors are designed to satisfy.

For even further energy conservation, a number of living units can be consolidated under one building envelope, thereby reducing the wall area exposed to the outside air. Heat transmitted through interior walls can be used in adjoining living units and not lost to the atmosphere. The temperature difference across these walls would be so small that heat transmission could be kept to a minimum, stabilizing interior temperatures and decreasing energy consumption.

The average person in the United States now uses more energy than the average citizen of any other country in the world. Up to 60 percent of the energy used in this country is wasted, partly because of poor planning of buildings and transportation systems. As the supplies of fossil fuels, on which we depend for heating, transportation, and electricity, diminish, increased prices will be paid. In some cases shortages of energy will be encountered. To ease the impact of these high prices and shortages, the American life-style will have to change. Greater respect for the natural environment will have to be cultivated along with a decrease in overall energy usage. Energy-conserving space planning should take place at all levels of design, from small homes to entire cities. Appreciation for the urban core instead of the suburban sprawl must become a reality.

People generate varying amounts of heat when engaged in different activities. Because of this, various rooms can be heated to different temperatures. Various tasks can also be performed at different temperatures; for example, sleeping areas can be kept cooler than living rooms or dining rooms, and bathrooms can be heated in the morning and evening and left relatively cool the rest of the day. Certain rooms may not be used for days at a time and should not be heated when unoccupied.

At times the heat given off by lights and people may exceed the heat loss of a building. When this occurs no additional heating is necessary, and excess heat can be used to warm other portions of the building. Any excess heat generated in one section should be captured and used as a supplementary heat source for other areas. Heat given off by washers, dryers, and other mechanical equipment can be stored for later use. The moisture from dryers can also be used to add humidity, and thus increase perceived temperature, to a space. In some large high-rise structures, the south side during the winter may be air-conditioned while the north side is being heated. Buildings should have controlled zones for heating and cooling, so that the excess heat from one area can be used to warm another. This would decrease the wasteful operation of systems.

The shape of a building is a factor to consider when planning the placement of insulation. The envelope of a multistory building has a higher percentage of wall area and a lower percentage of roof area than a sprawling single- or double-story shopping center. Additional roof insulation in the multistory building would not control as large a percentage of the total heat flow as it would in the roof of the shopping center.

In multistory structures wall insulation is a primary consideration. It should be continuous through the height of the building, including between stories. Too often insulation is provided only between the ceiling and floor of a given story and not between the ceiling and the floor above. These areas may comprise 50 percent of the wall area and should be insulated.

Overhangs, shading devices, and recessed windows in tall buildings decrease the internal heat gains yet allow for year-round natural illumination. Without these, the direct solar gains can be excessive, and energy-consuming mechanical equipment has to be provided to handle the heat. It may be practical in some cases to use heat pumps to distribute excess heat to areas throughout the building.

In many commercial buildings a system of heat pumps, water tanks for heat storage, and cooling towers would be practical. Using heat pumps serving various zones, heat can be extracted from areas with southern exposure, transferred to storage, and then distributed to areas requiring heat. The cooling towers could be used for evaporative cooling of the storage tank water during the hottest summer months.

Compartmentalization of interior space into logical thermal zones relative to both the external climate and the interior functions can conserve energy and make the passive attributes of the building work more effectively.

The location of rooms within a building should be determined by calculating the amount of heat that activities within the room will generate and the amount of heat that will be lost because of the room's location. Spaces that are not occupied a great deal of the time, such as corridors, closets, mechanical-equipment rooms, laundries, and garages, can be kept at lower temperatures. They can be located on the north side of a building to provide buffer zones between cold north winds and occupied living areas.

Rooms should be oriented to microclimatic conditions. During the day the north and east walls receive the least amount of radiation; they therefore remain the coolest. South walls receive radiation throughout the day, while west walls are exposed to intense late-afternoon summer sun. Walls with either of these last two orientations will be the hottest and may significantly contribute to heat gains.

Areas in which heat is generated, such as laundry rooms or mechanical rooms, should be located on the north or east sides of a building so that heat can either be vented to the outside in the summer or used to warm the inside in the winter. If vents are located on these cooler sides, the hot air is most easily vented. Using the hot air to preheat domestic water or to supplement other heat sources should be considered.

The north side of a building is a good location for a kitchen, hall, stairway, closet, or other areas that are not continually occupied. Internal heat is generated in kitchens by stoves, ovens, and other appliances, which warms the kitchen in the winter and may be used to initiate cooling inductive ventilation in the summer.

The south side of a building is a good location for living, family, and dining areas. These may be adjacent to greenhouses or contain some type of beneficial as well as decorative foliage. Heat gains and illumination to these areas are greater than to any others, making them ideal for continuous occupancy. Entries on the south are functionally compatible with these rooms.

Bathrooms can be placed at the interior of an arrangement of rooms, and the walls of the bathroom can be thermally insulated from adjacent spaces. Once heated, a closed bathroom will remain warm until the door to the outside is opened and outside air allowed to infiltrate. If equipped with its own clock thermostat, a bathroom can be warmer than surrounding rooms in the morning and evening and left relatively cool during the rest of the day.

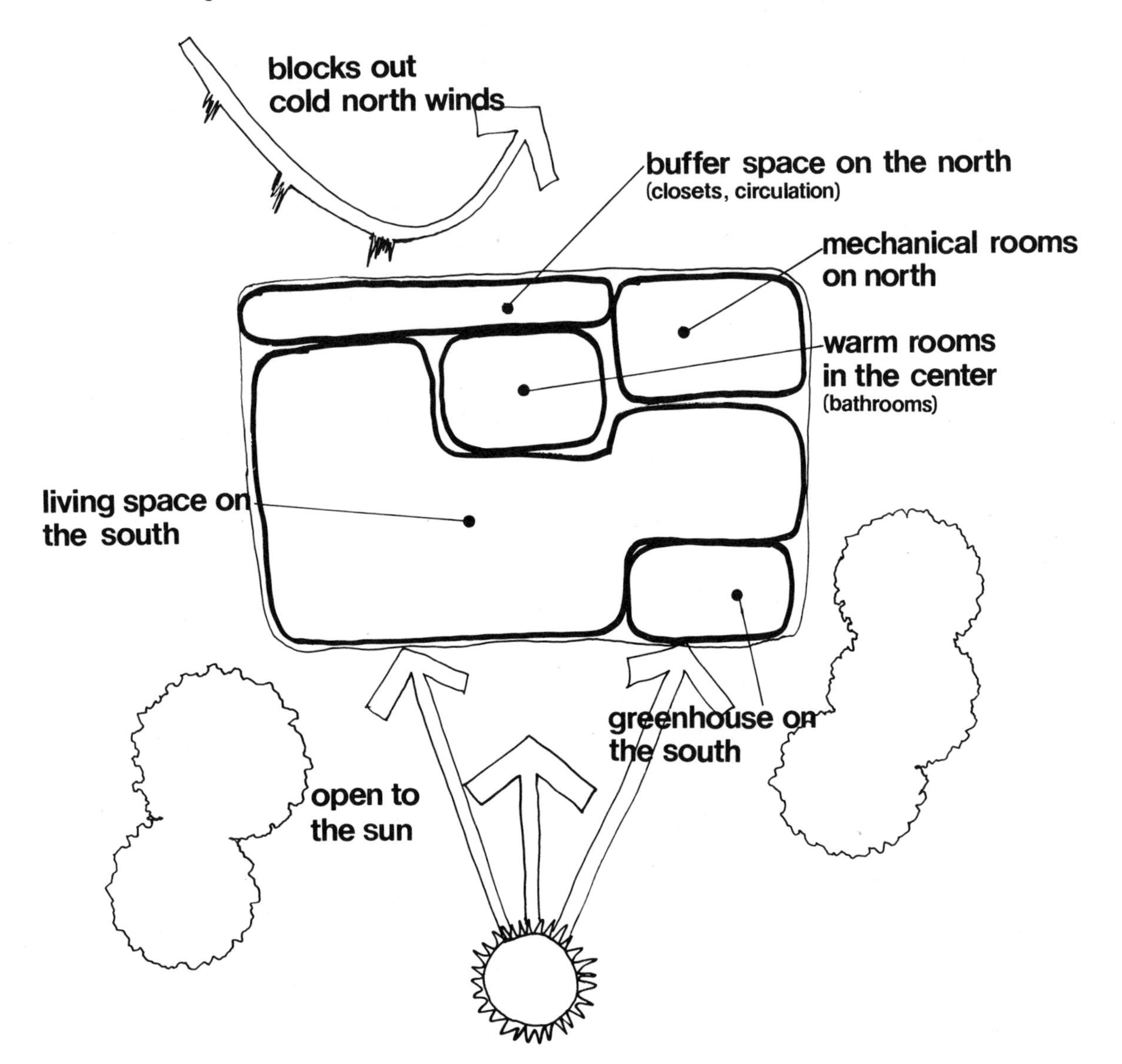

The position of interior rooms with respect to exterior spaces is important to optimizing energy conservation. Patios, gardens, trees, bushes, and ground cover play an important role in regulating exterior temperatures adjacent to buildings. They can add humidity to the air while lowering temperatures through evaporative cooling and shading. This conditioned air can be drawn into a house by inductive ventilation and used to lower indoor temperatures. Patios provide pleasant outdoor areas for relaxation, dining, and social activities, and their temperatures are influenced by nearby buildings as well. A massive wall will retain daytime heat and reradiate that heat at night, which can be a benefit, especially in the spring and fall, to nearby outdoor areas.

Space use and planning are extremely important in commercial structures. The effectiveness of leased spaces can be greatly improved by providing room layouts in which storage spaces to support activities have been carefully planned and in which rooms are arranged within the building according to function. This is true of low-rise, medium-rise, and high-rise structures; each has unique characteristics that must be considered as a consequence of their individual designs, shapes, volumes, dimensional ratios, and types of occupancy.

Support spaces for offices, reception areas, conference rooms, and retail spaces will increase the utility of each of these. As a consequence, tenants will need to lease a smaller total area, thus decreasing their costs and lowering vacancy rates. Energy conservation will result from maximum utilization of space, high population density, and the sharing of resources (heat) in properly zoned structures.

Life-styles, attitudes, and design philosophies will have to be altered if total energy consumption is going to be reduced. A simple procedure like turning off a light bulb in an unoccupied area is a start, but more drastic changes are necessary; for example, permanent improvements in the efficiency of the light bulb itself. Many of these will come about as more people realize that supplies and recoverable reserves of fossil fuels are limited. Changes in life-style do not necessarily mean a decrease in comfort. The suggestions presented here are intended to increase the comfort of all.

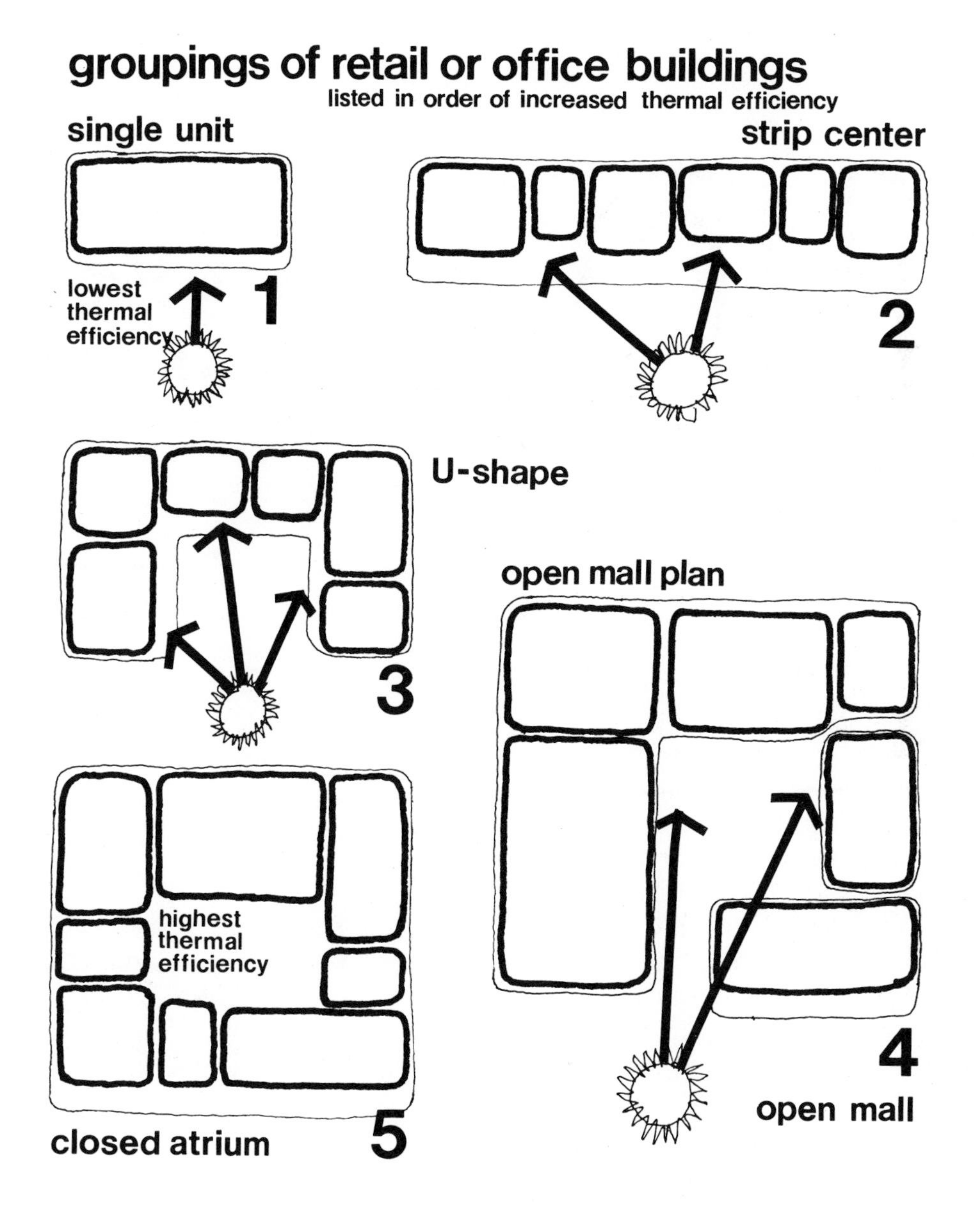

10 Use of Solar and Other Natural Energies

In order for the man-made environment to use the natural energy available to it most efficiently, it must be planned with consideration given to materials selection, site and building orientation, microclimatic conditions, landscaping, and many other factors. The environment-responsive building should have minimal negative impact on its site, maximize efficient use of all energy, maximize human effort, and promote human health and talents by providing a pleasing atmosphere in which to live or work.

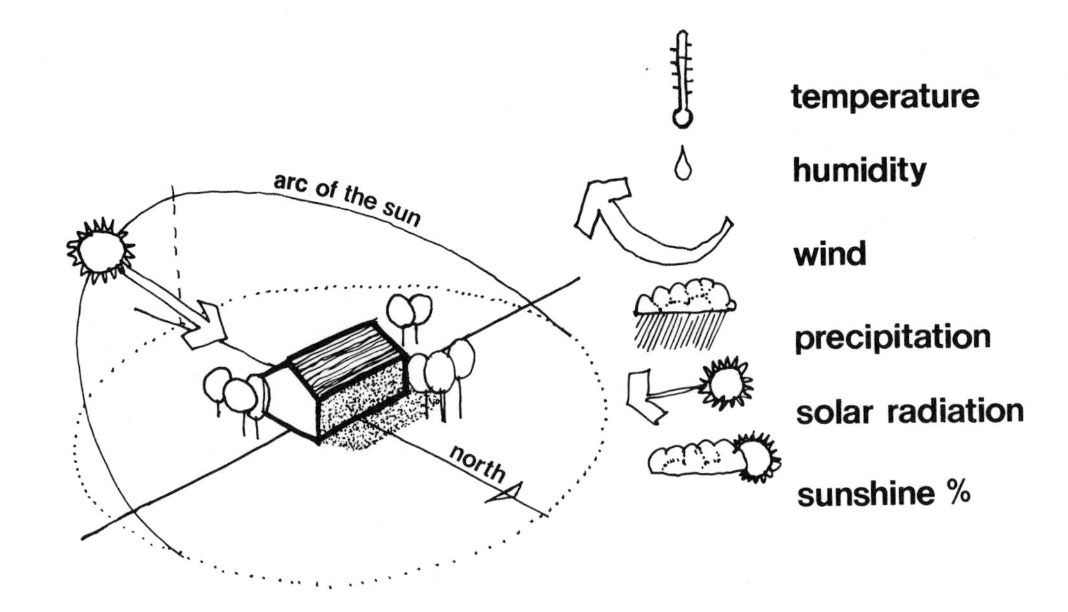

To incorporate alternative energy devices into buildings economically, energy-conserving measures must be taken to diminish the total energy usage. Energy conservation in some buildings can reduce energy demand by as much as 80 percent. Many different items relating to energy conservation should be considered and evaluated for possible use in all buildings. Areas in which energy-conserving practices can be employed have been divided into eleven categories and are presented in this chapter as follows:

- Site and Climate
- Entrances
- Walls and Roofs
- Windows and Other Openings
- Passive Solar Heating Strategies
- Passive Solar Subsystems/Energy Centralization
- Lighting
- Ventilation
- Passive Cooling Strategies
- Mechanical Systems
- Appliances

Each aspect of a building should be planned for its best utilization of all energy. Design ideas presented in this chapter include passive systems that use the natural energies available from the sun, wind, water, and earth.

Site and Climate

To consider the placement and orientation of any building on a site properly, its regional and microclimatic conditions must be evaluated. The microclimate is peculiar to each individual site and results from the influences of surrounding buildings and landscaping, topography, soil structure, and ground cover, as well as regional geographic influences.

Local climatic conditions must be evaluated in conjunction with regional weather data to determine the microclimate of a particular site. The regional data are usually available from the ASHRAE *Handbook* and the United States Department of Commerce, which maintains weather-monitoring stations throughout the country. Data that influence the design of a building and should be considered are:

- the daily and monthly average dry-bulb and wet-bulb temperatures
- the relative humidity (percentage of water vapor in the air relative to the maximum amount it can hold)
- the wind speed and direction—daily, monthly, and annually (a wind rose map indicates the velocity of the wind and the percentage of wind flowing in each compass direction)
- precipitation (the quantity, type, and duration of all snow, rain, hail, etc.)
- the incident solar radiation (usually measured in langleys, where one langley is defined as one calorie per square centimeter [3.69 Btu/square foot], on a horizontal surface)
- the seasonal percentages of sunshine (the percentage and probability of sunshine will help predict the monthly heating loads and storage capacities needed)

Plans for using alternative energy sources require that foresight be used in site selection and building placement. If an urban site directly north of a high-rise building is selected, it is unlikely that much direct solar radiation will be available at the site for collection.

In hilly or mountainous areas, a site with southern exposure is most advantageous for year-round control of indoor climatic conditions. This will have a favorable orientation for solar collection in winter, and the hill itself will offer partial protection from heat-robbing winter winds. Sites on the north sides of hills should be avoided, except in climates where heat gains during all seasons need to be controlled.

The proper orientation of a building on its site depends on the regional climate and the microclimate. To optimize energy use, it would be desirable for a building to have the greatest amount of area with southern exposure during the

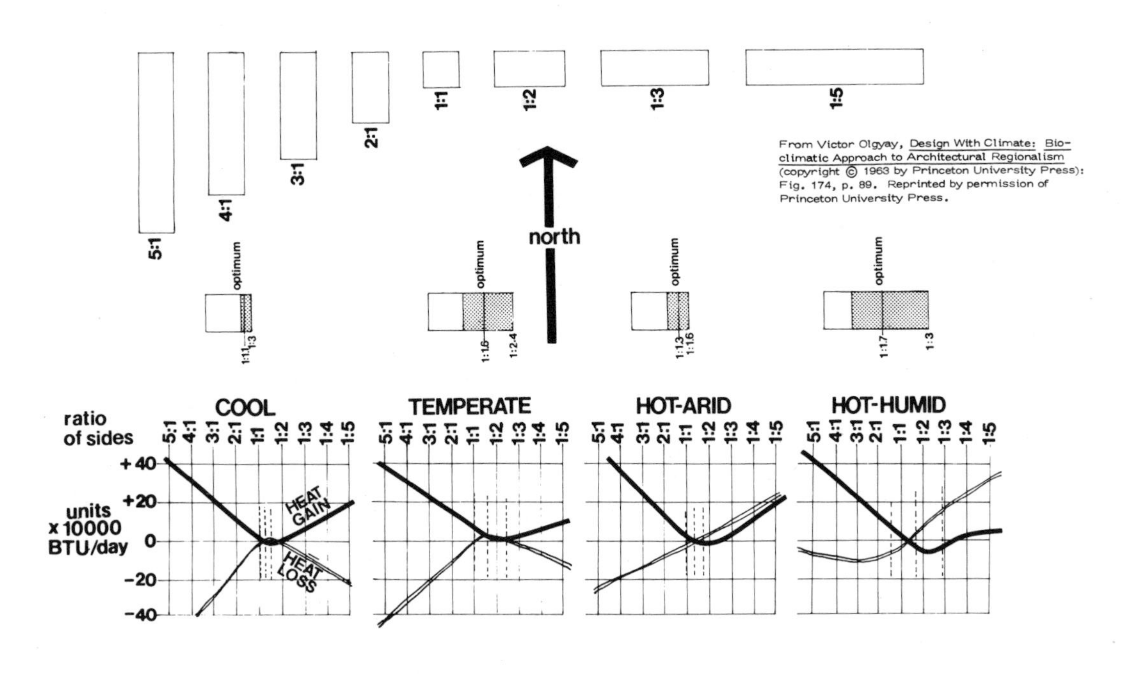

From Victor Olgyay, Design With Climate: Bioclimatic Approach to Architectural Regionalism (copyright © 1963 by Princeton University Press): Fig. 174, p. 89. Reprinted by permission of Princeton University Press.

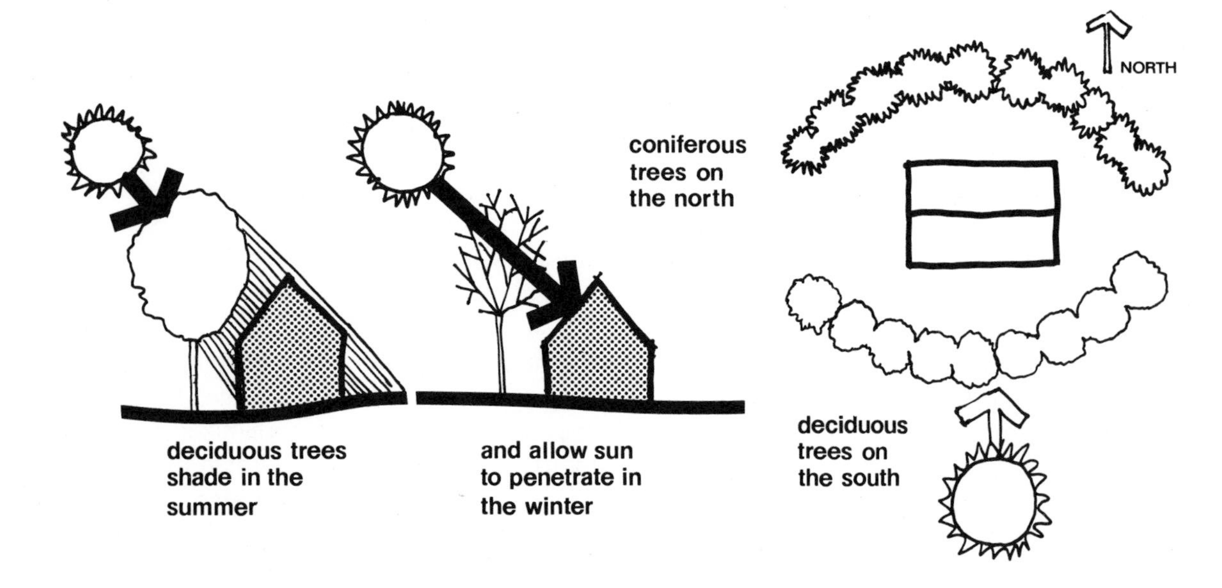

winter (to maximize solar heat gains when they are most necessary) but the least amount of area with this same exposure during the summer (to minimize solar heat gains when unnecessary). Because it is obviously difficult for the proportions of a building to change with the seasons, a compromise is needed to balance the annual heat gains against the annual heat losses, in determining the optimum building proportions based on thermal considerations only.

The accompanying chart (adapted from *Design with Climate*, by Victor Olgyay) shows various ratios of north-south dimension versus east-west dimension for different types of climates, with the optimum and an acceptable range of building proportions for each.

Changes in orientation can be effectively implemented by modifications in siting and design. Trees and shrubs can be used to reduce solar heat gains in the summer. Deciduous trees (trees that shed their leaves at the end of each growing season) provide shade in summer months and allow sun to pass through in the winter. Trees provide differing degrees of shade, depending on leaf structure and density. Many trees allow diffuse light to penetrate, permitting natural lighting levels to be maintained, while others are practically opaque.

Deciduous trees, although effective for providing shade, are not so valuable as windbreaks. Therefore, on the north or northwest side of a building (depending on the prevailing cold winter winds) coniferous trees should be used. Their use and type depend on the density of the branch structure, how close to the ground they grow, and their height at maturity. Their effectiveness as windbreaks is governed by the proximity of individual trees. It is necessary to direct the air flow over the trees instead of around them.

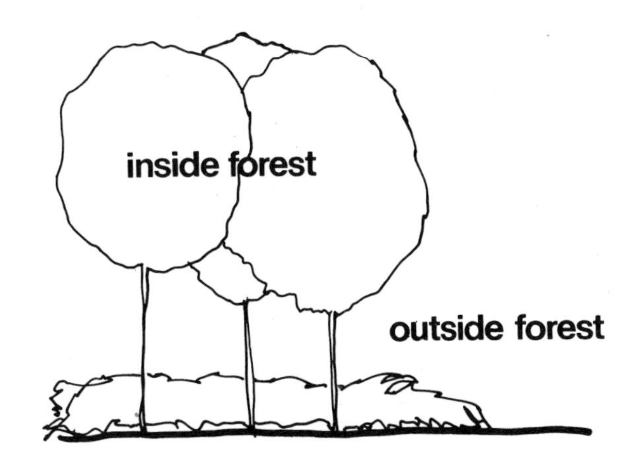

inside forest			outside forest		
ground temperature (F)	air temperature (F)	humidity	ground temperature (F)	air temperature (F)	humidity
60°	75-80°	77%	70°	90°	85%
45°	65-70°	75%	50°	75°	70
33°	58-60°	60	30°	60°	60
	44-48°			45°	
	29-31°			30°	
	17°			15°	

The dead air space behind trees can act as insulation space. With coniferous trees on the north and west and deciduous trees on the south and east sides of a building, maximum protection from the winter cold winds, minimum obstruction of the sun's warmth in the winter, and maximum shading from the sun in the summer can be attained.

The use of trees can also assist in the insulation of a building against both heat gain and heat loss. The accompanying chart is a comparison of temperatures and humidity inside and outside a wooded area. It is evident that trees produce conditions that moderate the extremes in both hot and cold temperatures.

Plant areas also help to purify the air. As air moves through the leaves of a plant, carbon dioxide in the air is exchanged for oxygen from the leaves. The plant

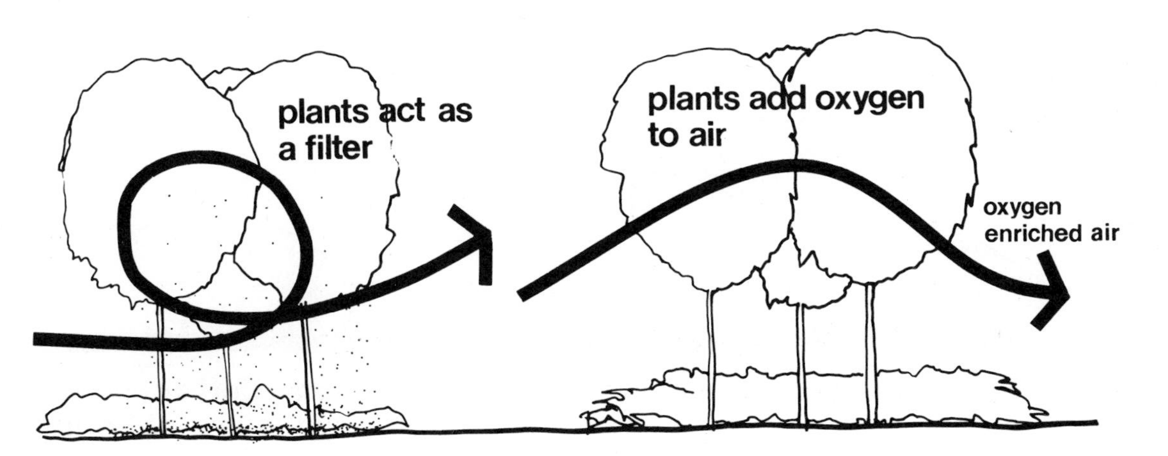

structure also causes air turbulence, permitting particles suspended in the air to fall to the ground.

Earth can be used to minimize the amount of exposed surface area of a building. Mounds of earth (berms) on the north side can considerably reduce the heat loss in that area. Prevailing winter winds (which usually come from the north or northwest) will carry away heat faster from an exposed north wall than from any other exposed building surface. Therefore, it is best to minimize the exposed wall surface area on the west and north sides.

Earth is effective as an insulator below the frostline. A mixture of mulch and soil can decrease the depth of the frostline because it is an insulator. An insulative ground cover, such as bark or leaves, also aids in diminishing cold penetration in the soil.

Berms can be useful in directing noise and snow away from a structure. Sound cannot penetrate the mass of a berm and is either absorbed or reflected by it. By tilting the berm's surface as shown, the sound is reflected upward away from the building. The proper positioning and forming of berms will direct winds, causing snowdrifts to form away from buildings and entrances.

One method for preventing excessive exposure of a building to the elements is to place it underground. In the opinion of Malcolm B. Wells, Massachusetts architect, underground architecture offers a more ecologically-sound alternative approach to building design. He has, in fact, completed several underground structures, a number of which also use solar heating. These buildings restore and intensify the natural landscaping of their area. Mr. Wells uses the accompanying chart, entitled "MAN'S WORKS compared to the miracle of WILDERNESS in fifteen ways . . . ," to evaluate buildings. A score from −100 to +100 can be attained in each of fifteen categories. If the total score is negative, the building is "dead"; if it is positive, the building is "alive."

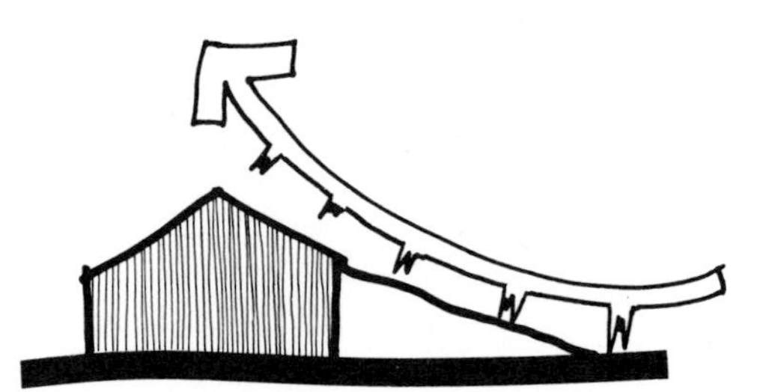

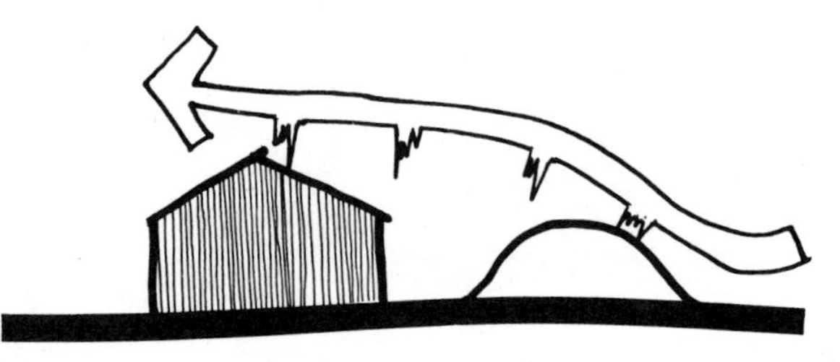

cold north winds directed over the buildings by berms

noise reflected upward by berms

snow drifts can be diverted by the use of berms

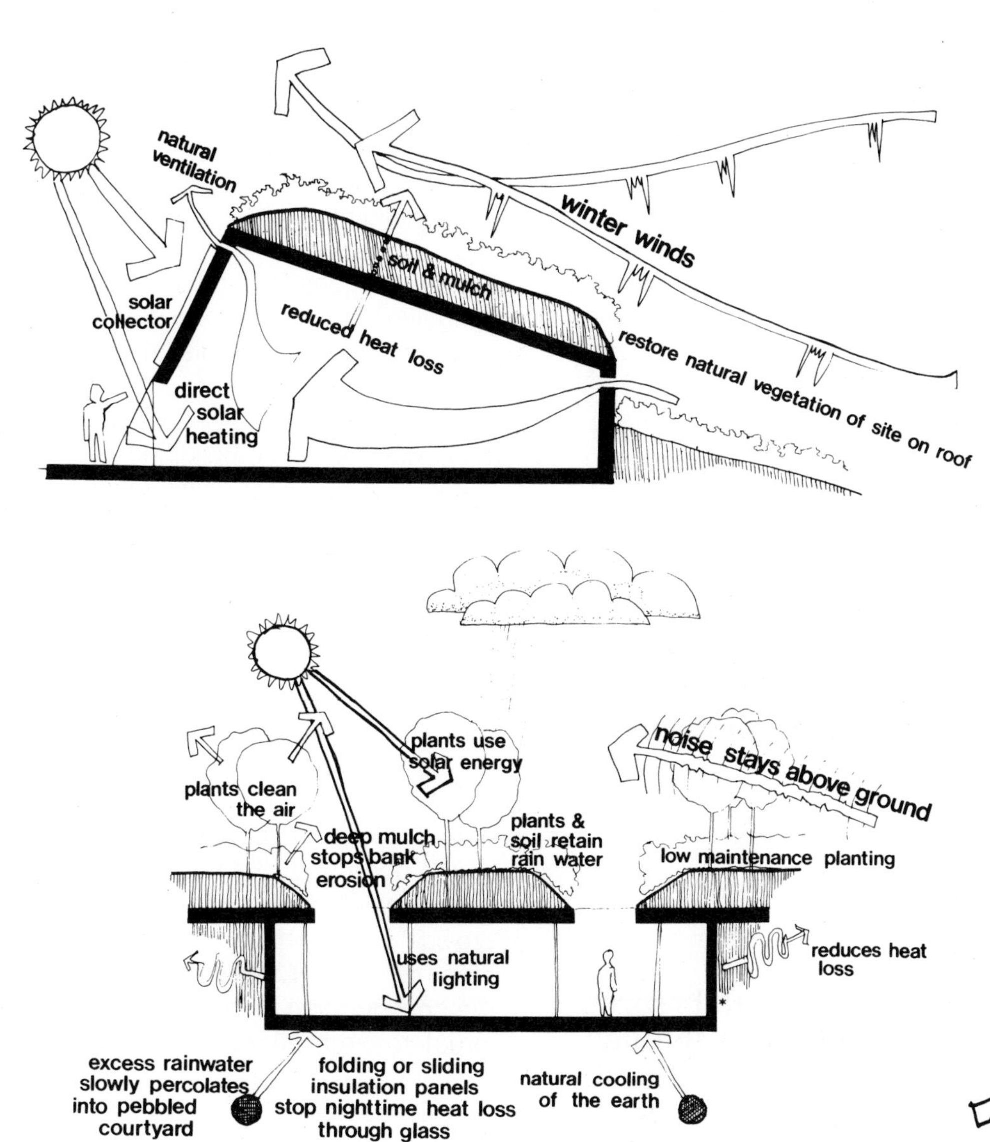

	-100	-75	-50	-25	0	+25	+50	+75	+100	
destroys pure air										creates pure air
destroys pure water										creates pure water
wastes rain water										stores rain water
produces no food										produces it s own food
destroys rich soil										creates rich soil
wastes solar energy										uses solar energy
wastes fossil fuels										stores solar energy
destroys silence										creates silence
dumps its wastes unused										consumes its own water
needs repair and cleaning										maintains itself
disregards nature's pace										matches nature's pace
destroys wildlife habitat										provides wildlife habitat
destroys human habitat										provides human habitat
intensifies climate and weather										moderates climate and weather
destroys beauty										beautiful

Fences, walls, and other external structures can be used in a variety of ways to deflect winter winds over or around structures, decreasing the heat loss of a building.

In the summer months, heat loss resulting from winds is not important, but heat gain is. Ventilation and circulation of the air within a building are of primary concern. During the summer, low-velocity winds generally come from the south and southeast. These air currents can be used to encourage summer ventilation. For maximum efficiency of air flow and heat gain/loss, a building in the northern latitudes should block out cold winter winds from the north and west and encourage the infiltration of cooling summer breezes from the south.

Earth structures built to the north side of a building can be used to reflect radiation into northern windows and against northern walls. White rock covering the earth will provide diffuse reflected radiation, while with a mirrored surface a predictable reflection is possible. These site modifications direct heat and light into alcoves that are in the building's shadow and would otherwise receive no direct heating or illumination.

Some suggestions for landscaping and site modifications to improve comfort and conserve energy are listed below.

- Plant deciduous trees and vines on the south side of the building to decrease summer heat gains yet allow the winter sun to enter. Conifers planted on the north side will act as an effective windbreak, decreasing winter heat loss.
- Fences of the proper design will provide privacy, security, sound control, and wind control.
- Where few or no openings are present on the north side of a house, the earth can be banked against the side to provide insulation. The height of the embankment will depend on the site, the configuration of the house, plus structural and waterproofing details.
- Berms strategically placed on a site will direct wind flow and control sound. These and embankments should be surfaced with ground cover.

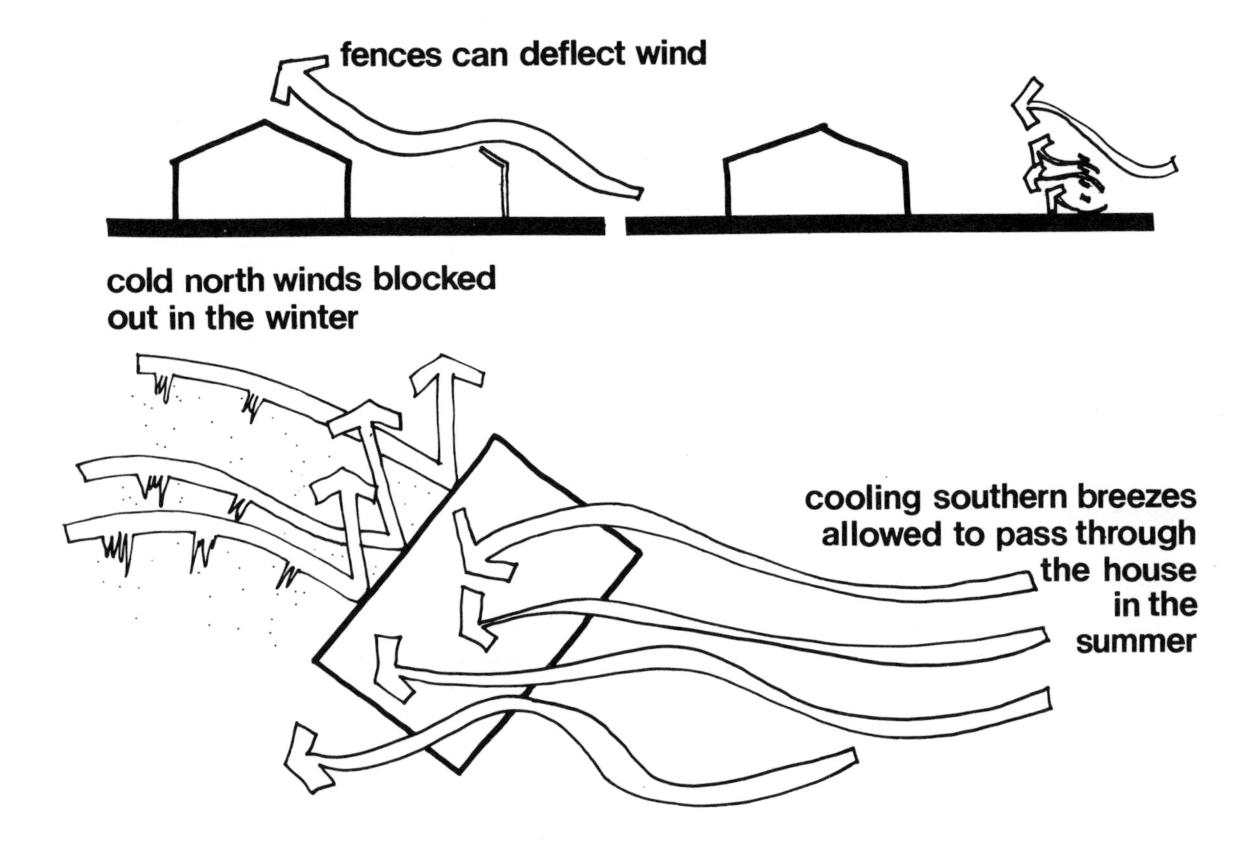

- Shrubs can be used effectively to produce cool ventilation-intake air, as a consequence of evaporation and the shade they cast near ground level.

- Patios, offering comfortable outdoor areas in which to spend leisure time, may also be used to temper indoor temperatures. Patios with white gravel on the south can be used in cold climates to reflect heat to interior spaces or occupants. Around patios, deep trenches filled with gravel and moistened with water will reduce temperatures through evaporation.

- Manure and natural compost from yard clippings are good fertilizers. Most commercial fertilizers have a petroleum base and therefore should be used only if compost and manure are not available.

Entrances

Controlling heat loss and gain through entrances is a major concern. All edges around entrances leak air and correspondingly lose or gain heat. To minimize these heat transfers, doors must seal as tightly as possible at the sides and at the bottom threshold. The edges must set square with a weatherstrip to prevent air movement. Weatherstripping can usually provide adequate protection in this area. Winds will carry away heat at a greater rate than still air outside a door.

To keep air as still as possible, wingwalls and landscaping can provide windbreaks. One of the major and immutable problems with an entrance is that each time it is opened a great quantity of cold or warm air enters the adjoining room. It is possible for the entire volume of air in a room to be exchanged when the adjoining door is opened. This problem can be intensified if the entrance is oriented to intercept the prevailing winter winds.

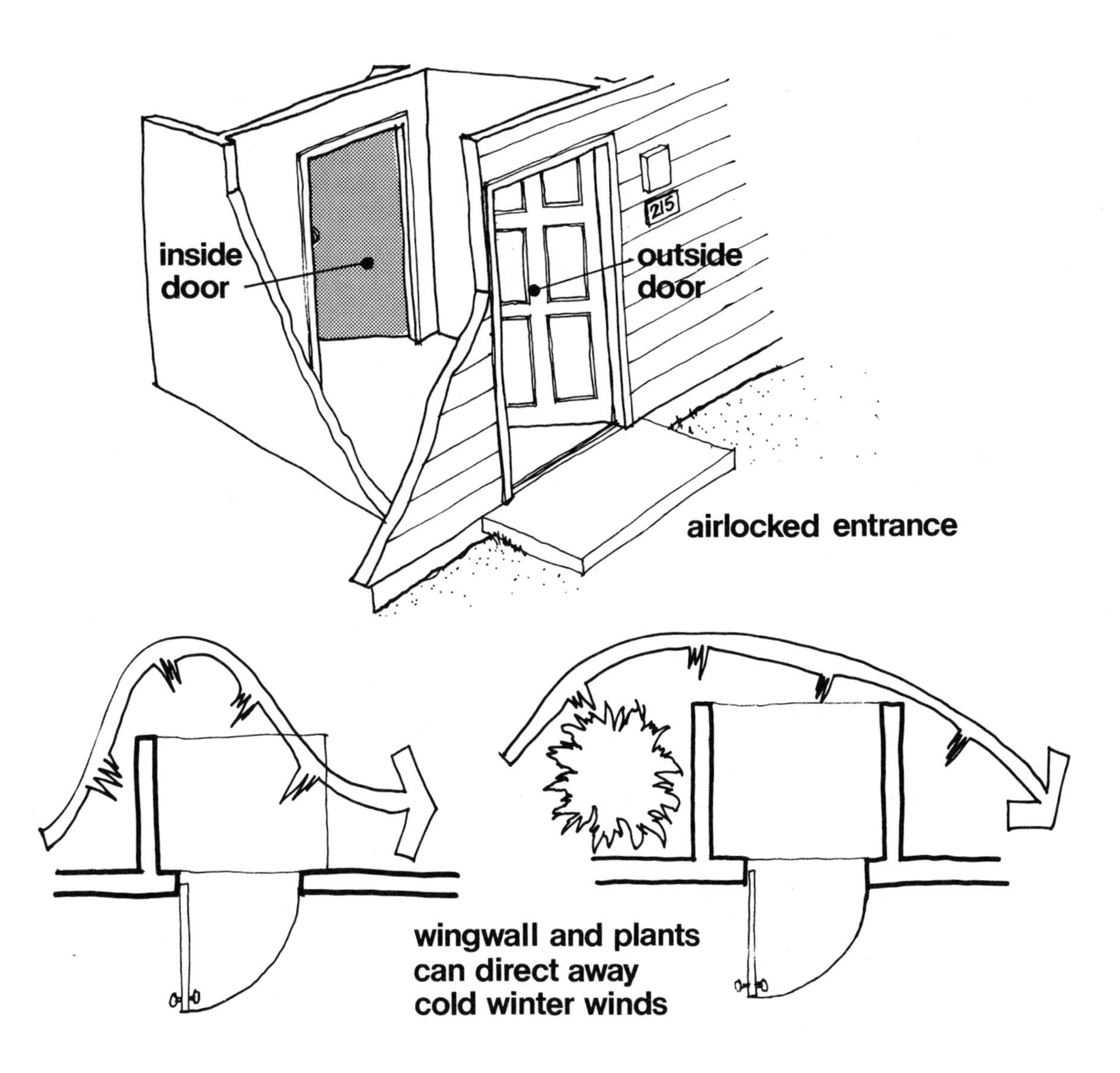

To prevent large heat losses caused by infiltration when a door is opened, airlocks can be provided. An airlock has two doors and an air space between; one opens to the exterior and one to the interior. When the exterior door is opened, only the air in the airlock can escape; therefore, only a small quantity of heat escapes. The exterior door is then closed before the interior door is opened, allowing only a small amount of the interior heat to escape to the airlock. This prevents the entire room from being infiltrated with hot or cold air. In general, airlocks are not heated, but they naturally maintain a temperature between that of the exterior and the interior temperatures.

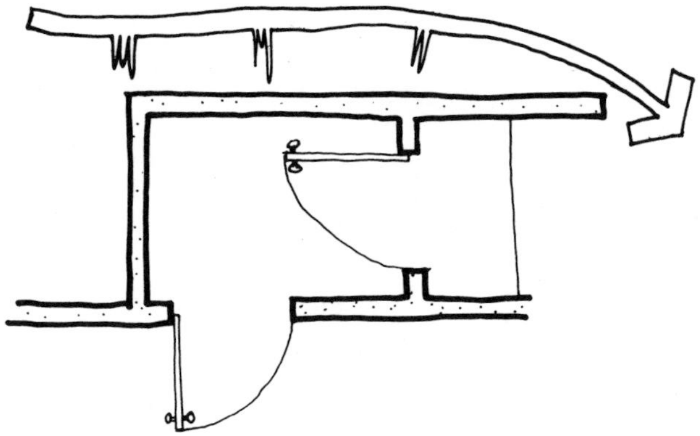

airlocks on the north need protection from cold north winds

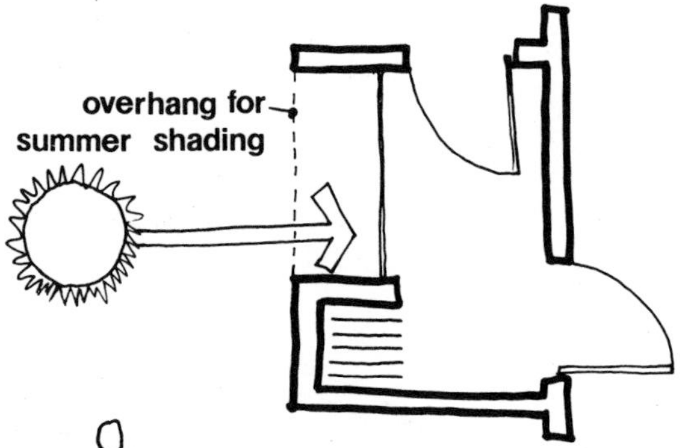

overhang for summer shading

south-facing airlocks can be warmed by the sun

vertical airlock

Airlocks work most effectively when placed on the south or east sides of a building. Wingwalls must be included with airlocks and be placed on the west and north sides to prevent prevailing winds from decreasing their effectiveness. Airlocks on the south can be warmed by the sun. With insulated glass on the south side and a floor constructed of a heat-retaining material, the airlocked unit can stay warm several hours after dark.

Airlocks can be used for more than just passageways. They can be used to store unheated items (sleds, snow shovels, umbrellas, etc.) or as areas to remove debris (mud, snow, etc.) before entering the building.

Another entrance that is effective for reducing heat loss is the vertical airlock. Since cold air will not rise and warm air will not fall, an entrance located below a building will allow the least possible amount of heat to escape. When the door is opened a small amount of warm air will exit and be replaced with cold air. The infiltrated air will not rise since it is colder (and heavier) than the air in the upper room. As the person who entered ascends into the heated space, he is greeted by the layer of warm air.

A revolving door works as an airlock because each chamber traps and releases only a small amount of heated air to the outside each time it completes a revolution. The edge, top, and bottom seals are of major importance to keep this type of door effective. One problem that is encountered is that as the tightness of the seal around each chamber is improved, the force needed to rotate the door and overcome the friction increases. This places a limit on how effective the door can be as a heat-loss barrier, but such doors are often successfully used in large building entries as windbreaks.

In snowy regions provision should be made to prevent snowdrifts from obstructing entrances. Reflective surfaces can be used above or around the door area to accomplish this. These should be calculated to allow entryways to be heated by the sun during the winter months and to be shielded from the sun during the summer. Care should also be taken to avoid directing the solar radiation into the eyes of people when they are exiting or entering. To absorb as much energy as possible, the reflection should be directed toward a flooring surface dark in color. Adequate drainage must also be provided so melted snow doesn't freeze in the entryway at night.

A garage can act as a climatic buffer when it is located to attenuate north and northwest cold winter winds or extended to protect an east or west entry. Radio-controlled garage openers conserve energy in cold climates by decreasing the time that the door needs to be open for passage of the vehicle.

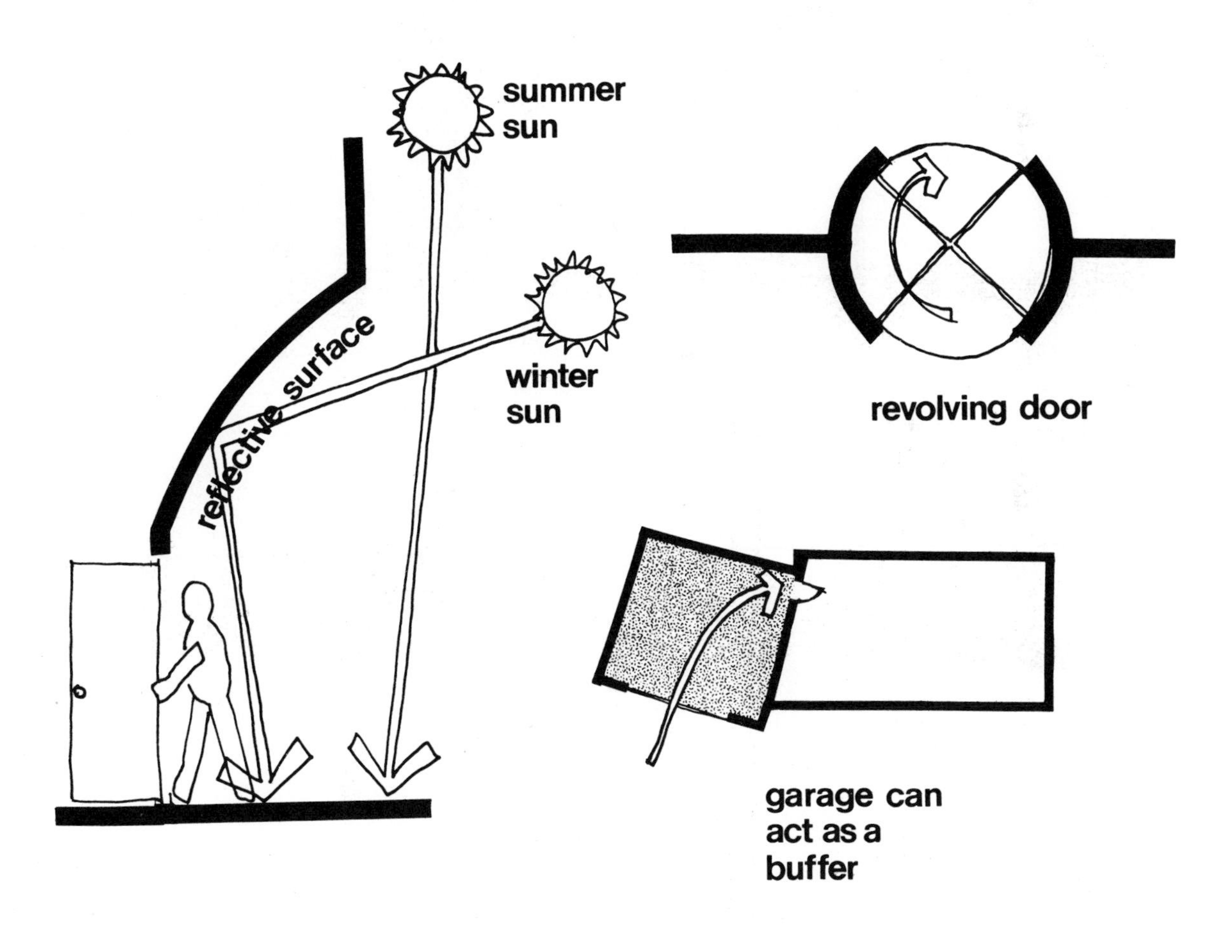

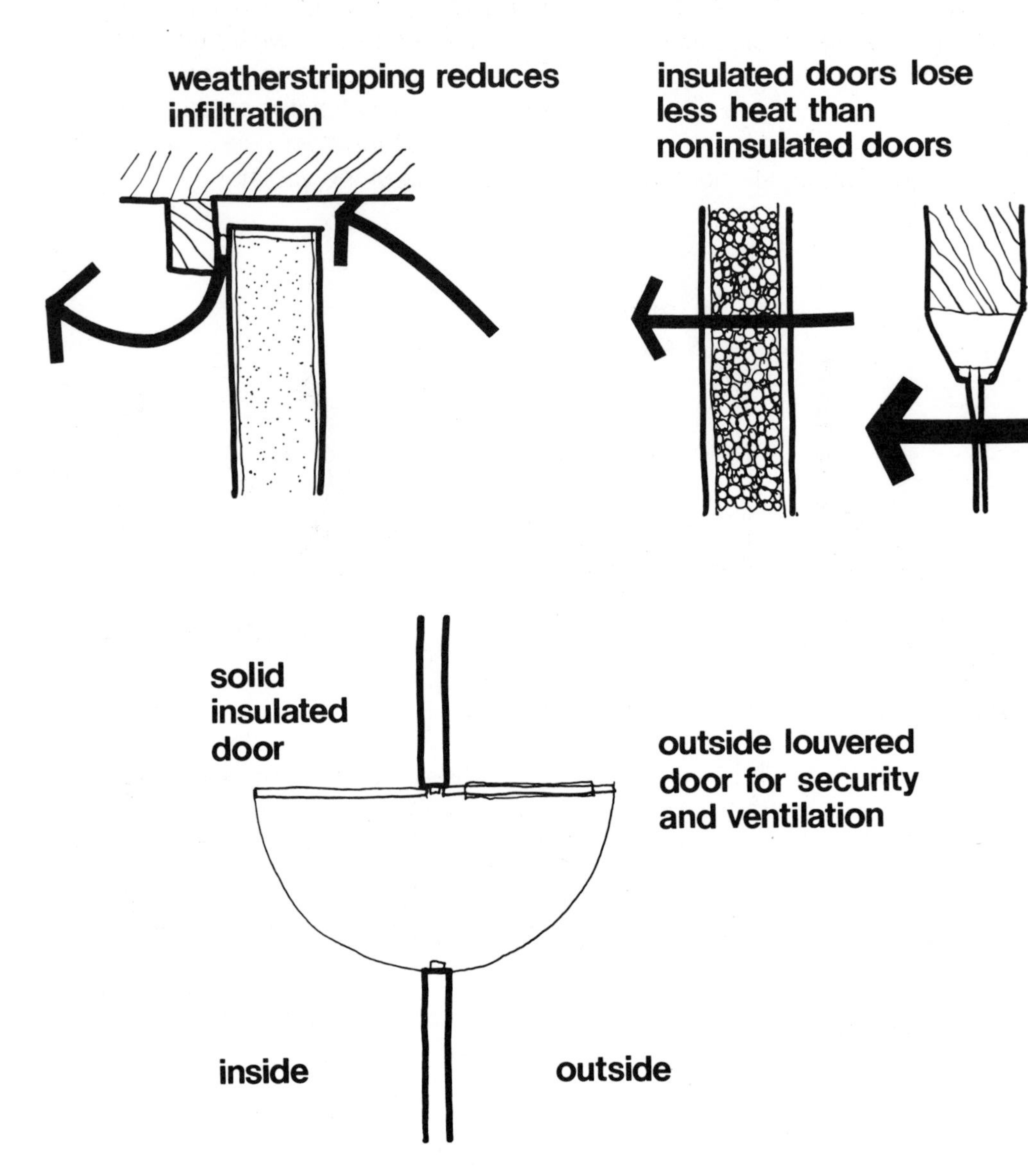

Some hints for improving the energy efficiency of entrances are presented below.

- Weatherstrip around all doors to decrease infiltration. Continuity is important, especially at corners.
- Use storm doors as additional insulation panels. Be certain these fit tightly enough to create a dead air space.
- Avoid using entrance doors against which the cold winter wind blows directly.
- Add a porch to a house to serve as an airlocked entry.
- Wingwalls can be added to shelter entrances from prevailing winds.
- A dark-colored south-facing door will absorb winter heat, but it may be too hot in the summer unless sheltered by overhangs from higher sun altitudes.
- During cold or hot weather avoid having doors open for extended lengths of time.
- Use insulated exterior doors.
- Avoid the use of mail slots that lose energy and allow infiltration.
- For ventilation, use a weatherstripped, louvered, and screened door with a lock for security. This should be designed so that an insulated panel can be added in winter.

Walls and Roofs

Walls and roofs form barriers that separate the inside of a building from the outside environment. These barriers resist the passage of most elements but are not impervious to the flow of air and moisture, with their accompanying heat. Methods for calculating the rate of heat transfer were presented in Chapter 8. Ideally, walls and roofs are designed and constructed to minimize winter heat loss and summer heat gain while maximizing winter heat gain and summer heat loss from interior spaces. Two important properties that regulate heat flow are the thermal resistance (insulation) and thermal mass (thermal inertia) of walls and roofs.

To minimize the rate at which heat is lost or gained through exterior surfaces, they should be constructed of materials having high thermal resistance. Generally, the higher the thermal resistance, the slower the heat flow through the wall or ceiling, with a corresponding decrease in interior heat loss or gain. One component contributing to the thermal-insulation value of an element is the layer of air on its inside and outside surfaces. The surface texture has an influence on the thickness and stability of these layers of air. Shake-shingle exteriors produce a thicker, more stable layer of air than metal or other smooth-surfaced materials.

Any material that is in contact with the inside and outside surfaces of a wall or roof aids in the transmittal of heat through it. Wall studs or floor and ceiling joists can be staggered to produce discontinuities in wall construction that act as thermal barriers. This system can also be used to thicken wall cavities to produce greater space for insulation without adding larger studs.

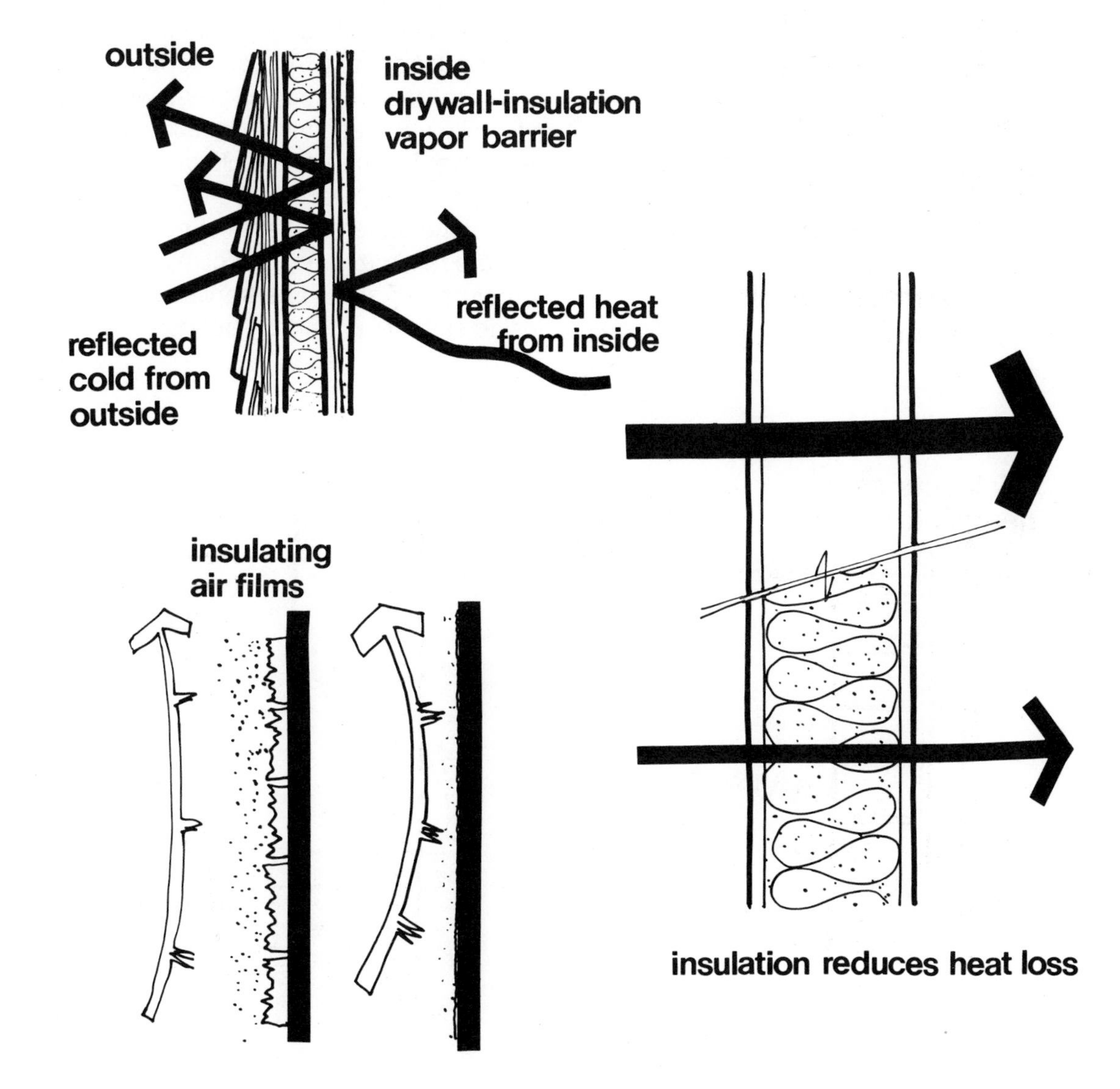

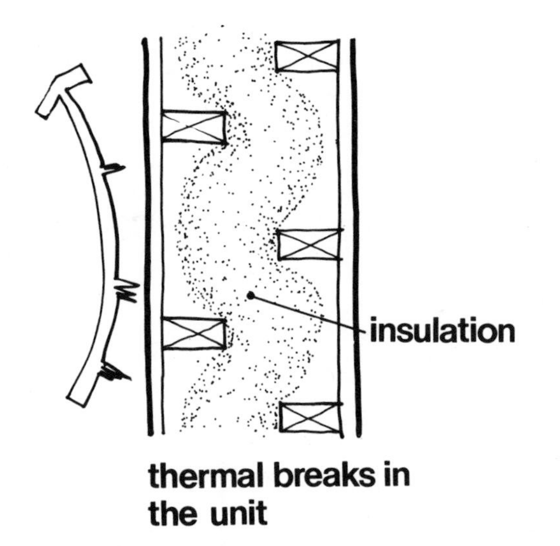

thermal breaks in the unit

Most building materials are porous and allow moisture-laden air to penetrate through them. Since latent heat is required to evaporate the moisture, a moisture loss also represents a heat loss. When the moisture condenses, it may cause discoloration, rotting, or peeling of exterior paint. A vapor barrier around all wall and ceiling areas will significantly decrease air penetration and heat loss. The barrier should be positioned on the "warm" side of the wall so that the adverse effects of condensation are minimized. In cold climates the vapor barrier should be applied near the inside surface, preferably between the drywall and studs. In hot, humid climates it should be applied under the exterior-skin material.

By virtue of their mass, all materials have the ability to store heat. The amount of heat that can be stored per unit mass varies from one material to the next. The thermal mass of building materials is a very important property that must be considered in the total design process.

Certain materials such as brick, block, stone, concrete, and adobe can store large amounts of heat. In their traditional position on the exterior of a building, these materials in summer will absorb heat for long periods of time before allowing it to flow to the interior. This produces the "time lag" effect and completely alters the heat flow, which is generally calculated using the U-value alone. Later in the day, when the outdoor air temperature is lower than the internal wall temperature, the heat stored in the walls will be conducted to the outside air. The thickness of the wall, the density of the material, and its heat capacity govern these effects.

Brick veneer can be used most effectively to moderate interior temperatures if it is applied to the inside rather than the outside of a building. In the winter months, the brick absorbs heat during the day or at anytime when the temperature of the air in the room is greater than

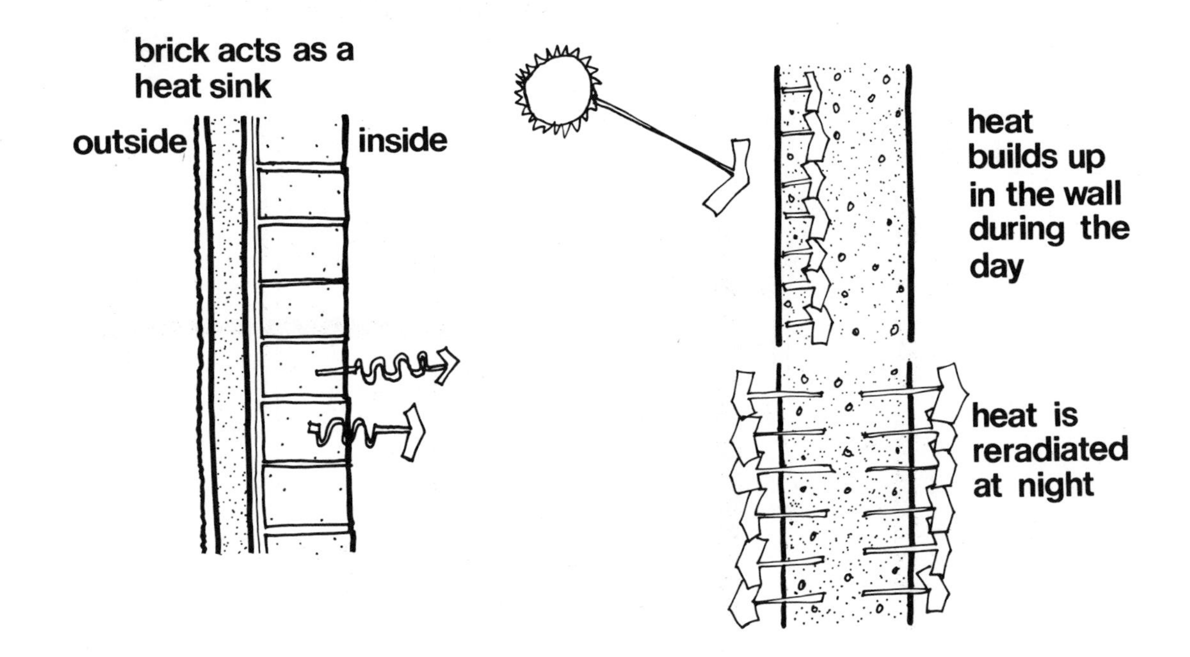

the temperature of the brick. When the air temperature drops so that the temperature of the brick is higher, the brick will give heat back to the room by radiation and convection, helping to maintain a stable temperature. Insulation with a reflective barrier should then be placed on the exterior of the wall where it is most effective. The summer operation of this system is similar to the process described above but produces an even greater time lag.

A house design by the French engineer Felix Trombé integrates the thermal storage capacity of concrete into a flat-plate collector. Components include a thick concrete wall facing south and painted black, plus a sheet of glass covering the wall to decrease convective heat losses and to form an air chamber between the wall and the glass. Vents at the top and bottom of the wall connect the air chamber to the adjacent room. As the sun warms the wall, the air next to it heats up and rises, passes through the top vent into the room, and allows cold room air to enter the bottom vent by displacement. The cold room air is warmed as it passes next to the wall and exits the collector at the top. The concrete wall also absorbs heat, which is conducted directly into the room.

By opening a vent between the glass and the outside, the collector works as a ventilator in summer. Air warmed by the sun flows up the collector and out the

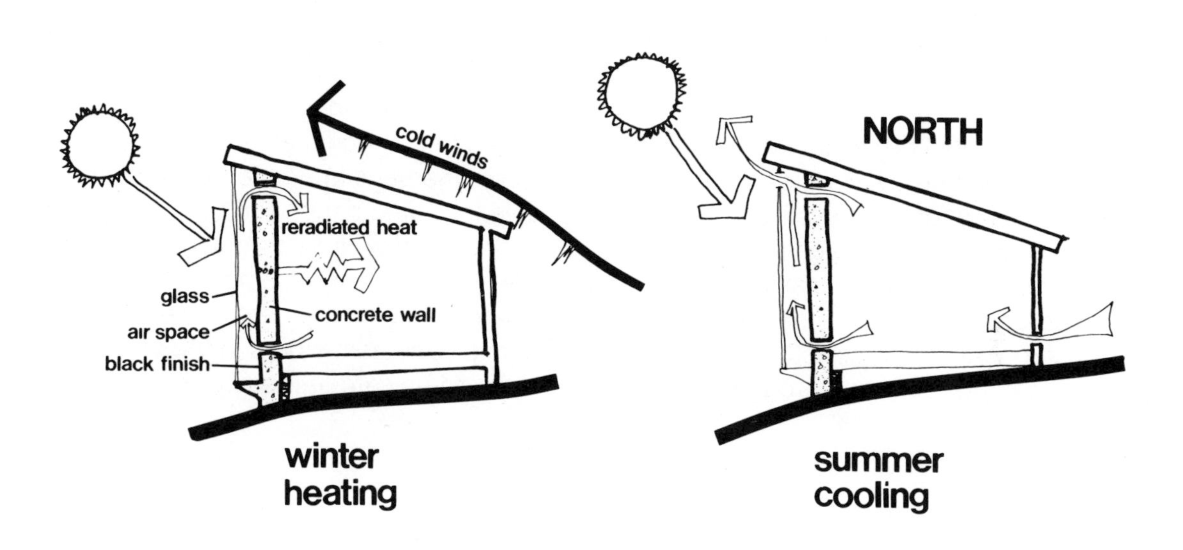

Trombé house

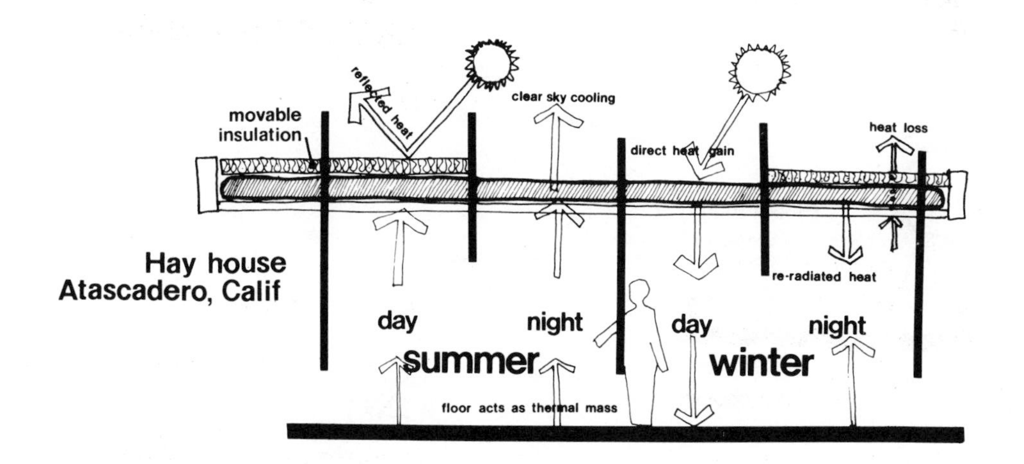

vent. Cool air enters the room by displacement through vents located on the north side of the building.

Roofs can also be designed to retain or reflect heat energy. Harold Hay has successfully tested a system that uses sealed clear-plastic bags filled with water, backed with black plastic, and located on a roof to store heat. The bags are supported by a structural corrugated-metal roof through which heat can be transmitted. During winter days the sun heats the bags. Most of the heat is stored in the water, and a portion is conducted through the roof and radiated to the room below. During winter nights the bags are covered with a movable insulated panel, and the heat stored during the day radiates downward into the space. During a summer day the insulation reflects the heat of the sun, and at night any heat stored in the water is radiated upward to the sky.

The presence of water on a roof can help moderate solar-heat gains into a building by acting as an evaporative coolant. Simple ponding on the roof can also cool its surface and may extract heat from the interior. In the winter, however, a roof pond may lose more heat by evaporation in twenty-four hours than it gains during eight hours of daylight.

Similarly, earth or sod on a roof helps to moderate heat loss and gain while sus-

roof treatments

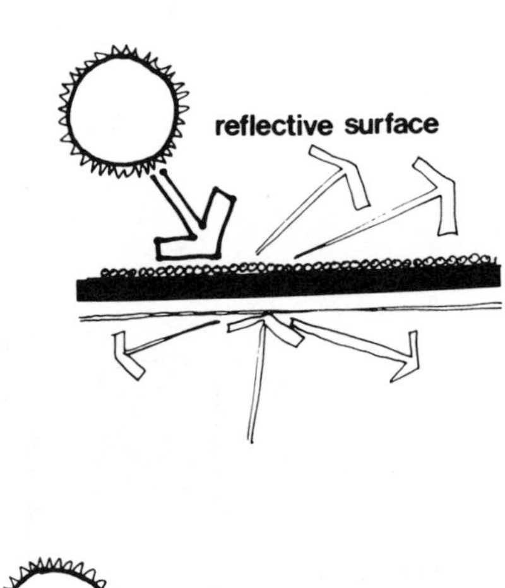

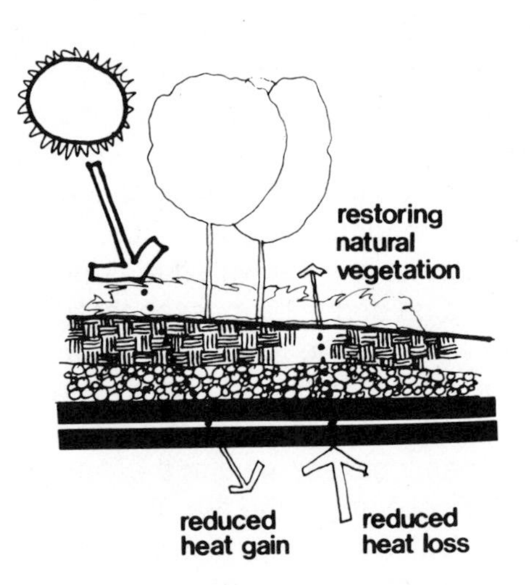

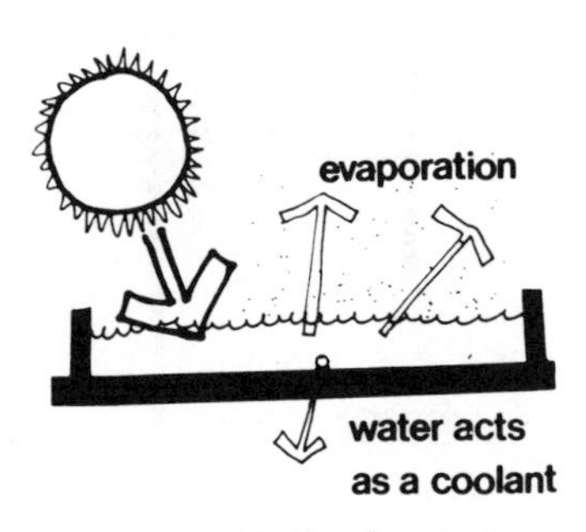

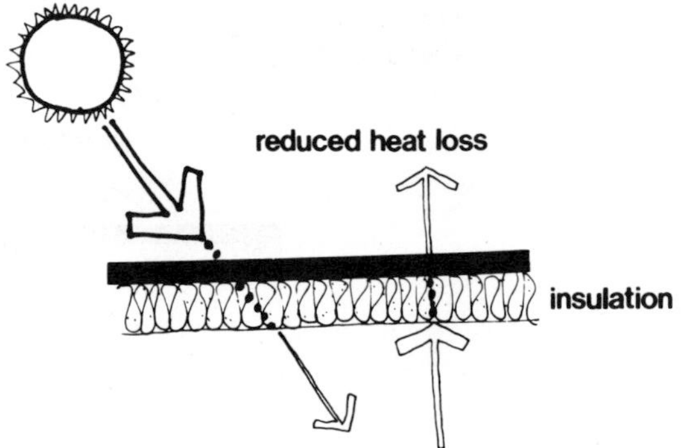

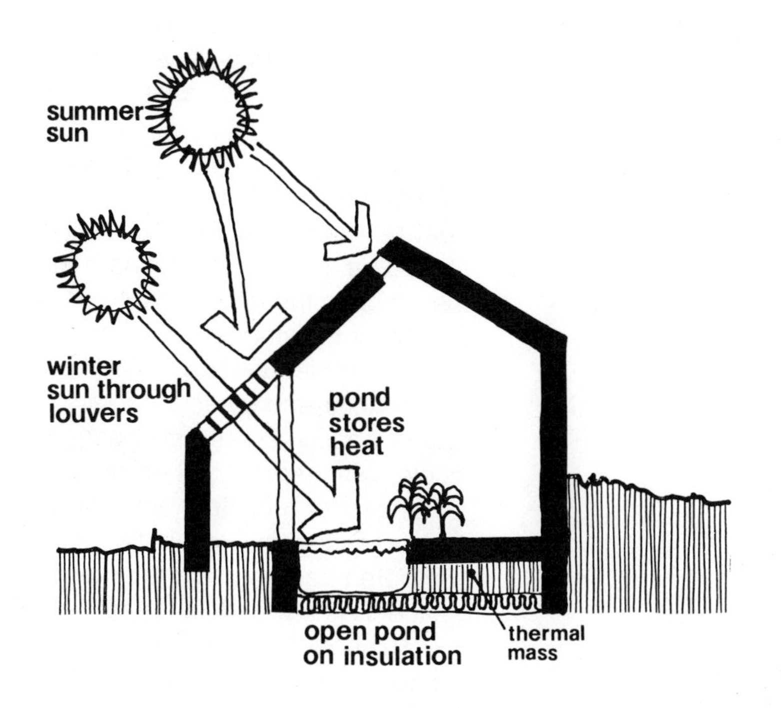

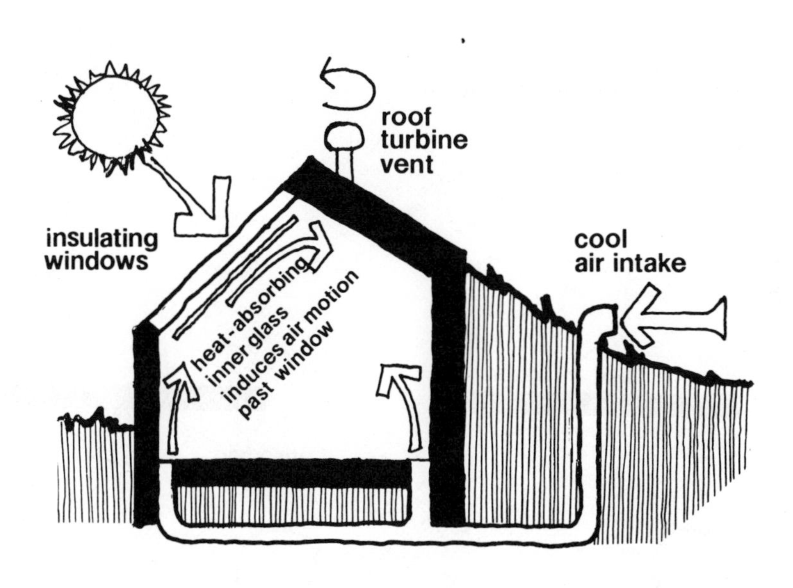

taining natural vegetation. This minimizes the negative impact of the building on the environment.

A mass of water may be incorporated into a building design as an open pool, swimming pool, closed tank, or storage wall to act as both a stabilizer of internal temperatures and a reservoir of thermal energy. With just a few degrees rise in temperature a large water mass can hold substantial amounts of useful energy. Solar, wind, and earth energies all work well in combination with water.

A relative humidity of 50 percent is regarded as physiologically ideal for most human beings but may not be high enough for some plants. Open bodies of indoor water, in supplementing ambient humidity, can cause problems by evaporating excessively at high room temperatures. This can be turned to advantage by lowered room temperature, in which human comfort is more easily sustained.

Earth temperatures in most geographic locations remain fairly cool and constant in temperature. During the warm and hot seasons of the year, air can be naturally circulated through underground areas to reduce its temperature and then be brought into a building to sustain indoor comfort. This action can be maintained by employing the sun to increase the upward convection of air through roof stacks.

Interior thermal masses of earth, concrete, masonry, water, phase-change materials or their combination with a well-insulated building envelope make the building itself a storage system of thermal energy. Moisture should be kept away from insulation, because water content will reduce or nullify its efficiency.

The properties of walls and roofs are affected by their surface colors, with dark tones absorbing more heat than light tones. Mirrored or white surfaces can reflect a great deal of radiant heat from any source, including the sun. Combinations of reflective and absorptive materials can be used to direct solar radiation, provide diffuse lighting, and effectively employ these thermal gains.

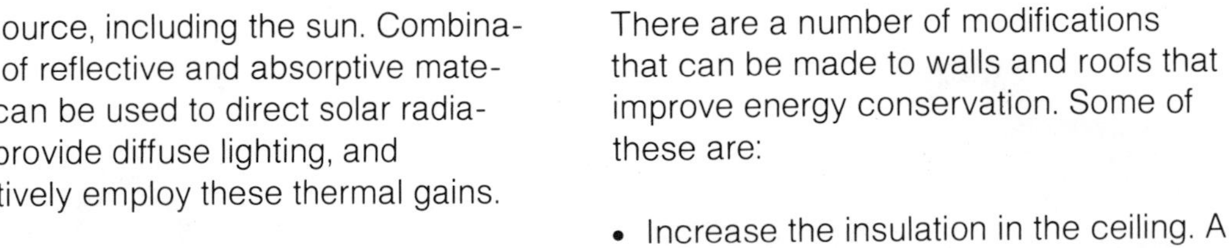

There are a number of modifications that can be made to walls and roofs that improve energy conservation. Some of these are:

- Increase the insulation in the ceiling. A heat-resistance value of 30 (R-30) should be a minimum standard for the insulation in the ceiling. (Note: This and other resistance values are federal recommendations. Cost-effective use of solar energy as a primary space-heating source may suggest, in cooler climates, far higher thermal resistances for roof, floor, and walls.)
- Increase the insulation in floors above ventilated basements and crawl spaces. A heat-resistance value of 19 (R-19) should be a minimum standard for insulation in the floor.
- If possible, increase the insulation in outside walls to a resistance value of 13 (R-13). Insulation in the upper portions of walls is most important.
- If possible, increase the mass of outer walls. An application of masonry veneer would increase the thermal inertia and time lag of the wall.
- Close attic and crawl-space vents in the winter if doing so will not interfere with combustion air to the furnace.
- When remodeling, place vapor barriers between the drywall and studs. This will decrease infiltration and interior moisture-vapor loss.

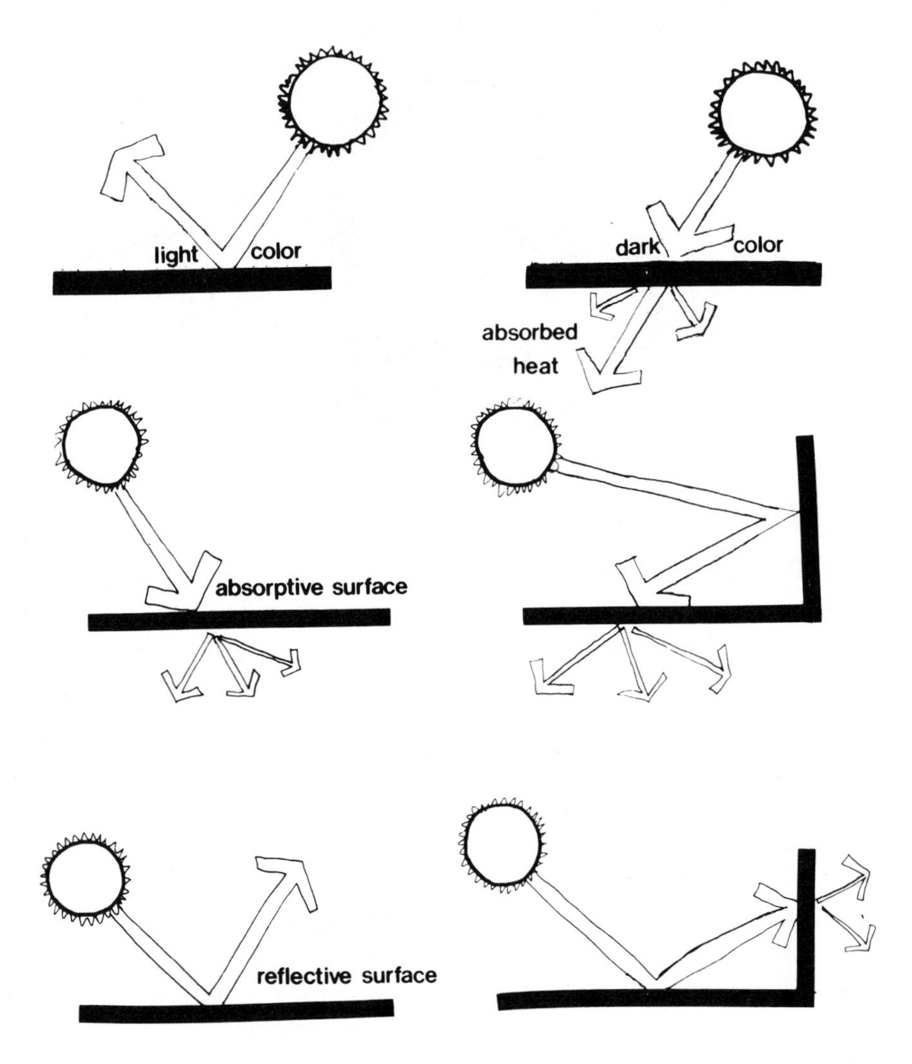

Windows

A great deal of heat is lost from a building through windows and other openings. Beneficial heat from the sun can also be gained through windows. To minimize energy loss through glazed areas, insulating glass that decreases heat conduction can be used. Insulating glass consists of two or more panes with a dead air space between. In the best prebuilt units, two panes are edge-welded after the air separating them is partially evacuated. Within limits, the insulating quality of the air increases as the distance between the panes increases. The optimum distance is 2 centimeters (3/4 inch); beyond that the velocity of convection currents formed between the panes of glass will transport large amounts of heat from one surface to the other. In some extremely cold climates, three panes of glass can be used, which produce two insulating air spaces.

The location of windows on a building will have an influence on their size, shape, and type. On the west and south sides of a building it may be advantageous to use glass with a reflective surface to shield out the intense summer sun. The illustrations show the approximate percentage of heat transmitted and absorbed by different glass types and configurations.

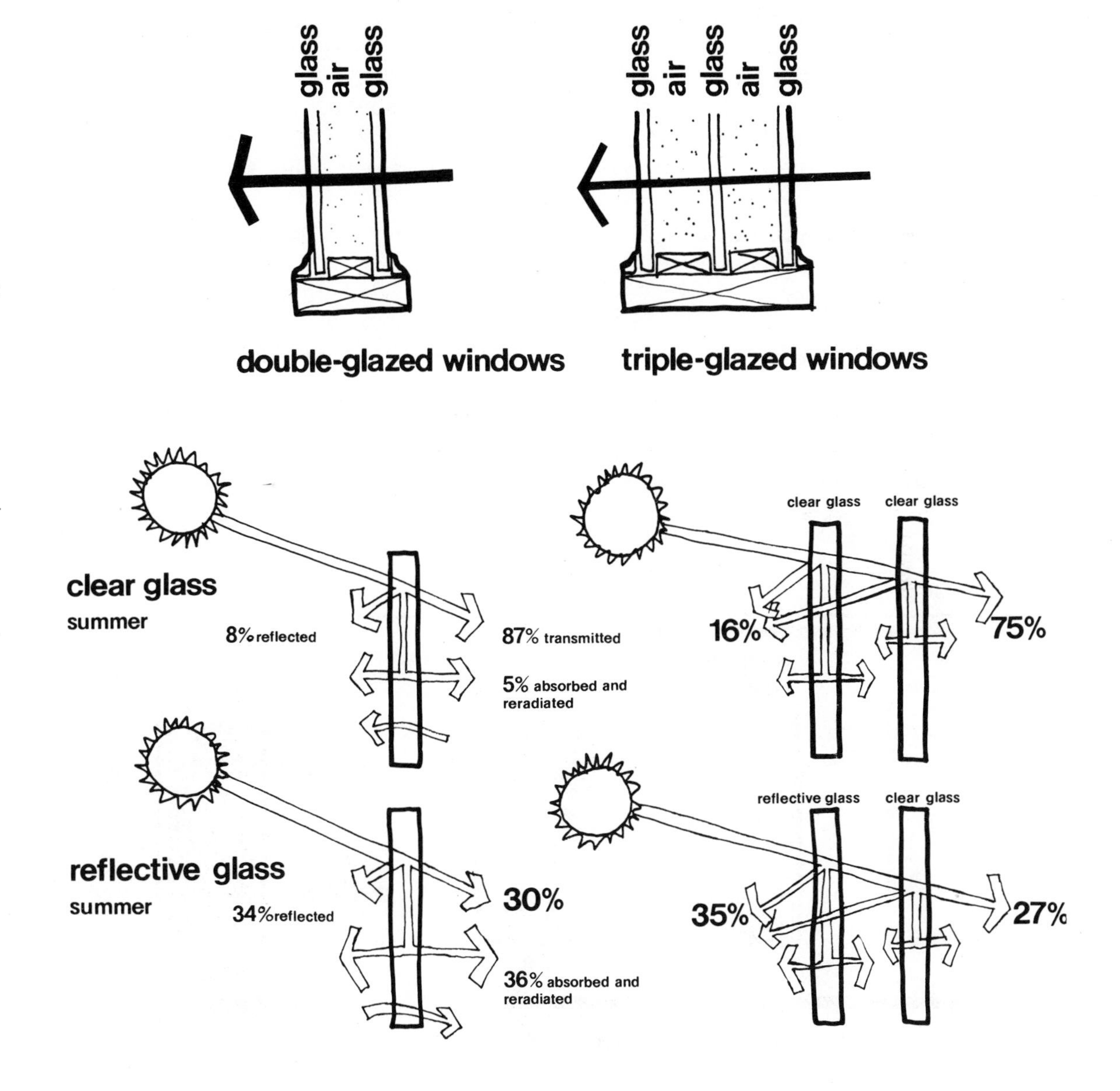

A true dead air space is a good insulator, but it loses its effectiveness when convection currents propagate in it. One interesting system that uses a material other than air was patented by Steve Baer of Zomeworks and is called Beadwall. This system has two panes of glass spaced several inches apart. During the day the space or chamber between the glass contains air only. At night, expanded polystyrene pellets are blown into the chamber, displacing the air and forming a heat barrier.

An impressive range of other forms of movable insulation that temporarily covers windows or skylights to reduce thermal losses or gains has been developed. These systems are installed either indoors or outdoors, and some feature automatic operation. During winter the insulation is closed to reduce heat loss at night or during cold, overcast days and is then opened to optimize daytime passive-solar gains and allow light penetration (except on north-side exposures) to interior spaces.

Special care in installing and operating movable insulation is essential. A tight fit to the opening is necessary; weatherstripping, weighted materials, magnetic stripping, pressure members, Velcro tape, or other closure mechanisms can be used to ensure a good seal. In many instances, architectural solutions and selective use of glazing methods that reduce dependency on movable insulation would be a better option, depending on climate extremes, specific exposures to sun and wind, and all other conditions affecting heat loss in cold weather or overheating in hot weather. Although movable insulation can control a major source of heat loss in buildings, it is a poor solution in terms of human presence, attention, and effort.

The following is a list of advantages and disadvantages of movable insulation.

Advantages:

- thermal protection of energy-vulnerable openings; retention of passive-solar gains
- more daylighting control
- possible reflective surface for artificial or solar light
- elements can be moved to correspond to solar position
- increased privacy and security

Disadvantages:

- care is needed to ensure tight seal
- loss of outdoor views
- manual operation requires time-of-day presence
- automatic operation invites problems
- barrier to fire safety
- initial high cost and need for maintenance
- need for stacking and storage space
- technological advances in glazing methods can make the systems obsolete
- lacks uniform gradation control of daylight

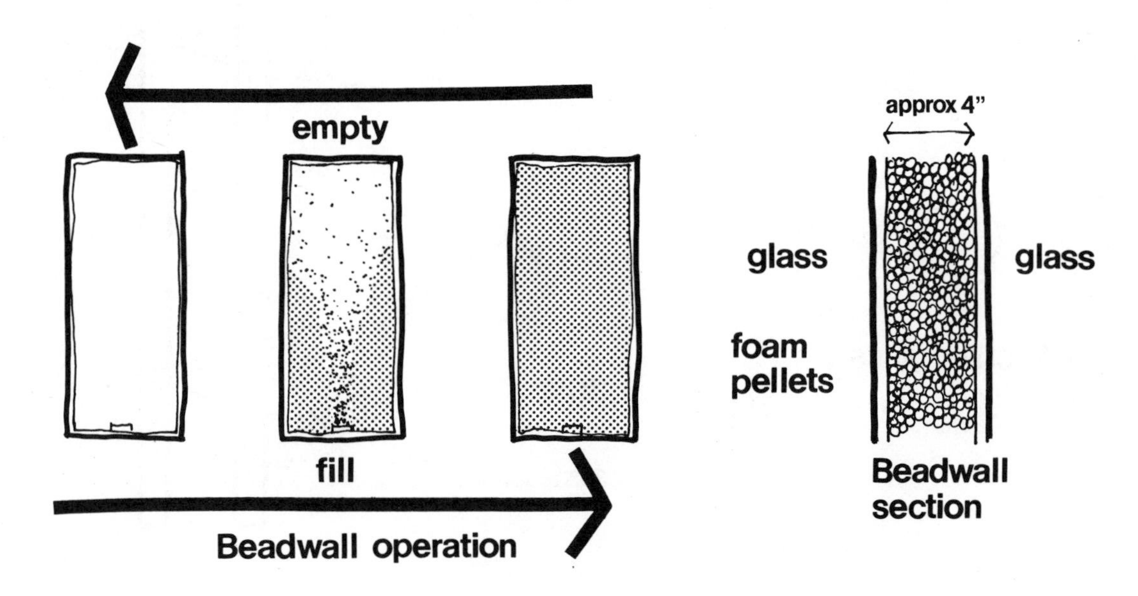

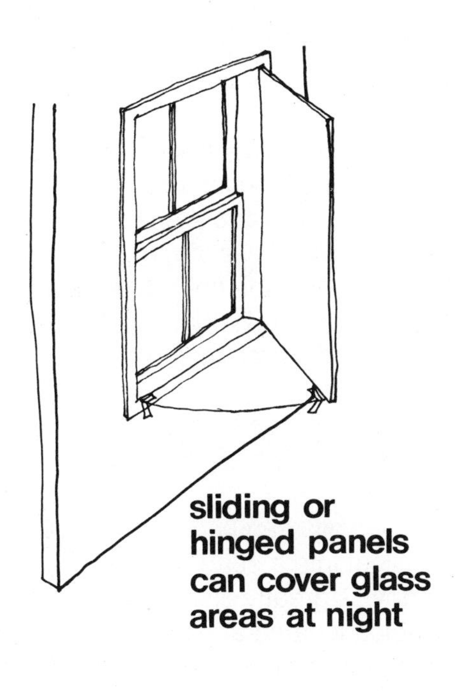

There are exceptions according to latitude, but the window area should generally be kept to a minimum on all sides of a building except the south, which should offer the maximum possible consistent with energy conservation. The winter sun should be used to heat a building during the day, and the windows should be sufficiently insulated to prevent heat loss at night. Window recesses or overhangs can be designed to allow the sun to enter a window in the winter months and shade the window from the intense summer sun.

While determining the size and placement of windows, many things should be considered. One of the more important variables is the position of the sun, which is continually changing, not only throughout each day but seasonally as well. Two angles are therefore needed to specify the position of the sun in the sky: one is the azimuth angle, which is a measurement in degrees of the sun's position in plan view (looking down from above); the other is the altitude, which is a measurement in degrees of the sun's elevation above the horizon. At a given

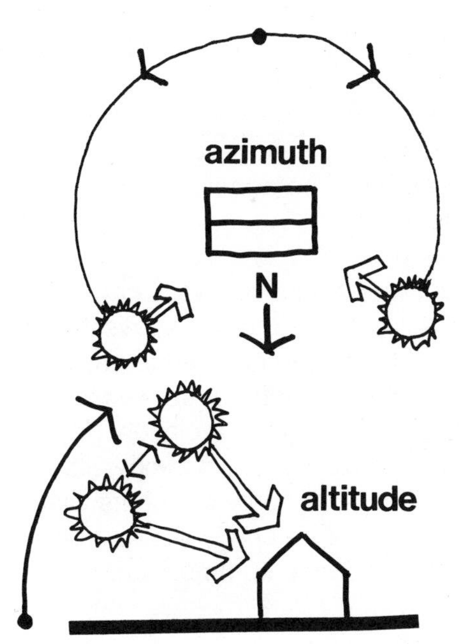

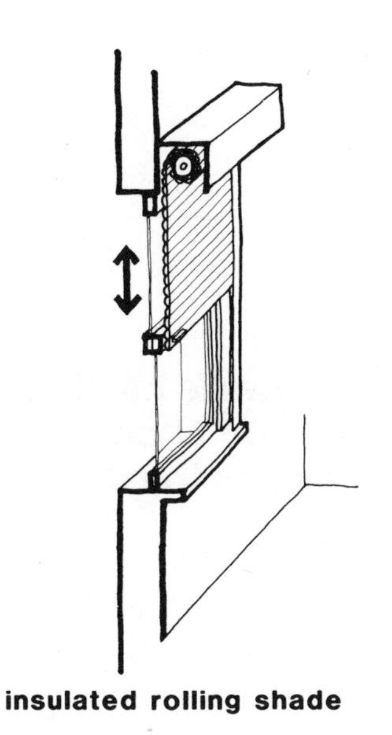

insulated rolling shade

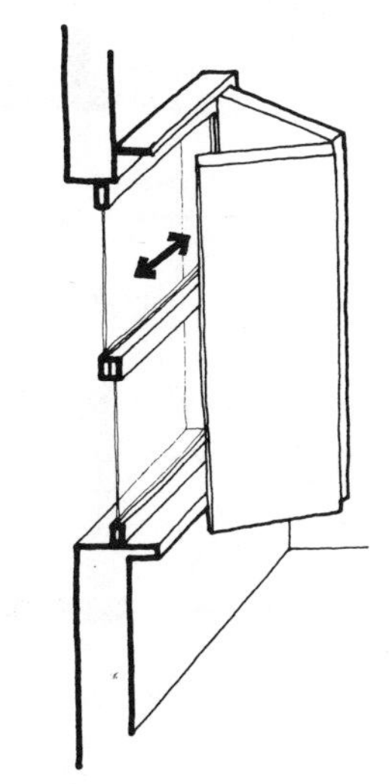

insulated folding shutter

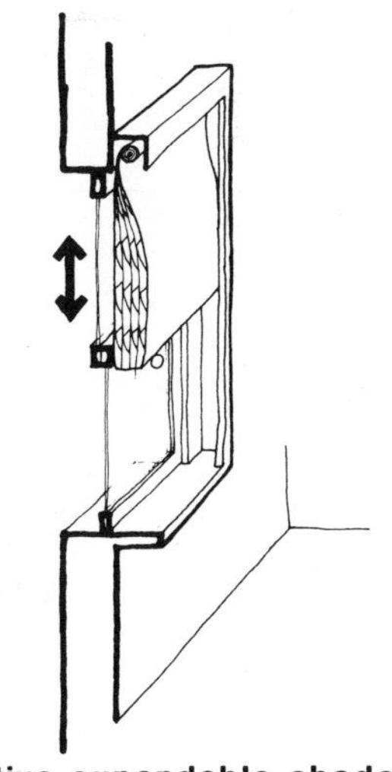

reflective expandable shade

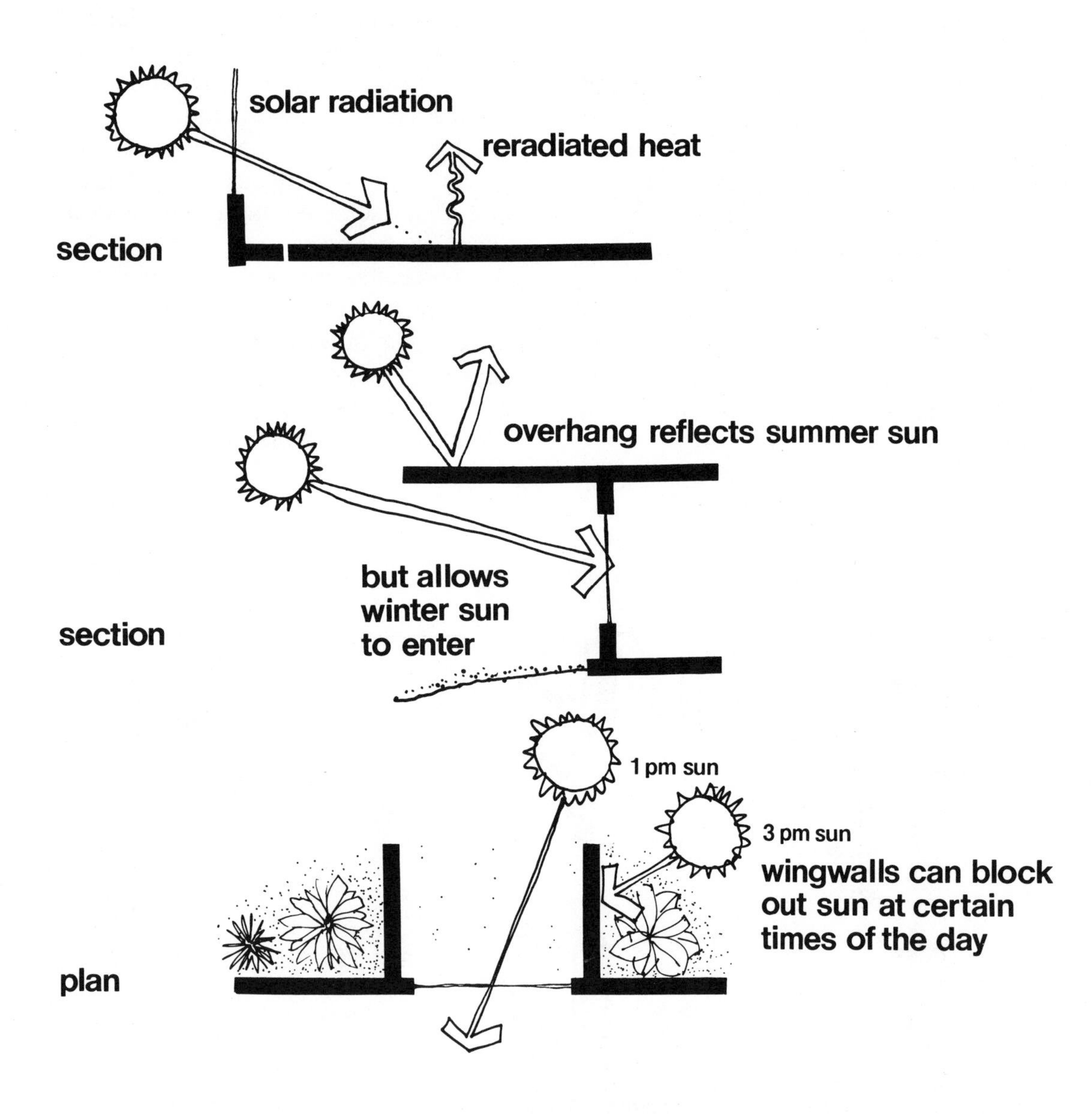

date and time, both the altitude and azimuth differ according to earth latitude.

Direct solar radiation can be intercepted and stored by the interior floors and walls of a room and then reradiated to the room. Windows that extend to the floor line allow the floor to be heated by direct radiation. The heat will be absorbed and stored in floors of massive construction (concrete, masonry, rock, etc.) and released when room temperatures start to fall.

Interior walls can be used to reflect diffuse light into a space. A long, narrow window is greatly enlarged in its effect by being located adjacent to a wall or ceiling of light color while softly diffusing the glare component of direct sunlight. Using the sun for natural diffuse lighting can also replace significant amounts of energy required for artificial lighting. In all buildings, natural lighting in some areas could be adequate for direct occupant use.

Overhangs and wingwalls can serve to reflect or block the sun's rays; the reflected light can be used to provide natural lighting to interior spaces that do not receive direct solar radiation. Proper design and placement of wingwalls and overhangs can provide rooms with warming winter sun while controlling the entry of intense summer sun.

To optimize the design of buildings to be energy conserving, windows should:

- have overhangs or wingwalls to block out the summer sun
- allow the winter sun to penetrate and warm interior spaces
- allow the heating of floors and walls, which can retain and reradiate heat
- be recessed, to trap heat and reduce convective losses and reduce summer heat gain through shading
- prevent heat loss at night and during days of limited sunshine
- provide natural lighting for interior spaces

Because gaps exist around movable sections even when they are closed, more heat is lost through operable windows (ones that open) than through fixed windows. To allow the least possible heat loss, the use of operable windows should be kept to a minimum, and air outside a window should be regulated for natural air circulation induced by the sun and wind. Insulated windows provide dead air spaces, which decrease the amount of heat reaching the outer glass surface.

Large glazed areas can be protected by buffer spaces. A greenhouse area on the south will act not only as an insulating buffer but as a source of heat for the building. As the sun heats the greenhouse, excess heat energy can be vented into the house during the winter and to the outside in the summer. At night the greenhouse acts as insulating air space for the southern windows. During especially cold periods, the greenhouse will be tempered by the heat loss from the house, tending to maintain safer temperatures for plants. Greenhouses, atriums, sunspaces, and other "double envelope" buffer spaces can also reduce the need for movable insulation and act as a shield to changes in temperature, wind, solar intensity, and atmospheric pollution.

Every room space can act as a thermal buffer to adjoining room spaces. Corridors, galleries, and other transient spaces can be effective thermal-energy barriers. Tight construction, foundation insulation, and use of insulated tight-fitting ventilators can increase the buffer effect of these spaces.

wingwalls and overhangs protect window areas

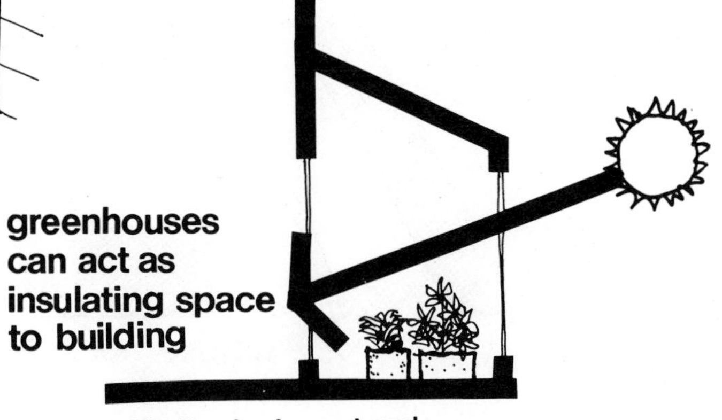

greenhouses can act as insulating space to building

infiltration is also reduced

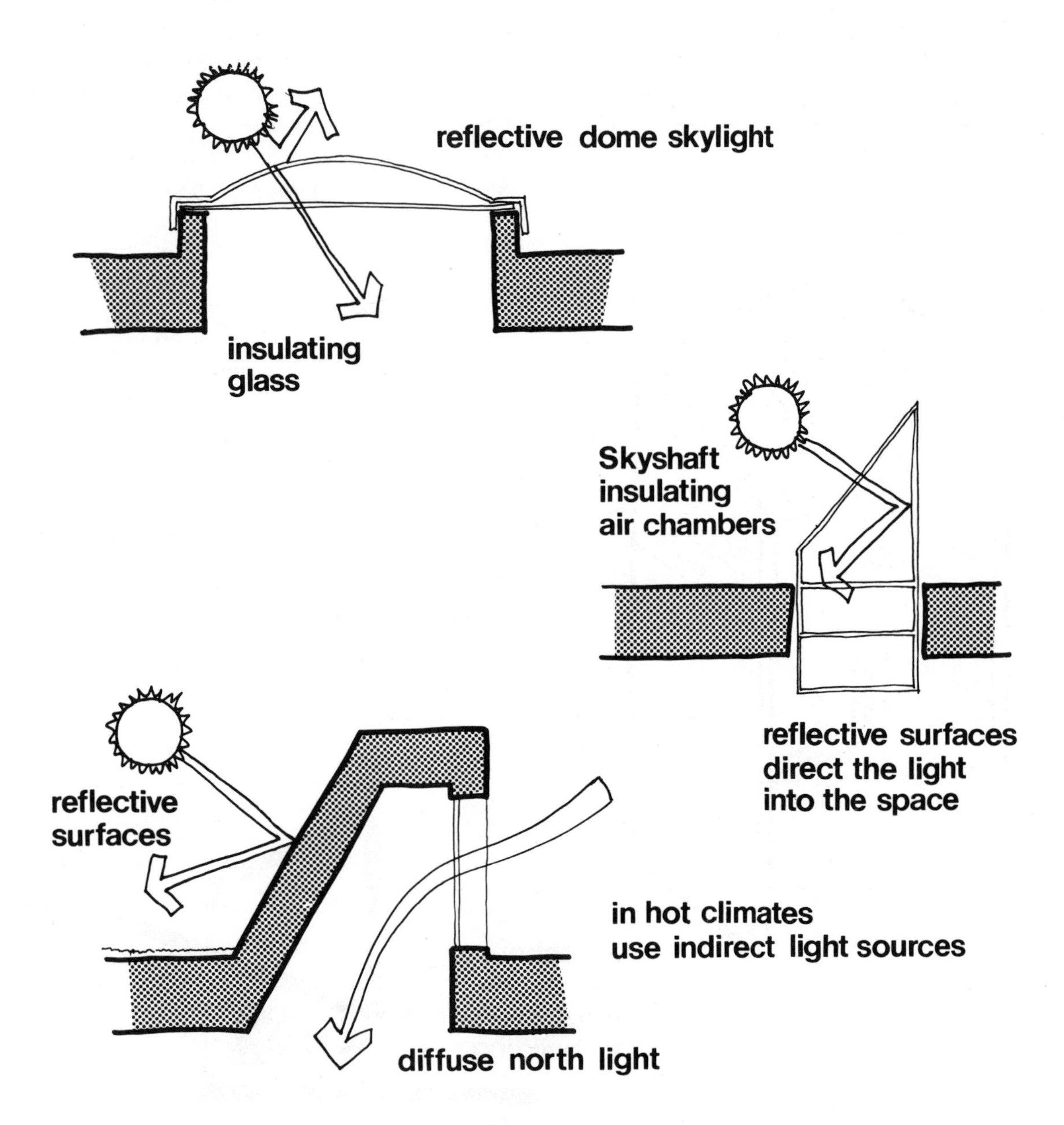

Skylights are another device that allow light to enter a space, but they can increase solar gains and overheat the interior during certain seasons. Normally, a great deal of heat is lost through skylights because the hottest air will be at the highest point in a room, which is generally the location of a skylight. Recent developments in double and triple layers of glass and plastic for skylights with insulating air spaces between the layers have decreased the heat-loss problems. The benefit of skylights is diminished if they are placed on a shaded north side of a building, a practice that should usually be avoided.

"Skyshafts," as developed by Richard L. Crowther, are composed of several sealed air chambers that are lined with white reflective sides and clear lenses at the top or horizontal positions. These provide an interior space with diffuse lighting.

In southern latitudes conventional skylights should be used sparingly because summer heat gains through them can be disastrous. A north-facing monitor window would be a more protected light source in such regions.

Skylights used in conjunction with sloping mirrored surfaces will illuminate and heat areas that do not generally receive direct radiation. This is especially important in providing light and warmth to rooms with northern exposures.

Another feature that will provide natural illumination to rooms not adjacent to the exterior of a building is an atrium—an open shaft or hole extending vertically the full height of the building, forming an interior courtyard at the bottom. The walls of the atrium may be surfaced with relatively large window areas to allow as much light as possible to enter adjoining spaces that without the atrium would be closed to daylight. Sliding doors of insulated glass can provide access to the courtyard. The effect that wind will have on the air in the atrium is controlled by the size, shape, and proportions of the atrium, the edge configuration between the roof and atrium walls, and the wind velocity. Ideally, wind will blow over the atrium, allowing the air against the windows to remain still, affecting a minimization of heat losses. This effect will also carry snow over the atrium and prevent its accumulation on the atrium floor. At times, vortices or other undesirable wind conditions may form in the opening to the atrium and disturb the air inside. These effects, however, are not easily predicted.

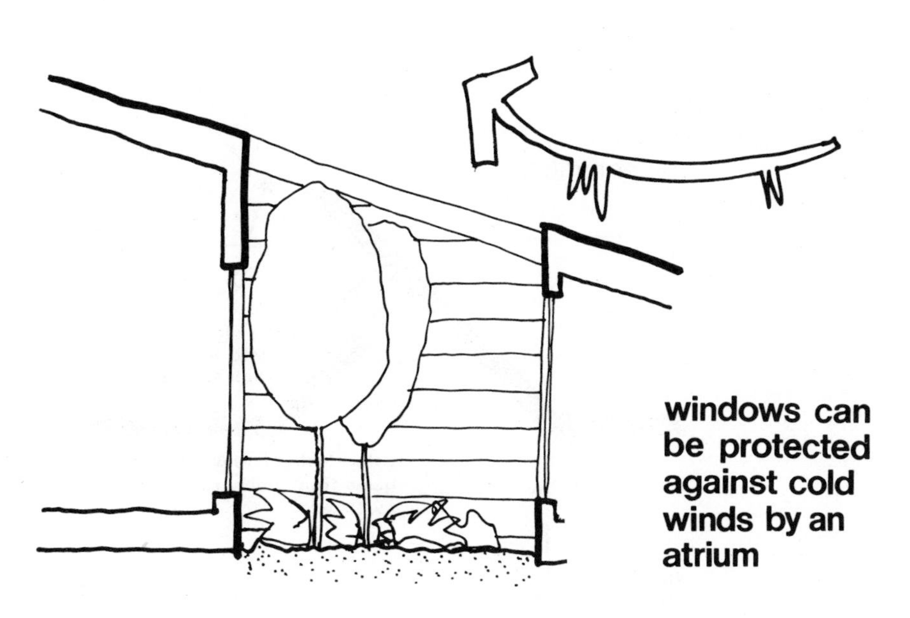

windows can be protected against cold winds by an atrium

Some suggestions for improving window areas for energy conservation are presented below.

- In the winter make sure all windows are closed tightly, and caulk or weatherstrip all edges not operable.
- Fasten storm windows of glass or plastic to exterior frame to increase thermal resistance. Metal frames must also be insulated, because they can conduct or convey large amounts of heat.
- Keep storm windows in place during the summer on all windows except those to be used for ventilating.
- Install draperies or shades on all windows and use them as movable insulating barriers to control year-round heat flow. These should fit tightly to sill or floor and be made of good thermal materials.
- Open windows on warm but not hot days to allow natural ventilation before resorting to powered air-conditioning.
- Attach external shading devices such as overhangs, sidewalls, recesses, awnings, or sunscreens to reduce solar-heat gain.
- Coat west-facing windows with reflective films or sunscreens, and/or install mirror-chromed venetian blinds inside.
- Replace metal window frames with wooden ones.

lower-level solar-inductive air convection through slots to upper building level

Passive Solar Heating Strategies

All buildings are solar collectors. Lacking energy-conscious design, they use the sun's energy inefficiently. Passive solar strategies aim to maximize occupant comfort and use with minimized utility demand.

We can optimize energy performance through its centralization. Subsystems can be tailored to specific occupancy needs, the best position for daylighting, amount of thermal mass, thermal flows, interface with the earth and atmosphere, and thermal-zone buffering. The design should be sensitive to both diurnal and seasonal change. As passive-solar architecture is more appropriately designed and fine-tuned for performance, it becomes more complex.

Various techniques have been devised to control the rate of heat flow from passive systems and thermal mass. Selective surfaces can limit passive solar-thermal loss. Selective-surface coatings, such as black chrome applied in a thin film, allow the sun's radiation to heat within a thermal mass but shield against energy loss from the mass. A diode is basically a fluid-type solar collector with one-directional flow to a vertical fluid reservoir with an intervening insulation panel; the fluid circulates by natural convection, and heat dissipates slowly

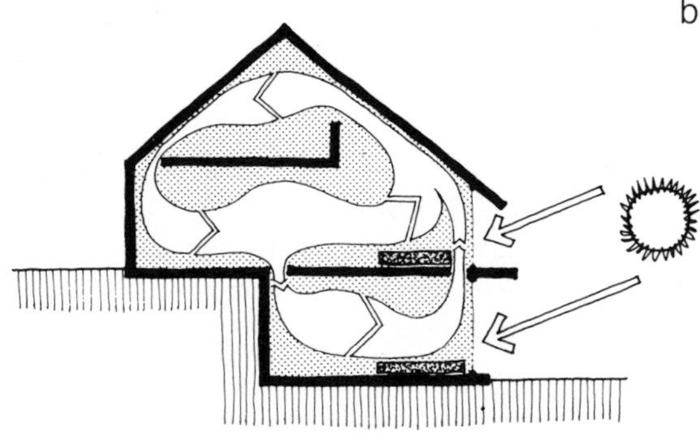

building acting as a natural convection solar-heat pump

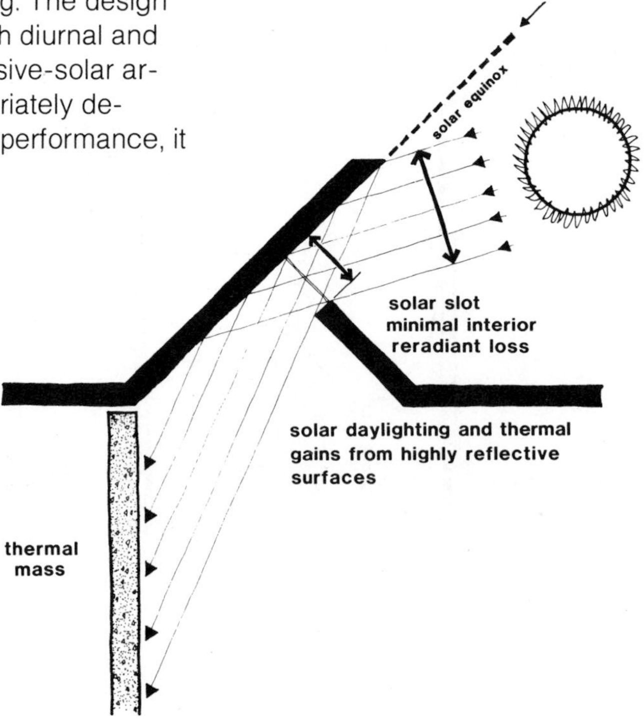

to interior space. Operable glass or metal louvers in front of thermal mass can help retain or dissipate the heat to interior space when needed. Hybrid procedures using fans for the relocating, remote storing, or distributing of solar thermal energy can improve performance. The sun's energy dissipates convectively throughout passive fluid storage at a lower temperature than for concrete, masonry or rock surface; thus, reradiant losses are lower.

The passive-solar system oriented due south or not more than 15 degrees to the east or west, depending upon the geographic location and site condition, most favors solar collection.

Daylighting

Daylighting is a key element of solar design. It provides useful interior light and a view of the great outdoors. Astute daylighting, considering every hour of the day through all seasons, can eliminate or drastically reduce daytime artificial lighting. Artificial lighting also increases air-conditioning demand above that of daylighting at equal levels of illumination.

Daylighting's potential negative aspect is great energy loss. Conventional skylights gain and lose energy excessively. In northern latitudes, south-facing reflective-slot clerestories and south exterior sidewall and compartmented Skyshafts provide daylight more energy efficiently than skylights. Northside triple glazing with Heat Mirror films and south-facing clerestory daylighting of the interior north walls are practical design strategies. It is critical that the design accommodate vision for interior tasks in both transient and nontransient space.

Glazing Systems

Glazing systems require careful thought. The use of single, double, triple, or even quadruple glazing should be weighed in terms of transmittance and insulative value. In a colder climate, east-facing glass might be double, with the outer glass clear and the inner heat-absorbing; the south-facing might be double and clear; the west-facing glass might be double and reflective; and the north-facing be triple or quadruple glazed.

Water-white (first choice) and low-iron (second choice) glazing transmits more sunlight than clear-float or window glass. The characteristics of glass blocks, diffusing and translucent glass, and plastic glazing should be checked with the manufacturer. High-transmittance or heat-reflective, clear mirror films interfere little with solar intensity but reduce heat loss.

Window mullions and divisions should be minimal in width and depth to avoid shadowing thermal mass. Without mullions, glass strong enough for the method of installation can be sealed edge to edge.

Holistic Energy Design

Every site has its particular microclimate. Passive systems and subsystems should relate to these microclimatic energy flows as well as those of the earth, architecture, interior, and time space use by the occupants. The ultimate design for each project should be based on the dynamics of human motion and use as well as the fixed elements, furniture, and furnishing of the architecture and interior.

Passive solar design requires a special sensitivity to holistic organic and inorganic energy flows. Ignoring applied-energy physics may sacrifice the great benefits of passive solar design to overheating or excessive energy loss. The design equation should also reflect human biophysics, behavior, and cultural viewpoint.

Health and vitality should be the result of a holistic design process. The sun's radiation is a biophysical necessity and, through thermal and daylight variation, a neuromental stimulant. Passive solar homes and buildings are more alive than traditional architectural shelter. Passive solar energy is economic and desirable not only for space heating but also for air tempering, humidification, biophysical regeneration, plant growing, and the simple enjoyment of basking in the sun.

Primary Space
Primary space is where sunlight first enters the building.

Secondary Space
Secondary space is where solar radiation or its thermal effects penetrate through or are received from primary space. Passive solar gains are augmented by diffuse irradiance from the skyvault and reflection from ground, vegetation, and neighboring structures. Sunlight heats primary interior spaces that buffer secondary spaces that benefit by direct and indirect solar radiation.

Direct Gain
Solar radiation that enters the interior space directly is converted to heat as it is absorbed by or reflected from surfaces. A south-facing window accepts sunlight, which is absorbed by thermal mass, converted to heat, or reflected to other interior surfaces.

Indirect Gain
Solar radiation is converted to heat at the surfaces of a thermal mass or primary space, which then gives up its heat to inner living space. Thermal lag, radiation, conductivity, and convection all play a part in releasing indirect gain energy from thermal mass and isolated gain space.

Isolated Gain
Solar radiation enters a space and is contained as direct or indirect heat gain. This space, isolated from other interior spaces, can be an attached greenhouse, atrium, sunroom, or thermal envelope.

Passive Solar Subsystems/Energy Centralization

A single passive solar system may be suitable for a small dwelling. However, commercial or institutional projects with increased occupant densities and thermal loads from lighting and equipment always require year-round cooling and ventilating. Solar thermal subsystems are most appropriate in these large buildings and for moderate- to large-sized residences.

Passive, active, and hybrid subsystems should be tailored to meet specific occupant space needs. These micro-energy systems have distinctive properties and performance characteristics. Subsystem energy can be used, stored, or relocated for heating, cooling, daylighting, natural convective distribution, humidification, air tempering, and other specific applications.

Architectural form should relate to daily and seasonal use. Internal heat, off-peak utility power, and stored solar energy in commercial and institutional buildings can provide heat for early morning, cold weather start-up or for unoccupied periods.

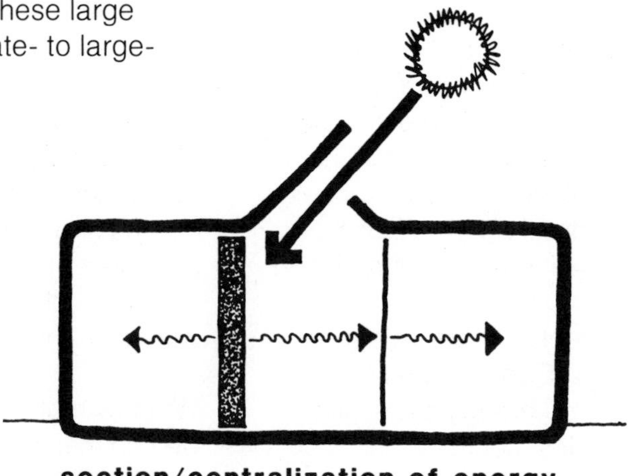

section/centralization of energy

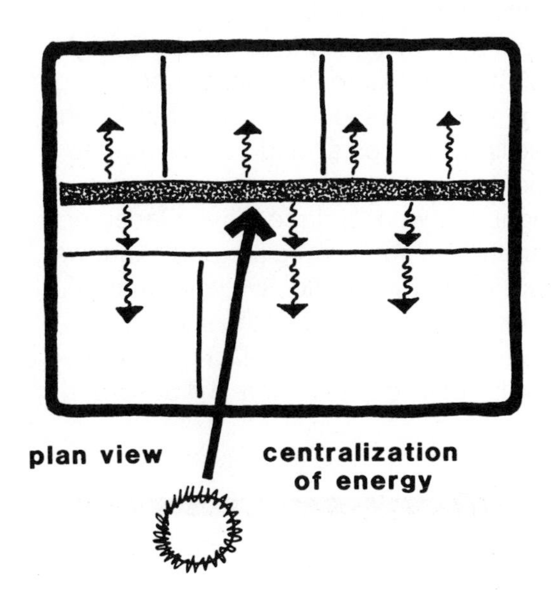

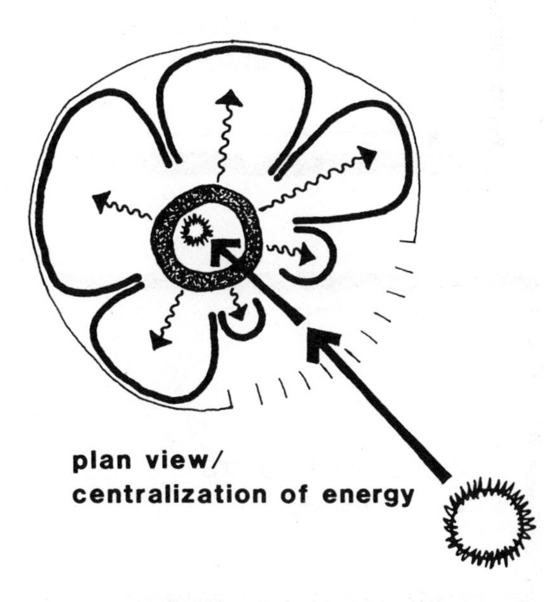

Energy centralization combined with synergistically integrated subsystems provides an opportunity to most effectively use the total energies of a building. Centralized energy can give surrounding spaces the most thermal energy "mileage" before dissipating. Energy consolidation in the lower levels of buildings with two or more stories can provide thermal flow to subsequent interior spaces and upper floors. Small peripheral spaces with low ceilings opening into a high-ceilinged area can expedite the natural convective movement of warm air to a central collection space. Energy centralized in core interior spaces can also be mechanically distributed to where it is most needed or to a place of thermal storage.

Lighting

The proper lighting of buildings is of major importance, accounting for up to 50 percent of the total energy expended in some nonresidential buildings. The heat produced by lights in a large building can be a substantial amount and may be useful either for heating or as an additional load against cooling.

Lighting inevitably affects the biophysical responses of man. Certain wavelengths of light help the eye maintain proper levels of the chemicals necessary for vision. Lighting can also have the psychological effect of altering a person's mood, attitude, or efficiency.

Artificial light is generally produced by converting electrical energy to light energy, using a lamp. Some light is produced by the direct burning of fuels, as exemplified by a candle or kerosene lamp. For the most part, the light produced in this way is used to create an atmosphere rather than for illumination.

A measure of the intensity of light is the number of "lumens" it produces, and the efficiency of a light source is expressed in terms of the number of lumens of light it emits per watt of electrical power consumed. Energy conservation in lighting can be accomplished in a number of ways; the simplest is to change from one light source to another. Different light sources have different electrical conversion efficiences, light output, lamp life, and color temperature, expressed in degrees Kelvin. Following are some typical efficiences for different types of lamps.

Lamp Type	Output
Incandescent	19 lumens/watt
Mercury vapor	57 lumens/watt
Fluorescent	75 lumens/watt
High-pressure sodium (yellow)	125 lumens/watt
Low-pressure sodium (orange)	183 lumens/watt

By using the most efficient lamp that will provide the necessary light intensity and color, electricity consumption can be kept to a minimum. As the power rating of a lamp decreases, so does its efficiency; hence a 15-watt lamp will produce fewer lumens per watt than one rated at 100 watts. Operating a lamp at a lower voltage than specified will decrease its light output but will disproportionately prolong its life. The light output can also be affected by the cleanliness of the bulb or tube, as well as the type of maintenance program for a building and its light fixtures.

Light output, along with fixture design and surface reflectivity of walls and ceilings, affects the amount of useful light striking a surface.

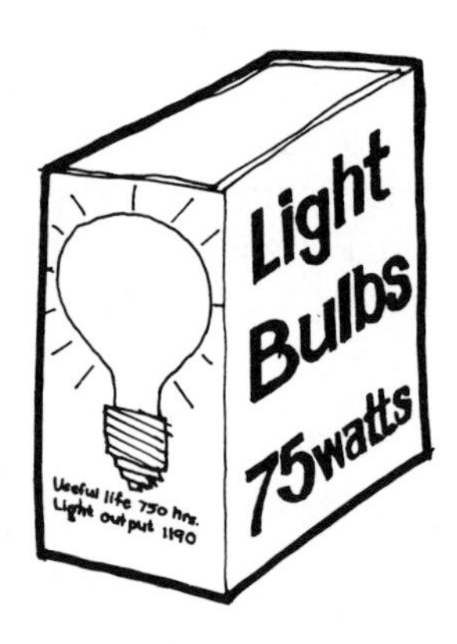

light bulb packages usually include the useful light in hours and the average light output in lumens

Another important consideration in the selection of a lamp type is the color it emits. Lamps do not equally produce all the colors of the spectrum; their spectral content varies greatly from one type to another. Incandescent bulbs give off a warm reddish light. The three types of fluorescent bulbs—cool white, warm white, and daylight—have unique spectral qualities. The color content of cool white is predominantly yellow and green, while warm white is red-orange. The color emitted by daylight tubes varies from one manufacturer to the next, but all attempt to simulate natural outdoor light. Some manufacturers have, in fact, developed fluorescent lamps showing 91 percent chromaticity, against a natural daylight index of 100 percent. By contrast, average cool white fluorescent lamps yield an index of 68 percent.

The life of a lamp is expressed as the number of hours it will provide useful light (figured as a statistical average). This can be an important consideration when specifying a type of lamp. On large projects the replacement of burned out lamps can be a major long-term expense in terms of labor and the cost of replacement bulbs. Fluorescent and sodium lamps offer up to 36,000 hours of life, or between ten and fifty times the longevity of incandescent lamps.

Other factors that influence the quality of light at the work surface are the design of the light fixture (luminaire) containing the lamp and the reflectivity of walls and ceiling. Fixtures should be designed to deliver the maximum possible useful light to a work surface. Their placement should be such that glare and shadows are minimized. Reflective walls and ceilings can help to lower the lighting requirements of a space. Dark-colored walls, ceilings, and floors will absorb

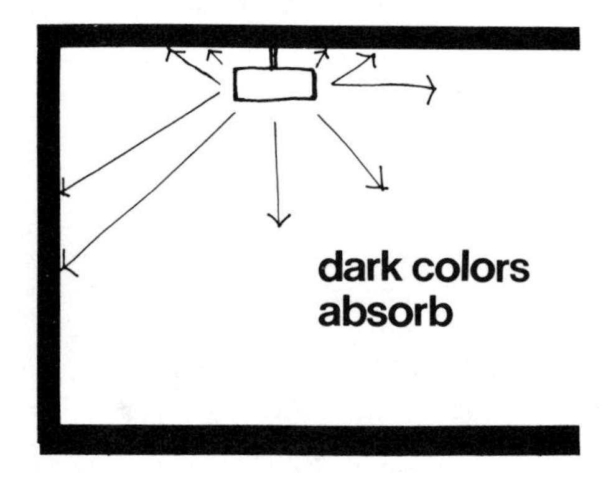

dark colors absorb

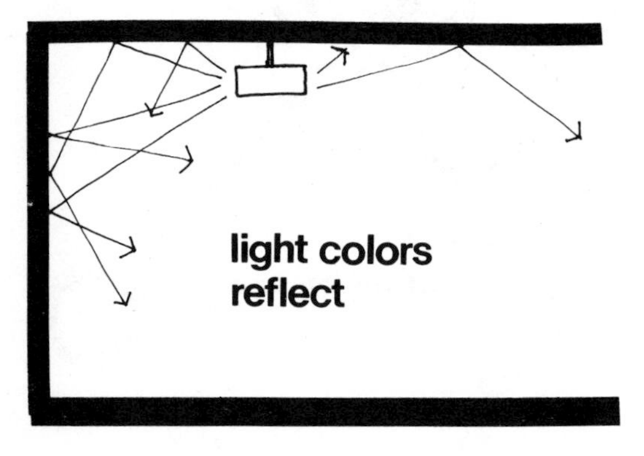

light colors reflect

light that could be reflected and used for illumination of a task area.

Existing light fixtures can be relamped to great advantage and with no wiring alterations in the following categories:

Intermittent or Task Lighting

Incandescent (tungsten filament) lamps should contain krypton gas, which can increase lamp life by a factor of five, at lower wattage-per-lumen output.

General Indoor Lighting

Incandescent (tungsten filament) lamps should contain krypton gas, as should fluorescent lamps, which in this case lengthens lamp life while maintaining the same lumen output at a 9 percent lower wattage level. High-pressure sodium, even more economical, may be used where color rendering is not critical.

Outdoor Area Lighting

For small areas, self-ballasted mercury vapor, high-pressure sodium, or krypton/tungsten incandescents can be used; all of these offer longevity and lumen-output superiority over incandescents. Large areas are best lighted by low-pressure, low-wattage sodium (LPS), which offers outstanding lumen/watt output and lamp life. Numerous large American cities have recently made partial LPS street lighting or freeway installations, recognizable as a warm yellow-orange glow. This color rendering produces monochromatic night vision and high object resolution; that is, all objects that under mercury-vapor lighting appear black or indigo blue are instead highly visible as shades of gray. The improvement is immediately noticeable in cities using red fire trucks, for example, since these no longer appear black at night.

Task Lighting

Lighting levels should meet, not exceed, the requirements of the tasks or activities to be performed in a space. In the past, general overall lighting of high brightness was utilized to provide equal intensity levels over an entire room, but much of the light was not productively utilized. The amount and quality of light required to perform a given task varies from one activity to the next. The concept of task lighting involves providing an area with an amount and quality of illumination sufficient to perform the task for which that area is intended. Quality of light is extremely important and, in many cases, is a more imperative criterion than the footcandle level (quantity). Task lighting makes more sense than equal lighting throughout a building. Lighting fixtures over individual desks use much less energy than is used illuminating an entire office area. Since light intensity varies as an inverse function of distance from the source, the closer a fixture is located to the task area, the less light it must emit to provide the necessary illumination.

Photocells can be used to switch on artificial lighting when natural lighting is inadequate or to prevent outdoor lighting from coming on until it is dark. They can also be used to activate shading devices when sunlight is too intense.

provide task lighting instead of general illumination

Continuing research and improvements on gases, circuitry, luminaire design, materials, optics, and heat recovery are leading toward increased lighting efficiency, extended lamp life, and improved color rendering.

In some buildings all of the lights in a large room have to be activated in order to turn on the lighting fixture over one desk. If the light fixtures could be turned on independently, a great deal of energy could be saved. It is not always practical for each light to be independent. Energy could be saved if zones, composed of a small number of fixtures, could be turned on at once.

Listed below are measures that can improve the efficiency of the lighting in an existing building, contribute to energy conservation and a reduced electricity bill, yet maintain minimum lighting standards necessary for seeing and safety.

- Use natural daylighting to replace or supplement artificial lighting whenever possible; it produces less comparative heat than artificial lighting and is more beneficial to human health.

- Turn off all lights when they are not needed. One 100-watt incandescent bulb burning for ten hours requires 11,600 Btus (12.2×10^6 joules) of energy to be produced at the generating plant.

- Use fluorescent lamps in suitable areas since they produce more lumens per watt and have a longer life than incandescent bulbs.

- Keep lamps and lighting fixtures clean. If they are dirty, light output can be significantly reduced.

- Light-colored walls, rugs, draperies, and upholstery will reflect more light than dark colors and reduce the wattage of artificial lighting fixtures required.

- Consider installing solid-state dimmer switches when replacing light switches. They reduce energy consumption by permitting lamps to be operated at reduced power levels.

- For any application it is better to use one large bulb instead of two equivalent small ones. As an example, one 150-watt bulb delivers 2,880 lumens while two 75-watt bulbs deliver only 2,380 lumens.

- A three-way bulb should be at the lowest setting for watching TV. The higher settings can be used for tasks such as reading, writing, and sewing.

- Do not turn off any fluorescent lamp that will be needed twice or more during a fifteen-minute period; the shortened lamp life caused by switching will more than offset electricity saved in this manner. Fluorescent lamps are "ignited" by bursts of extremely high voltage, and the lamp and ballast (transformer) life is affected according to the number of "starts." Incandescent lamps, by contrast, deteriorate according to total operation time, with the number of starts being less important. These lamps should be turned off whenever unused.

Ventilation

As air moves or circulates, it gains heat from objects warmer than itself and loses heat to colder ones. Wherever two air masses of differing temperature mix, however, the larger will prevail in its effects. During cold weather it is therefore important to prevent air from carrying heat out of a building, while during the summer maximum ventilation is encouraged, unless mechanical air-conditioning is used or excessive outdoor air pollution prevails.

Tight seals on all the windows and doors of a building minimize the infiltration of cold air and the loss of warm air. Weatherstripped doors and inoperable (fixed) windows also prevent excessive heat losses caused by infiltration. Cracks in the construction around windows, doors, and other openings can leak air, particularly in masonry buildings. (See Chapter 8 for methods for calculating heat loss caused by infiltration.) Caulking and sealants can be used to stop infiltration heat losses around joints and openings, while vapor barriers prevent air and moisture penetration through walls, floors, and ceilings. Any exhaust fan, bathroom vent, or other opening for air circulation should be provided with a damper to minimize heat loss when not in operation.

Vents that are used year round, such as direct-ducted bathroom fans and kitchen exhaust fans, can remove undesirably large amounts of heat from a space. Ductless filtered fans are available that remove odors from room air and then return it into the room directly. Air is pulled into one such unit by a fan and then filtered through a natural citrus-based chemical that absorbs odors caused by bacteria and mildew. The air and its heat are then returned to the room. Exhaust fans can also be equipped with heat exchangers that remove heat from the air before it is exhausted to the exterior.

Any fans or vents not equipped with dampers allow heated air to escape almost continuously. All equipment of this type should be fitted with dampers that are designed to seal as tightly as possible to minimize heat loss. Dampers should be insulated and spring-loaded toward their interior (room) side, or otherwise controllable. In some applications they may be remotely located and require motorized control.

During the winter, heated air will leak out of a building through its attic vents. Doors or shutters can be attached to attic vents and be closed in the winter and opened in the summer. Cracks around hatch doors or stairways leading to an attic are pathways through which heated air travels on its way out of the building and should be sealed with weatherstripping or caulking. Access hatches themselves should be fully weatherstripped, with insulation equal to that of the surrounding attic.

Controlled infiltration can be accomplished by making homes and buildings very tight and then providing for the intake of outdoor air through sunspaces, greenhouses, other noncritical temperature spaces, or underground ducts. This is a more rational approach than trying to deal with an architectural energy sieve.

For energy conservation, ventilation should be minimized during cold winter months and encouraged during warm months. However, in tight, thermally efficient buildings, indoor air quality is a critical issue. Adequate ventilation, systemic purification, and recirculation tend to offset the health hazard of indoor pollution of tightly contained air with contaminants from smoking materials, finishes, cleaning agents, fireplaces, stoves, radioactivity, and equipment.

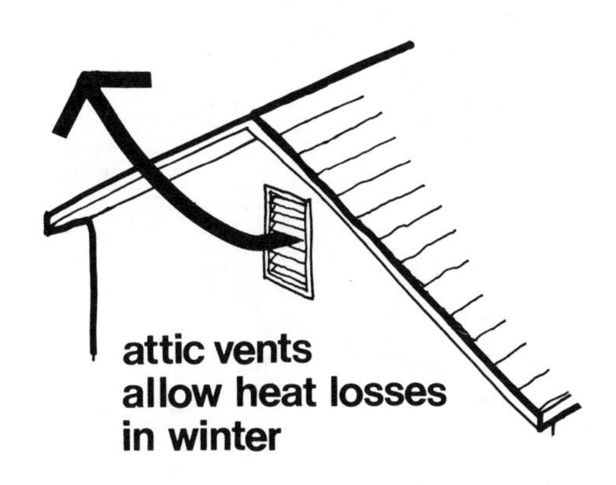
attic vents allow heat losses in winter

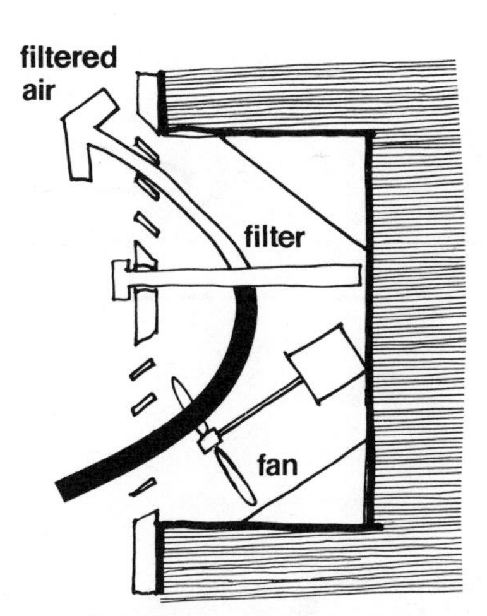

ductless fan does not exhaust heat from rooms

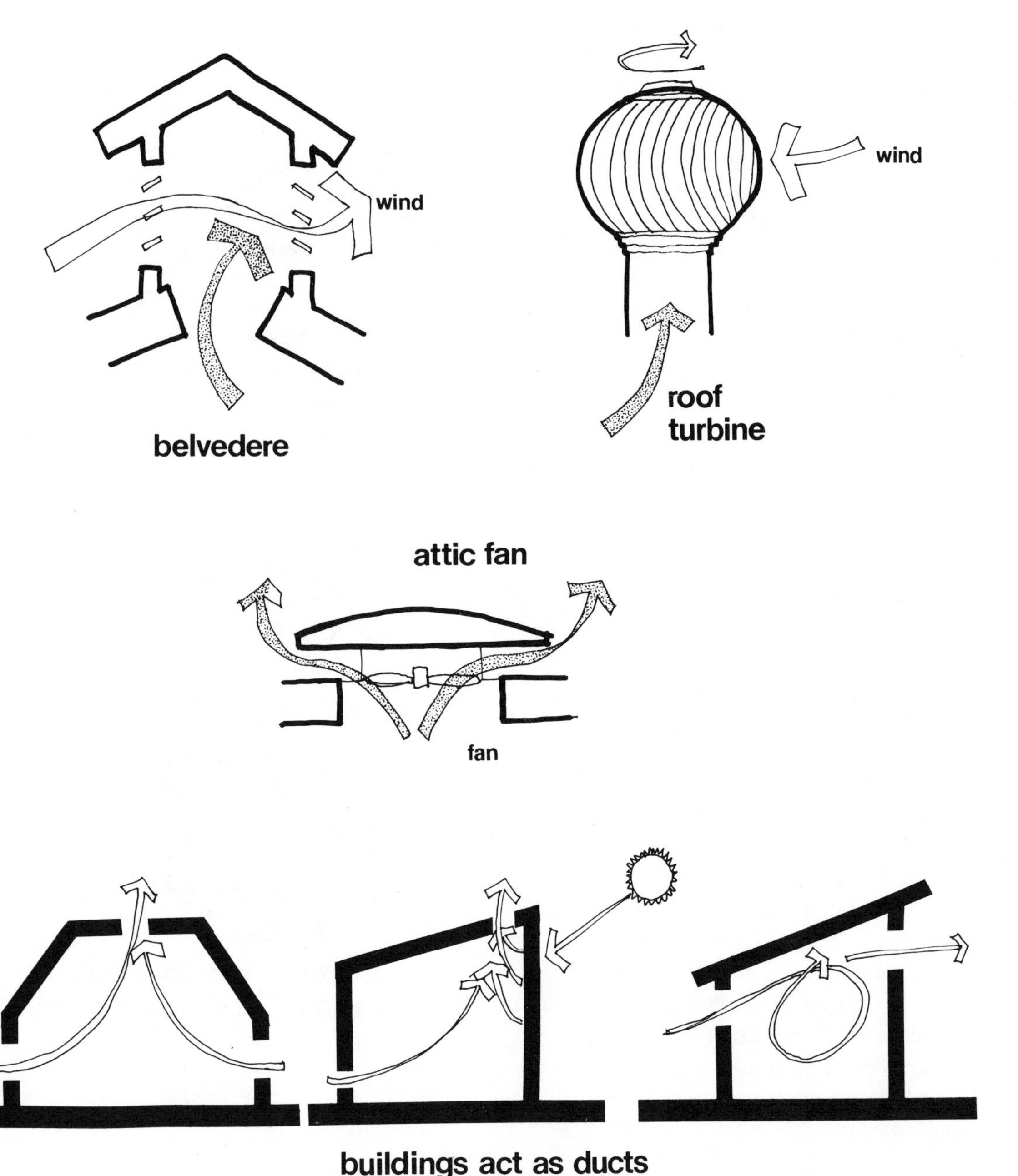

The air required for ventilation varies depending on the activities performed in a space and whether or not smoking is allowed. If smoking is permitted, up to nine times more air must be brought through a space hourly to maintain necessary purity levels than if smoking were prohibited. This can mean oversized mechanical equipment to move and purify the air. A more reasonable alternative might be confinement of smoking to certain small areas of any building. Interior building materials deteriorate at a greater rate in an environment where smoking is allowed. The air pollutants from smoking can even cause disorders in sensitive electronic equipment such as computers and activate smoke-detection alarms.

Outgassing of architectural materials and finishes, furnishings, floor coverings, clothing, appliances and other consumer products, food preparation, plants, cleaning agents, radon gas, and radioactive and allergenic biologic materials, particularly molds, spores and bacteria, can all add to indoor air pollution.

Polluted outdoor air is a poor source for indoor ventilation. Nevertheless, building codes require prescribed levels of air ventilation. Most filters fail to remove gaseous pollutants. Activated charcoal and negative ionization, used with efficient filtration materials, are more effective than conventional filters. Air-to-air heat exchangers or other thermal air-transfer systems can be used to conserve indoor heat and still ventilate a tight building.

Intake air should be between 12.8° and 21°C (55 to 70 F). Where outdoor air is too cold to vent directly to the interior, preheating, or "air tempering," can be provided by the latent heat of sunspaces, greenhouses, atriums, basements, patios, crawl spaces, building envelopes, solar collectors, and heat pumps or other mechanical equipment. The earth can temper temperature extremes in winter and summer. Outdoor air should be properly filtered as well as air tempered.

Large indoor plants can act as an air-filtration system, absorbing carbon dioxide and expelling oxygen into the atmosphere as well as providing humidity.

Natural ventilation of a building will initiate passive cooling during the summer. The exhaust of excess warm air and the intake of cool air help to lower interior temperatures. To facilitate the exit of warm air, dampered vents should be located at high points in a building. If natural thermal-pressure differentials do not produce sufficient flow velocities, fans, turbines, or plenums can be used to accelerate them. Attic fans will, if needed, more quickly remove hot air that accumulates during the day. Daytime venting can also be accomplished with a solar photovoltaic fan-powered exhaust. Nocturnal cooling with powered pressurization can be more effective than with powered exhaust. Operable vents connecting the building space with the attic provide a pathway for the upward flow of hot air. The motion of outside air can often be used to induce interior air movement without the aid of fans. Wind can be used to power a turbine, or it can be directed in such a way that pressure changes result that move inside air. Venturi roof caps can be more effective than wind turbines.

Homes and buildings have traditionally depended on open windows for ventilation and cooling. Windows sized and placed for view and daylight, however, are seldom best for ventilation. Bringing in wind, with its dust, and rain, with its moisture, can be better averted with close-to-the-floor, air-intake vents. Security is another advantage of using relatively small, insulated, operable vents and fixed windows. Air-intake vents and roof-exhaust vents must be opened simultaneously to function properly.

Buildings and their interiors act as large-scale ducts and should be designed with options for unobstructed movement of air. The location of properly sized openings to permit natural ventilation between rooms, corridors, and floors of buildings is critical to indoor air flow.

Doors and operable interior openings act as dampers to air movement. Time-of-day human-use factors as well as cross-ventilation should be carefully considered. In effect, the building itself performs as a heat pump. Air contained in southern rooms and glazed areas is heated by the sun, producing warm air currents and positive-pressure areas within the building. This warm air rises to the ceiling, and cool air moves across the floor to replace it. If the building has been designed to function as an air duct, natural upward air flows will be initiated, inducing compensatory outdoor air intake from properly sized low-level wall or floor apertures.

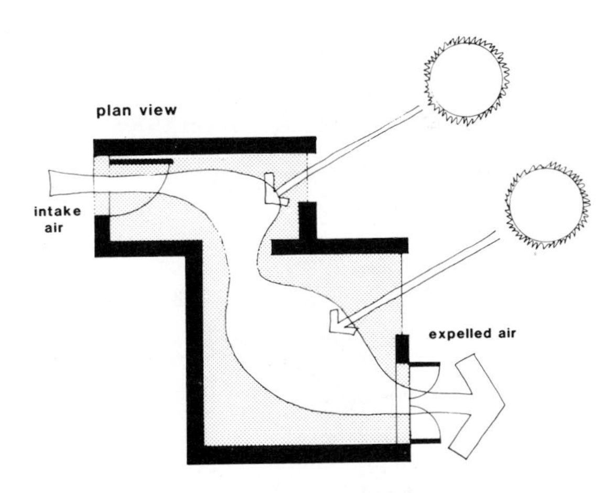

solar radiation provides daylight and acts as a driving force to inductively convect interior air

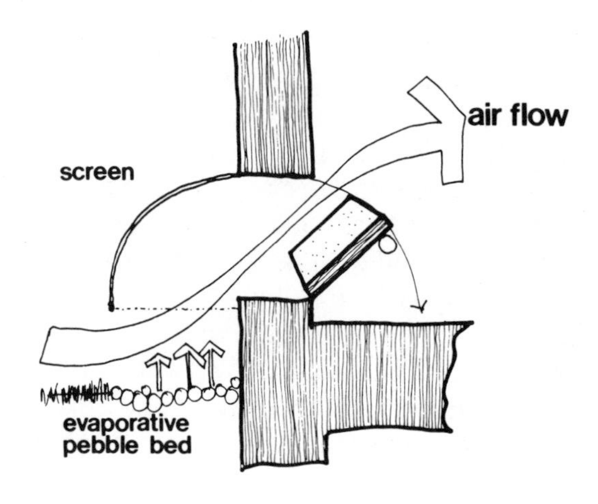

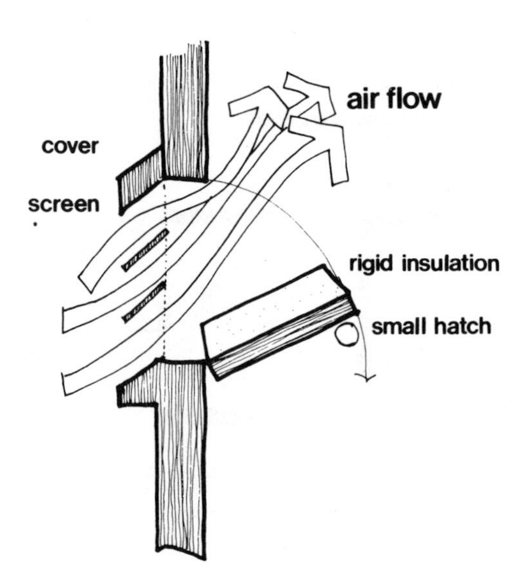

Unfortunately, conventional air filters will not remove pollution from incoming outdoor air or operate efficiently with naturally induced air flow. Hooded and louvered air intakes can reduce dirt and dust intrusions. Powered-fan intake of outdoor air through high-efficiency filters effectively captures atmospheric particles while increasing indoor cooling and ventilation when outdoor temperatures are 12.8° to 22° C (55 to 72 F). Also, negative ionization ahead of the filter can reduce the air pollution entering buildings.

The sun's heat can be used in other ways to induce natural air movement in a building. If a space is not properly ventilated, summertime heat will eventually build up. Solar-heated plenums are effective passive systems for ventilating a building. The sun heats the plenum and raises the temperature of air behind a cover. Warm air rises up out of the plenum, allowing cooler air to enter the building by displacement. The plenum can be equipped to induce air movement even after sunset by incorporating heat-storage material into the plenum volume, for daytime heat absorption and retention and for dispersal of heat at night. Solar plenums are most effective on west exposures, followed by south. In these locations, they make use of the day's highest ambient temperature, maximum solar angle of incidence, and latest radiation period before nightfall.

Air surrounding trees, shrubs, or ground cover is cooler than air away from vegetation and should be used for summer ventilation. Air for ventilation and cooling can also be extracted from basements, underground tubes, winter ice-storage bins, low-temperature chemical-storage, or subsurface water-to-air heat pumps. Conventional air-conditioning powered by photovoltaic cells or lithium-bromide absorption can be mechanically employed for cooling.

In some locales, evaporative cooling, which cools the air by adding moisture to it, can be effective. Evaporative cooling works well where there is an appreciable difference between wet-bulb and dry-bulb temperatures. Intake vents can be positioned near ground level over a bed of pebbles, with water running over the pebbles and the intake air drawn off immediately above. This form of cooling is not effective in humid climates because the wet-bulb and dry-bulb temperatures are nearly the same.

In high humidity areas, buildings can be cooled using desiccant (drying agent) interception of moisture in the air. Ceiling fans or even photovoltaic-cell-with-battery-powered fans can direct wind flow through large louvered grilles to interiors. Buildings in semitropical areas located on islands or shorelines can benefit from tradewind air flow through protective louvers and screened areas.

Some suggestions to improve the energy efficiency of existing buildings, relative to ventilation, are listed below.

- Use insulating batts, available with foil vapor-barrier backing, to decrease heat losses and infiltration.
- Install recirculating fans with charcoal or citrus-base filters in kitchens and bathrooms; they decrease the amount of heated air piped directly out of the building in winter.
- When needed, use bath and kitchen ventilating fans that exhaust to the outside only.
- In the winter, be sure that fireplace dampers are closed except when the fire is going.
- Modify the fireplace to deliver more heat to the building. Heat-exchange methods are noted in Chapter 8.
- On warm days, open windows or vents for ventilation instead of using air-conditioning or fans.
- Install controllable vents between the various levels of a building. Basement air can be drawn up for cooling. Vents can be closed and heat retained at the level at which activities are being performed.
- Install a turbine vent at the highest point of any pitched attic roof to draw hot air out of the attic during summer. The same vent should be tightly closed during winter. Eave venting or other vents are required to provide incoming displacement air.
- In the summer, use the fireplace flue as a vertical stack to initiate passive ventilation.
- Provide vents and a stack to allow natural ventilation of hot air in greenhouses.
- Heat-transfer wheels, heat pipes, and other heat exchangers can conserve energy by transferring exhaust air heat to incoming cooler exterior air.
- Outdoor air is often polluted. Closed-cycle air-quality reconstitution through filtration or by use of plants conserves the need to temper varying temperatures and eliminate the pollution of outdoor air. Indoor smoking usually renders the indoor air quality below that of polluted outdoor air. The health, efficiency, and quality of the environment are threatened by smoking and air pollution.

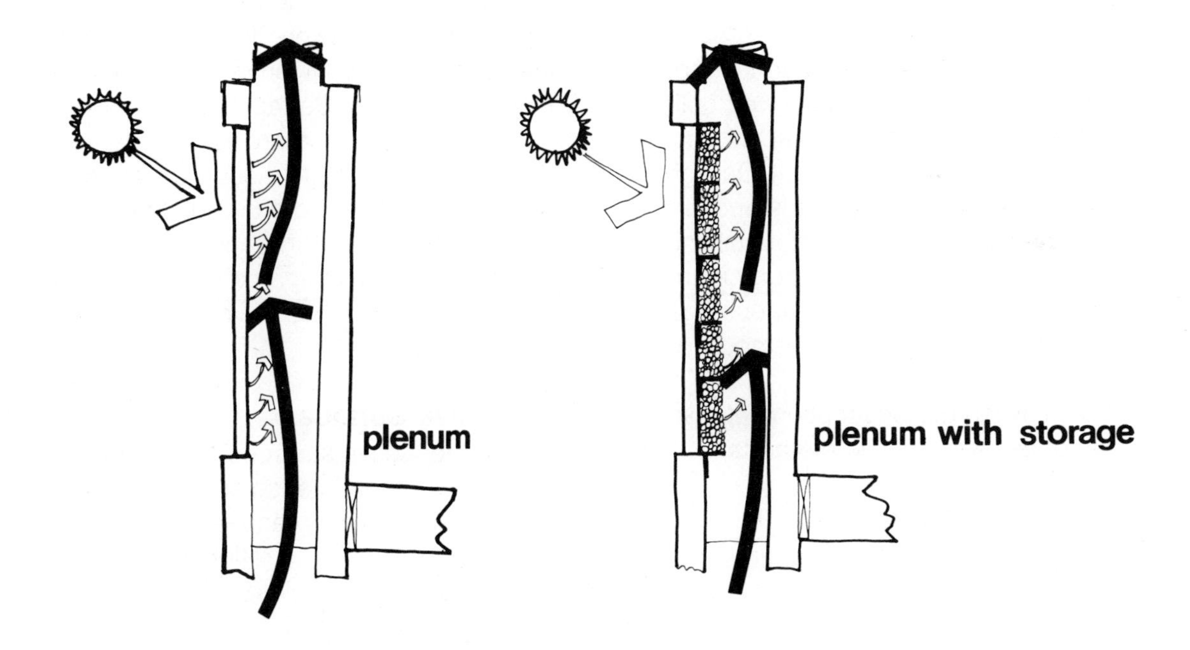

Passive Cooling Strategies

Many of the design ideas presented earlier in this chapter incorporate passive cooling strategies and techniques. Passive and hybrid cooling systems can control temperature and provide comfort in well-designed residences in many different climates with significant energy savings. Small-scale commercial and institutional buildings not partitioned into numerous offices can often replace a measure of mechanical air-conditioning with passive cooling.

Manual operation of vents and higher internal heat loads make passive cooling systems in large commercial and institutional buildings less practical. Daylighting can reduce cooling loads and electricity demand in commercial buildings. Also successful are natural and induced ventilation, night flushing, various shading devices, and evaporative cooling.

The following are descriptions of a variety of passive cooling methods.

Evaporative Cooling

Adding moisture to cool the air works best in hot, arid locations. Air can be directed through or over pools, fountains, ponds, water mists and sprays (such as from a lawn sprinkler), plants, or porous surfaces, such as walls. Roof sprays or ponds, or water trickling over roofs or against windows (thin-film surface cooling), can also be used. Outside air can circulate through an evaporative cooler to a heat exchanger. Interior air circulates through the heat exchanger and is cooled, but the moisture remains in the outside air, which is exhausted.

Shading

Devices can be designed to allow both summer shading and winter solar gains. Architectural shading devices (brises-soleil or sunbreakers) offer many possibilities for visual expression. Appropriate overhangs, wingwalls, awnings, canopies, balconies, porches, recessed glazing and doors, grillwork, fins, blinds, and internal and external louvers and shutters can be used. Certain kinds of fabrics and nettings also provide shade, as do trees and vegetation such as vines and shrubs. For a more complete treatment of shading, refer to Victor and Aladar Olgyay's *Design with Climate* and *Solar Control & Shading Devices* (see Bibliography).

Reflective Surfaces

Exterior reflective surfaces can reduce the impact of the intense summer sun and correspondingly reduce cooling loads. Light-colored or reflective roofs and walls, reflective glass, sunscreens, films, foils, metallized fabrics, and mirror-chromed blinds can be applied to a building for this purpose. Reflective surfaces can also direct light onto solar collectors (for higher efficiency) or into a space (for increased daylighting). Reflective glass or film has one drawback: Since it transmits less light and absorbs more heat than clear glass, partially shaded glass can break in the face of thermal stresses. Interior reflectivity and secondary daylighting can be optimized by careful attention to the reflectivity of interior surfaces and the size and placement of interior openings.

Natural and Inductive Ventilation

Natural convective air movement and induced "stack action" aid passive cooling. Solar chimneys and plenums, absorptive sunscreens, solar absorptive glass, and centralization of exhaust from low to high ceiling spaces can effectively produce this kind of air movement. A deeper-tone (transparent gray or bronze) inside glass with clear outside glass will induce air movement and, for west windows, lessen solar intensity and glare. Either a dust filter or access for cleaning is advisable with this system. Other passive ventilation and convection techniques include louvered or grilled doors; attic cross-ventilation with louvers or ridge vents; negative-pressure, trailing-edge or venturi roof exhaust (both wind assisted); diagonal cross-ventilation through building openings; nocturnal cooling (night flushing); strategically located air intakes (in lieu of windows); as well as interior door, stairway, and other interior openings; convective cooling through light troffers and window louvers; thermal mass cooling; and the use of furniture built to assist ventilation with wire, perforated metal, wicker, reed, rattan, or webbed chairs and other seating pieces. Body-heat-conductive furniture made of marble, stone, concrete, or other massive materials can also provide passive cooling.

Earth Cooling

During summer months, earth temperatures are substantially lower than air temperatures. Caves, tunnels, caverns, and mines are notably cool. Numerous passive techniques take advantage of the earth's cooling mass and draw hot summer air through earth spaces to cool building interiors.

Concrete or waterproofed, corrugated-iron retaining walls in contact with the earth can cool underground areas and excavated spaces. Subsurface ducts or treated metal, concrete, or plastic pipes can be used to cool intake air. If all or most of a building is buried, living areas, basements, crawl spaces, and building envelopes can benefit by contact cooling.

Evaporation of the earth's moisture or moist intake air can cool a building. Condensation must be drained away from subsurface tubes or other implements, however, to avoid mildew and bacteria growth.

Migrating from warm to cooler portions of a building can be effective. In the early 1900s, some families had summertime kitchens and dining space in their basements. On screened sleeping porches, cooler nighttime air and breezes made slumber more pleasant.

Additional Passive Cooling Methods

Other methods include airtight construction with adequate insulation. An open-construction sun-shield roof can shade a building's regular roof in desert or hot climates. The intervening air layer and movement reduces radiant impacts.

Dehumidification cooling uses a desiccant material to remove air moisture in hot, humid areas. Night sky cooling can be effective where the difference between the clear sky temperature and the ambient air temperature near the earth causes radiant heat loss. This principle was used in the Mideast and China to make ice. Also, subsurface water can circulate to and cool a thermal mass that in turn cools the air through evaporation, convection, and radiation to the night sky.

Mechanical Cooling

Mechanical cooling systems have been developed, such as air-to-air or water-to-air heat pumps or lithium-bromide solar-absorption cooling, that are generally more energy-efficient than traditional refrigeration methods (standard air-conditioning). Economizer cycles can bring in cool outdoor air, reducing energy consumption. Whole-house fans (although sometimes quite noisy) and ceiling fans exhaust hot air and provide air movement. Photovoltaic-cell-powered ventilation fans increase energy savings but are useful only during daylight hours. Relatively small pressurization fans can be effectively used for nocturnal cooling, but they must be located so that the building's occupants are not disturbed. Mechanical exhaust and ventilation systems require strategically located air-intake openings to function properly.

Mechanical

Mechanical systems use energy to heat, cool, and ventilate the interior environment of a building. In the heating mode, fuel or electricity is used to warm the air or fluid that circulates in a mechanical system. Exhaust gases from furnaces or boilers can carry as much as 50 percent of the system's heat energy up and out the flue. This need not be wasted; most heat can be extracted from the exhaust gases and used elsewhere in the building. To accomplish this, heat exchangers have to be added to the exhaust ducts; they may contain a fluid or air that is pumped through a series of tubes. As hot exhaust gases rise through the flue, they flow around the tubes, eventually giving up their heat to the cooler fluid inside. The exhaust gases still exit from the flue but only after giving up some of their heat in this manner. A series of heat exchangers could be used in a fireplace flue to effectively capture the energy given off during the burning process.

The successful operation of many of the systems that have been presented depends upon the transference of heat from one fluid to another or from one location to another. Heat pumps can perform this function, but they require an input of work and rely upon moving parts to accomplish their function. Two other types of devices that transfer heat are heat exchangers and heat pipes.

Some heat exchangers have moving parts while others do not. A heat wheel is one that slowly rotates to transfer heat, and it must be powered by a motor. In the winter it extracts heat from exhaust air to temper incoming cold air. In the summer, heat from incoming warm air is transferred to existing cool air, precooling the entering air.

The wheel, packed with aluminum or stainless steel wool, rotates between an exit duct and an entry duct, picking up heat on the warm side and rejecting it on the cold side. These have been designed to transfer not only sensible heat but latent heat of evaporation. Moisture is absorbed from the humid airstream by a desiccant, lithium chloride, and is released to the dry airstream.

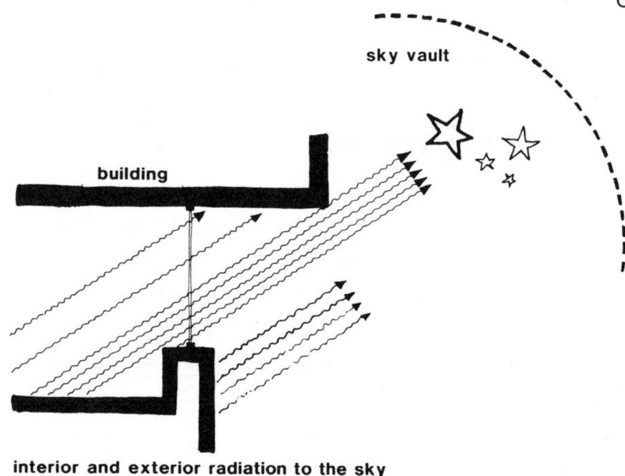

interior and exterior radiation to the sky

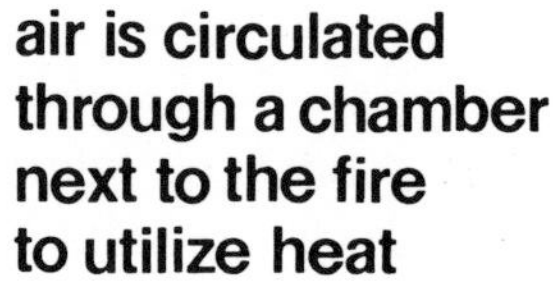

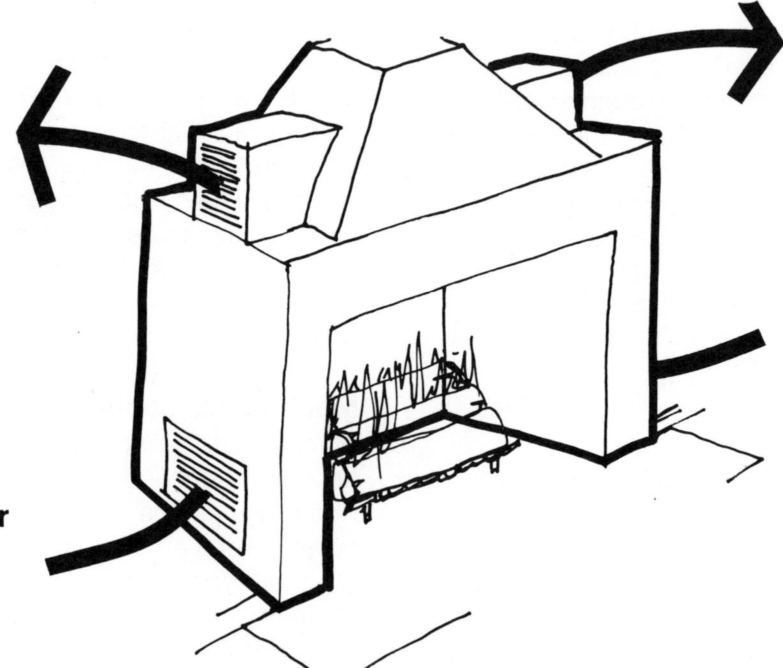

The most common heat exchangers have no moving parts and transfer sensible heat only. These are exemplified by automobile radiators. A hot fluid, air or liquid, is piped into the heat exchanger where it flows into a network of small tubes, greatly increasing its surface area. A cooler fluid circulates on the opposite side of the tubes, and heat is transferred to it by conduction. The rate of heat flow depends on the relative temperatures of the two fluids, their heat capacities, flow velocities, surface areas, plus the thickness and type of material out of which the heat exchanger is constructed. Transfers of heat from air to liquid or from air to air are generally less efficient than transfers from liquid to air or from liquid to liquid.

A device that is considerably more efficient for air-to-air heat transfers is the heat pipe. This simple object is composed of a sealed tube and contains a working fluid and a wick. When one end of the tube is heated, the working fluid evaporates, absorbing large amounts of heat, and moves to the cooler end of the tube. There it condenses, giving up the heat, and flows back through the wick to start the cycle over. A pipe one inch in diameter and two feet long can transfer 3,661 watts (12,500 Btu/hr) at 982° C (1,800 F) with only a 10° C (18 F) temperature drop from one end to the other.

The heat pipe was invented in the 1940s by Richard S. Glauger, a General Motors engineer, who used it in refrigerators. It has been used in many of the most recent spacecraft, but has had only limited use in homes and buildings. Principal among the reasons why it is not used more in homes and buildings is its high cost. It is hoped that the cost will decrease when these devices are manufactured in quantity.

Heat exchangers can be used in several different ways to prevent "waste" heat from leaving an interior space. Fluid exchangers can remove heat from water in shower and tub drains, dishwashers, and water or refrigerator condensors. This heat can be used to preheat domestic water or to supply heat for a space heat pump. Heat exchangers are available in a variety of sizes and configurations, but their purpose remains the same, to transfer heat from one medium to another.

A heat pump is a device that is able to produce more energy as heat than is contained in the fuel with which it is operated. This may seem to be a contradiction of the laws of thermodynamics, but it is not, for the energy input comes from two sources—the operational fuel (mechanical energy) and the natural (ambient) energy of the environment. Its name derives from the ability to "pump" converted ambient energy into a space to be heated or pump "heat" out of a space to be cooled.

The operation of a heat pump takes advantage of the principles of latent heat. A working fluid, a refrigerant, circulates throughout the heat pump system, which gives off heat when it condenses and picks up heat when it evaporates. The heat content of the refrigerant is affected more by changes of state than by changes of temperature.

In a normal cycle, the refrigerant is compressed to a relatively high pressure, producing a temperature increase. Mechanical energy is required to operate the compressor to raise the internal energy absorbed by the refrigerant at low ambient temperature to a temperature level useful for space heating. For Freon, the most common refrigerant, this temperature reaches approximately 60° C (140 F). Hot, gaseous refrigerant is pumped through a heat exchanger where it loses some of its thermal energy, cools, and condenses to a liquid. This is called a condensation heating cycle; while changing from a gas to a liquid, latent heat is given off.

After it has condensed, the fluid is still at a high temperature and pressure. By using an expansion valve to reduce the pressure suddenly, the temperature can be reduced below −17.8° C (0 F). The low-pressure refrigerant is then a mixture of liquid and gas and is piped to a second heat exchanger where it takes on heat energy and vaporizes. This second unit is called an evaporator. Energy can be taken from "cold" environments since the refrigerant temperature is below −17.8° C (0 F). At this point the cycle starts over with the gaseous refrigerant being condensed.

If the hot gas is pumped to a condenser located in a building, the system is operating in the heating mode. If it is pumped to a condenser located outside the building, it is operating in the cooling mode.

The advantages of this system are that it is reversible and that the heat-energy output at the condenser is greater than the mechanical energy input to the compressor. The second energy input to the system comes when the low-pressure, low-temperature liquid refrigerant travels through the evaporator in which it vaporizes and takes on ambient heat from the outside. Heat pumps are energy-conserving because the heat-energy output is generally 2.5 to 5.0 times greater than the fuel input. As earlier noted, COP expresses this ratio.

Heat pumps that draw their heat of evaporation from air and release heat through a condenser to air at another location are called air-to-air heat pumps. Heat can also be drawn from water and released to air, or vice versa, by air-to-water heat pumps. The ground can also be a heat source or sink and used in a ground-to-air or ground-to-water heat pump system. Each of these has advantages and disadvantages that should be considered for specific applications. Manufacturers of heat pumps should be contacted for this information.

Indoor annual temperature equalization can technically be accomplished by the seasonal storage of the winter's cold and the summer's heat. Natural ice, passively frozen in winter, can be used for summertime cooling. Summer heat can be stored in appropriate salts for direct or heat-pump-assisted winter heating.

The cost-effectiveness of seasonal storage depends on space requirements and a practical tie-in to a building's mechanical systems. Adequate insulation is necessary to sustain cooling and heating for such long periods of time. The earth itself can be used as the storage reservoir. For many years, underground water has been used as a summer and

air-to-air heat pump system

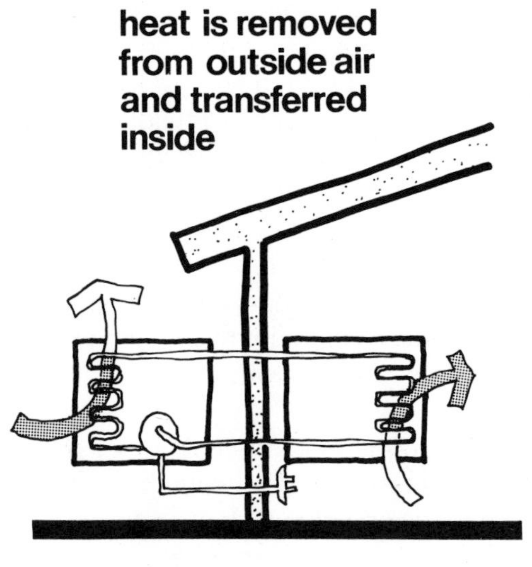

winter

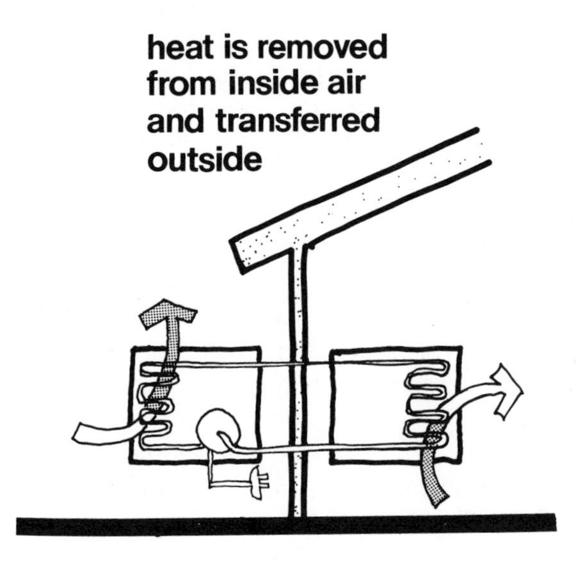

summer

winter temperature system by employing the coefficient of performance (COP) advantage of a water-to-air or water-to-water heat pump.

It is likely that seasonal energy storage will become more common as research into these techniques is developed and energy costs continue to escalate.

Because of the numerous adjustments and modifications that can be made to the mechanical systems of homes and buildings, these suggestions for energy conservation are divided into groups of listings for each of the different types of systems.

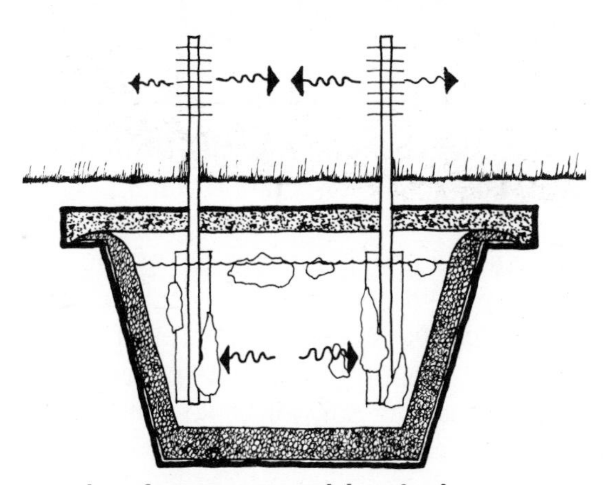

Space Heating

- Lower thermostat setting to 20° C (68 F) during the day and 15.5° C (60 F) at night. A clock thermostat automatically lowers the setting at night and then raises it again in the morning. One of these may be a good investment.
- Have the furnace and entire heating system serviced once a year, preferably in the fall to keep it operating at top efficiency.
- Clean or replace the filter in a forced-air heating system once a month. The fan will operate less efficiently if it doesn't have to force the air through a dirty filter.
- Dust or vacuum radiator surfaces frequently. Dust and dirt will block the heat, reducing efficiency.
- Consider installing a humidifier. Moist air helps you feel as comfortable at 21.1° C (70 F) with 35 percent humidity as you would at 23.3° C (74 F) with very low humidity.
- Make sure draperies, carpet, and upholstered furniture do not obstruct heat outlets or radiators.
- Do not heat to comfortable temperatures rooms that aren't being used. Close the doors leading to these areas and restrict the flow of heat through registers or radiators, or allow for cooler zones in large buildings.
- During the day keep draperies and shades open in sunny windows; close them at night, because the radiative heat gain will nearly always exceed the conveyed heat lost to outdoor ambient air.
- Move leisure furniture away from exterior walls to avoid cold drafts and the consequent lower perceived temperatures.
- Keep fireplace dampers closed when the fire is not burning.
- Provide combustion air from the outside directly to the fireplace and furnace. When heated indoor air is not used for combustion, fuel bills can be significantly reduced.
- Turn the thermostat down to 12.8° C (55 F) when the house or building will be unoccupied for a day or longer.
- For comfort at lower indoor temperatures, the best insulation is warm clothing. Clothes made from natural fibers—such as wool and cotton—may be warmer than ones made from a similar thickness of synthetic materials.
- Insulate all ducts that carry heated air and pipes that carry heated water.

Space Cooling

- Set air-conditioner thermostats no lower than 25.5° C (78 F). If humidity is controlled, this is a comfortable temperature.
- Run air conditioners only on very hot days and set the fan speed on high. In very humid weather, set the fan speed on low to provide less cooling but more moisture removal.
- In dry climates, consider installing an evaporative cooler. These devices generally use less energy than conventional air conditioners and cool an incoming air stream by extracting heat from it to evaporate water.
- Have the air-conditioning system serviced once a year, preferably in the spring, to keep it operating at top efficiency.
- Clean or replace air-conditioner filters once a month. Clean filters will decrease the amount of electricity used by the fan.
- Do not air-condition unoccupied rooms. Close them off from the rest of the home or building.
- Minimize the use of lights, appliances, sound equipment, and televisions. These all generate heat, which increases cooling loads.
- Shade windows with overhangs, wingwalls, or awnings, and deflect direct sunlight with light-colored shading devices, preferably mounted on the outside of the building and having an air space for convection between them and the exterior wall.
- Consider installing vents and exhaust fans to remove heat and moisture from attics, kitchens, and laundry areas and exhaust it directly to the outside.
- Consider installing a wind-powered roof ventilator along the roof ridge and louvered grilles between the attic and ceiling, so that cool air entering basement vents will be induced by the exit of the hot attic air to flow upward through the building.
- Locate outdoor cooling components behind opaque foliage, where they will be shaded from the sun.
- Insulate all ducts used to distribute cool air.

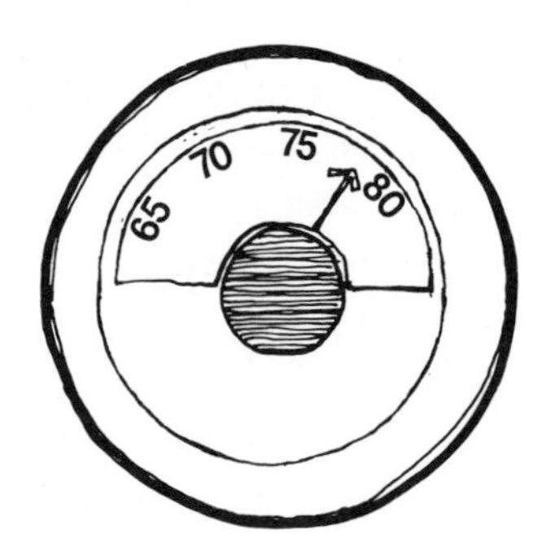

set air-conditioning thermostat at 78° or higher

Plumbing

- Repair leaky faucets as quickly as possible. A hot-water faucet leaking one drop a second wastes 2,650 liters (700 gallons) of hot water per year.
- Keep the temperature control on the hot-water heater set no higher than 60° C (140 F). Temperatures higher than this waste energy and shorten the life of a glass-lined tank.
- Minimize water and energy consumption by running clothes washers and dishwashers only with full loads.
- Make sure hot-water storage tanks and hot-water pipes are well insulated to decrease heat loss.
- Take more showers than tub baths, because showers use less hot water.
- Consider installing a flow restrictor in the showerhead that limits the flow to four gallons of water per minute while allowing an essentially unchanged water-spray velocity.
- Place a brick, or quart jar filled with water, in the tank of the water closet to decrease the amount of water used each time it is flushed. Do not decrease the volume too much because plumbing problems may result.
- Do as much cleaning as possible, including the car, with cold water. This will save the energy used to heat water.

Appliances

The analysis of homes and other buildings for optimized energy conservation requires that appliances be treated as part of the unified building system. Appliances are devices that perform work, provide heating or cooling, and require energy to accomplish their functions. Because they require energy they generate heat that is transmitted by conduction, convection, and radiation to surrounding air. This heat can either contribute to space heating in the winter or contribute to space cooling loads in the summer.

The value of an appliance can be assessed in terms of a time-based energy audit of alternative methods for accomplishing the same end result. For example, a healthy human adult is able to open a can using a mechanical can opener in about the same length of time it takes to open a similar can using an electrical can opener. Much less energy is required to accomplish this task by hand than by the electrical appliance, and since the times involved are similar, the choice should be to not use the appliance.

Variable conditions may enter into the judgment concerning the value of an appliance. For example, another choice may be whether or not to dry clothes in an automatic dryer. If the weather is sunny and dry, the clothes can be hung outside and will dry naturally, or a well-ventilated indoor space using sunshine or internal heat should suffice.

Electrical-utility rate structures may have a bearing on the utilization of an appliance and at what time during the day it should be used. Some companies charge a lowered rate for electricity consumed during off-peak periods and tailor their rate structure to the maximum amount of electricity used at any one time. Local utility companies should be consulted to find out their regulations for determining rates.

It may be desirable to take advantage of off-peak rates, while keeping the usage at any one time to a minimum. This would mean that certain appliances may have to be used in the early morning or late evening hours. At times, appliances may have to be switched off to keep total usage to a minimum.

One potential use of off-peak electricity is as an auxiliary heating source in solar-heated buildings. The heat must be stored so that it is available during peak hours. This can be accomplished by lacing an insulated gravel-storage bin with heating coils, which heat the gravel during off-peak hours. When heat is required, air can be drawn through the bin where it will absorb heat (or a series of electric hot-water tanks can be phased to receive off-peak electric heat) and then be delivered to the building.

The wise use of appliances is necessary for energy conservation in homes and buildings. Appliances not only consume energy directly but may contribute to increased cooling loads and further energy consumption. Below is a list of suggestions for the energy-wise use of appliances.

- Turn off all appliances (television, radio, phonograph, lights, etc.) when they are not being used.
- Make sure all appliances are operating at maximum efficiency by having them serviced periodically.
- Check refrigerator and freezer door seals to ensure that they are airtight and that their condensing coils are clean, allowing good air flow. Open the doors as little as possible.
- Operate heat-producing appliances in the early morning or late evening when air temperatures are cooler and heavy demands are not being made on electrical distribution systems.
- "Instant on" television sets use energy even when the screen is dark. This waste can be eliminated by plugging the set into an outlet that is controlled by a wall switch and turning the set on and off with the switch or by installing an additional on-off switch on the set itself or in the power cord.

- Check the ENERGYGUIDE label on appliances before buying. These labels can be used to compare the energy efficiencies and estimated annual operating costs of major home appliances.
- Match appliance size to the size of the job, especially major appliances. Instead of cooking one potato or meat dish, cook an entire oven meal. For small meals a "toaster oven" may be most efficient.
- Use flat-bottom pans that completely cover the burner or heating element of the stove. More heat will enter the pot and less is lost to the surrounding air.
- Maintain electrical tools in top operating shape. Keep them clean, sharp, and properly lubricated.
- When purchasing a water heater, match its size to the needs it serves. Oversized water heaters use more energy than necessary.
- Use a small appliance in place of a major appliance whenever possible.
- Do not preheat appliances longer than necessary.
- Do not overload any electrical circuits; this results in reduced energy efficiency.
- Use the waste heat from major appliances, other equipment, and lighting fixtures to assist with cold weather space heating. Vent this heat to the exterior during warm or hot weather.
- Consider providing electronic ignition devices for gas-fired equipment. Pilot lights that burn constantly waste energy.

PASSIVE SOLAR STRATEGIES direct, indirect, isolated systems

plan view

solar gain

thermal mass Trombé wall

thermal mass sidewalls

thermal mass pylons

thermal mass pylons

sunspace

south sunspace with thermal mass

sunspace

east sunspace with thermal mass

sunspace

west sunspace with thermal mass

thermal mass fireplace

thermal water wall

thermal water tube sidewalls

thermal water tube pylons

thermal water tube pylons

thermal eutectic salt wall

thermal eutectic salt tube sidewalls

thermal eutectic salt tube pylons

thermal eutectic salt tube pylons

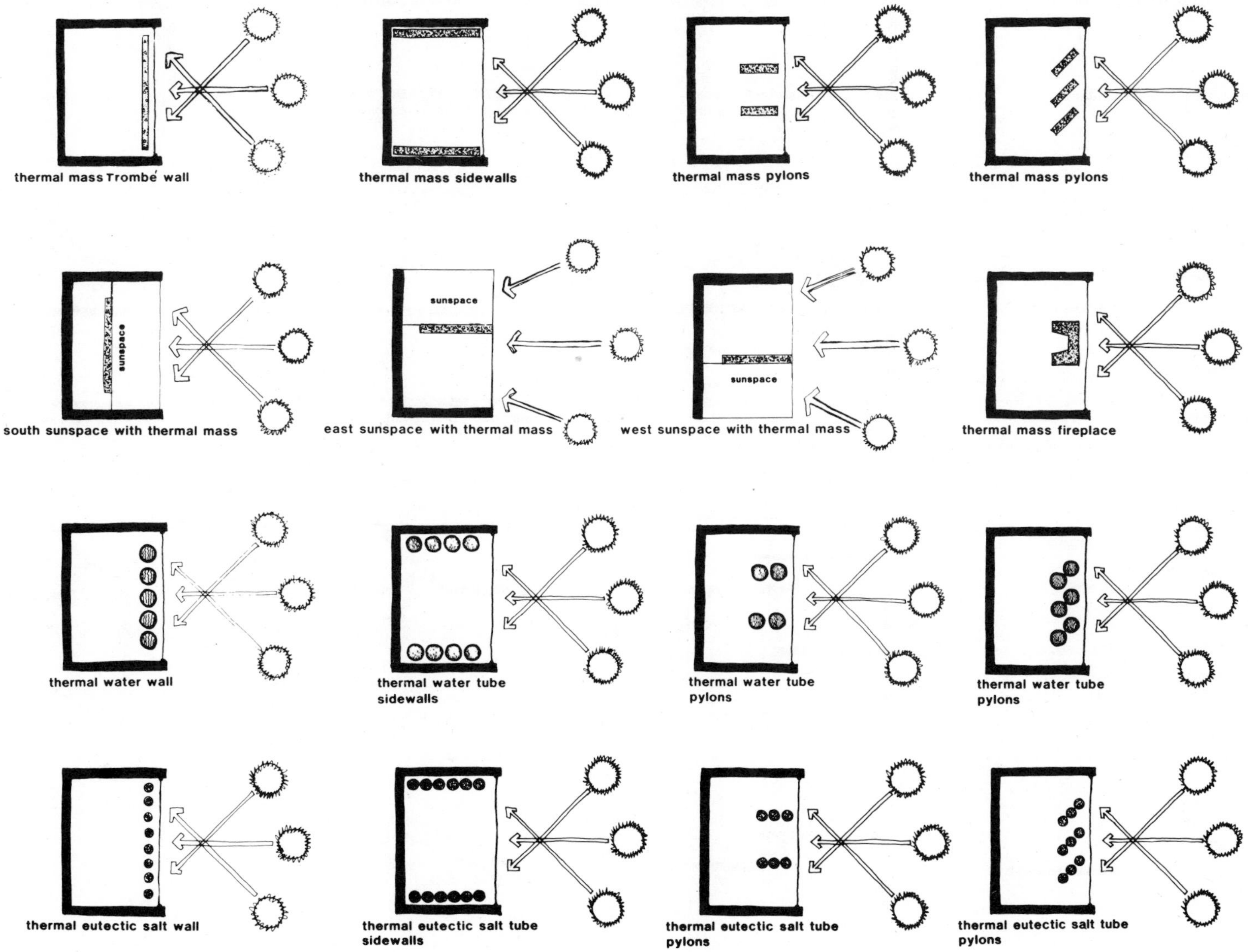

passive solar heating / concrete, masonry, gravel mass

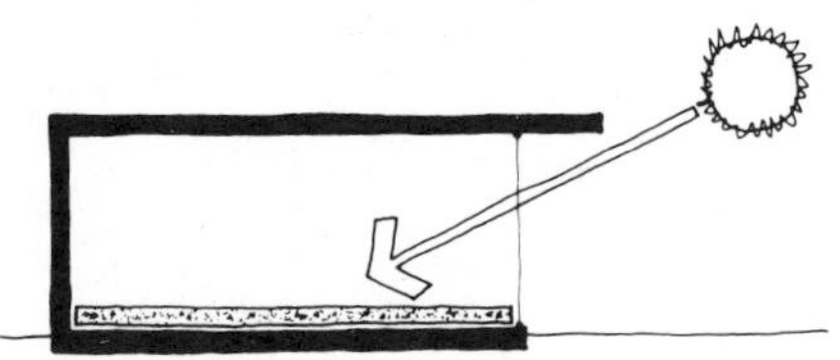
thermal mass floor

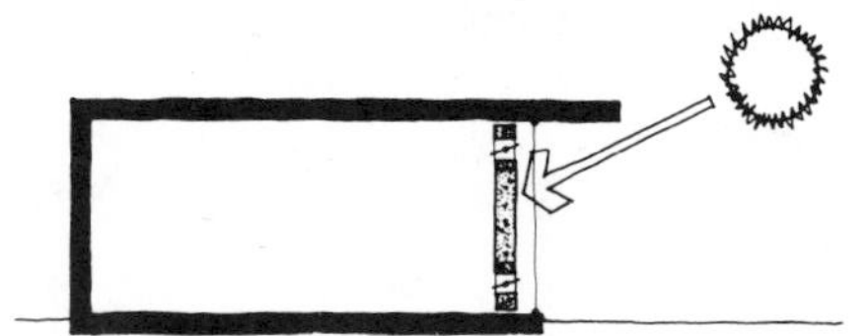
thermal mass Trombé wall

thermal mass solar bunker

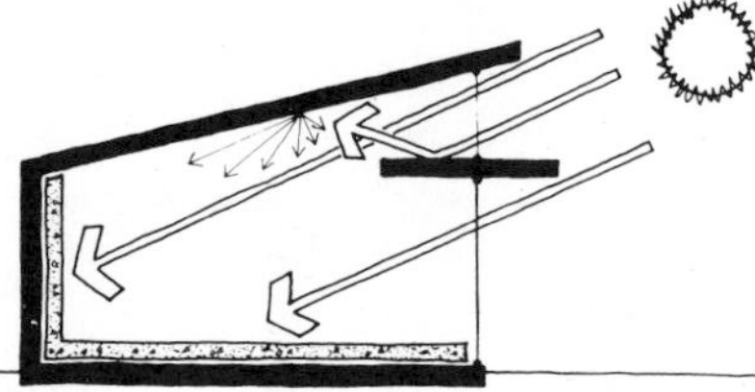
thermal mass wall and floor
with solar reflective light shelf

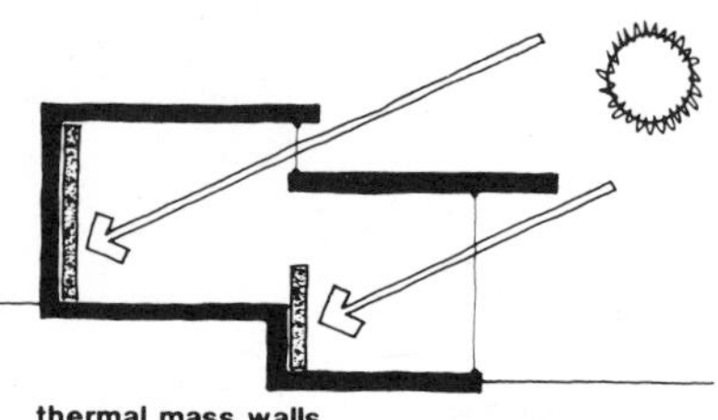
thermal mass walls
solar clerestory

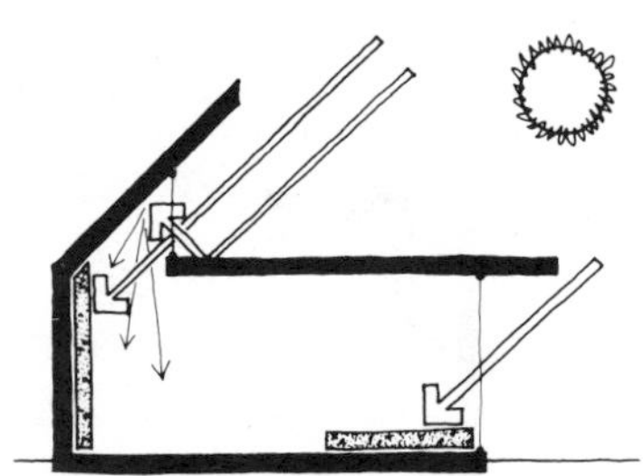
thermal mass wall and floor
with equinox position solar
reflective clerestory

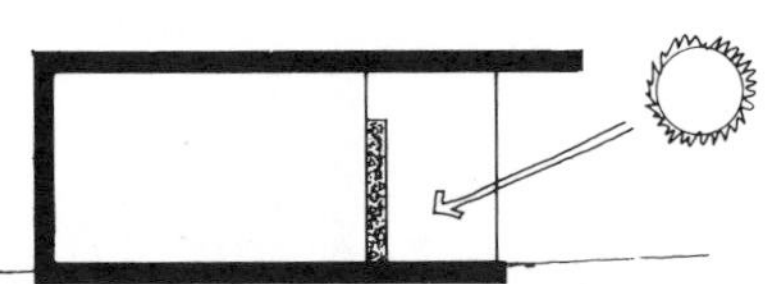
sunspace thermal mass
thermal lag

sunspace thermal mass
secondary thermal gains

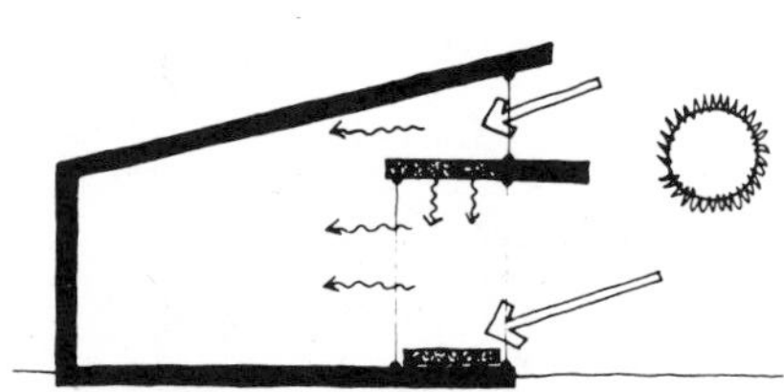
sunspace thermal mass
secondary thermal gains

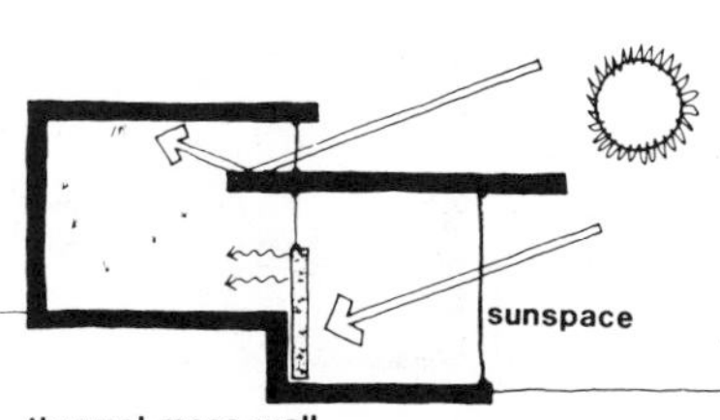

thermal mass wall
thermal lag/reflectivity/secondary gain

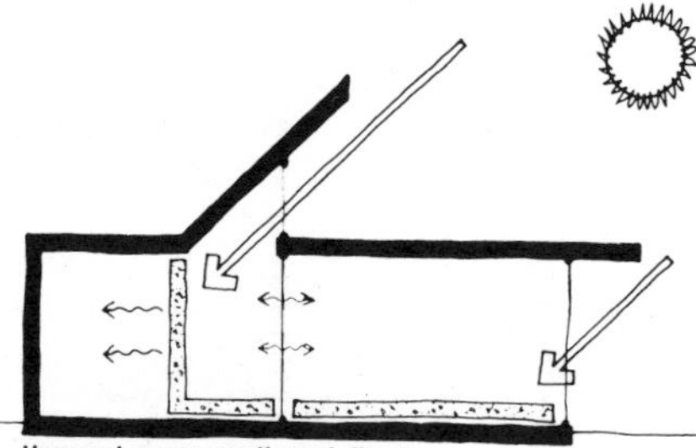
thermal mass wall and floor
thermal lag/secondary gain

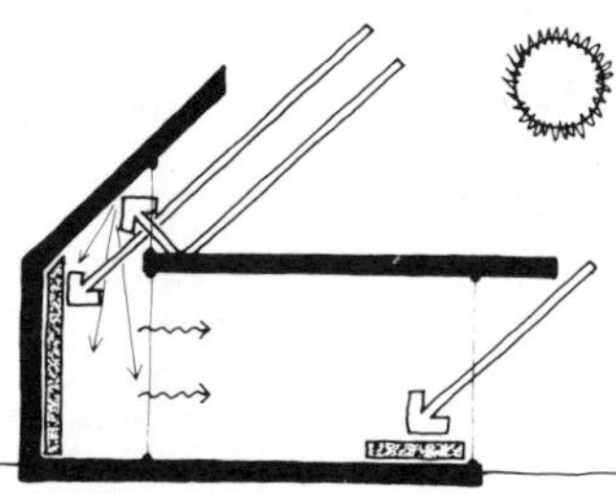
thermal mass wall and floor
secondary thermal conductivity

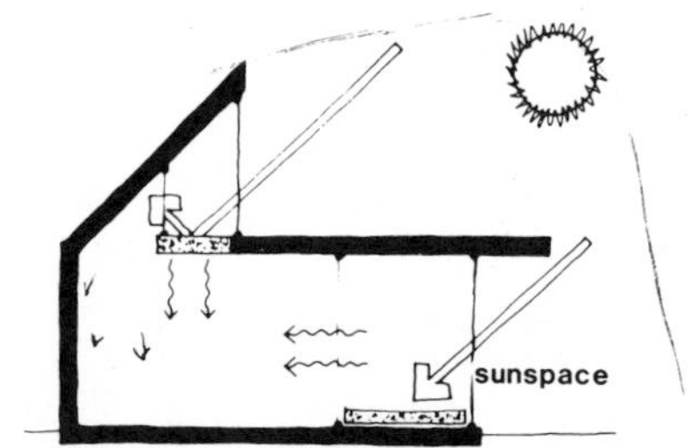

thermal mass ceiling and floor
thermal lag/secondary gain

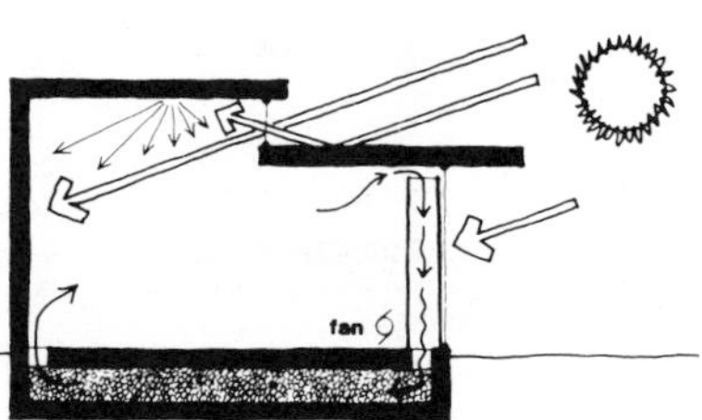

solar air plenum
low-temperature gravel storage

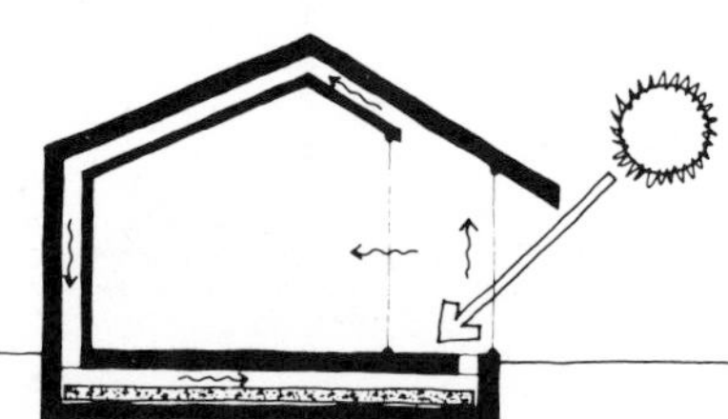
solar thermal envelope with sunspace
under-floor thermal storage
secondary thermal gain

passive solar heating / concrete, masonry, gravel mass

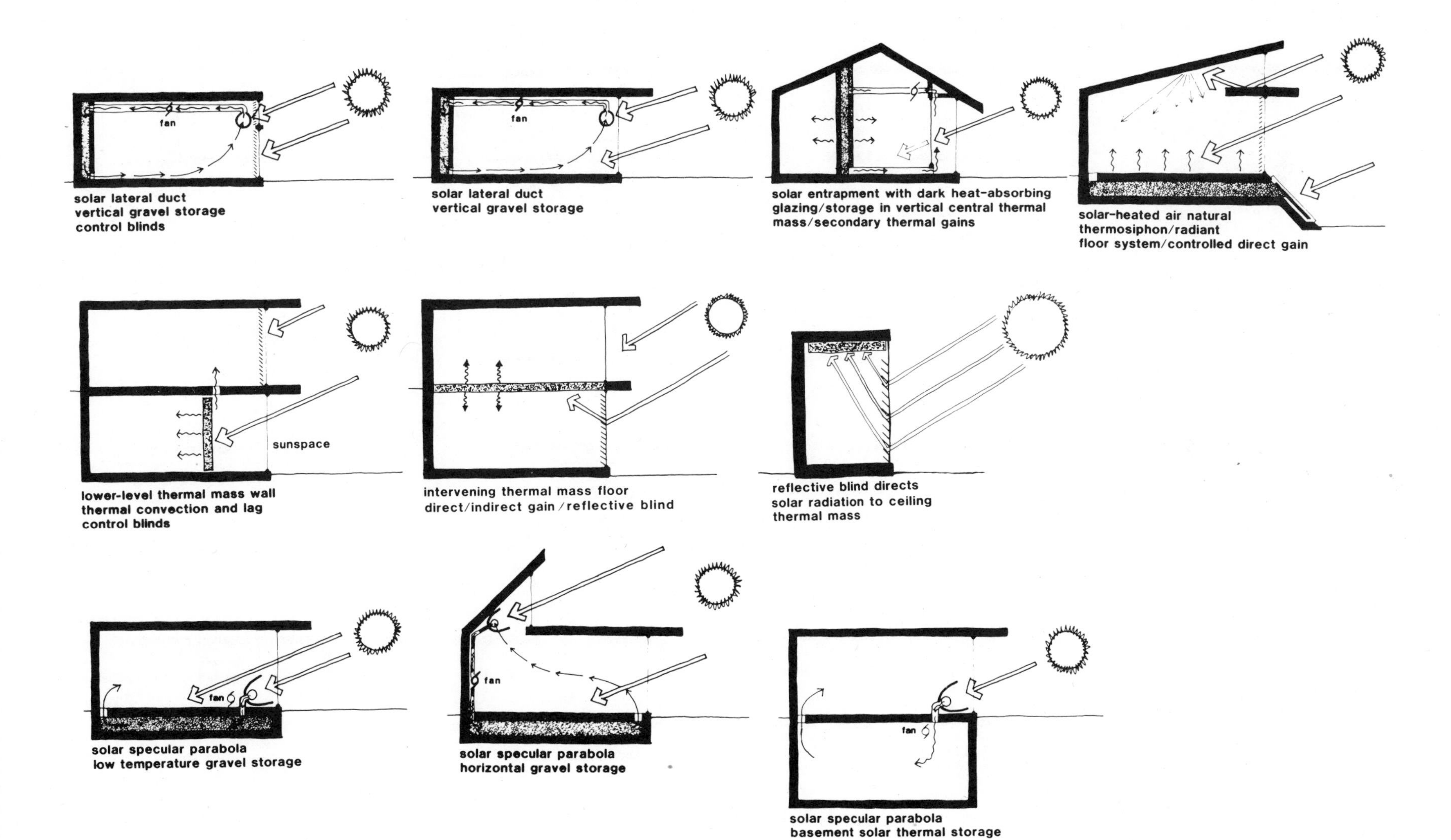

passive solar heating / water and fluids mass

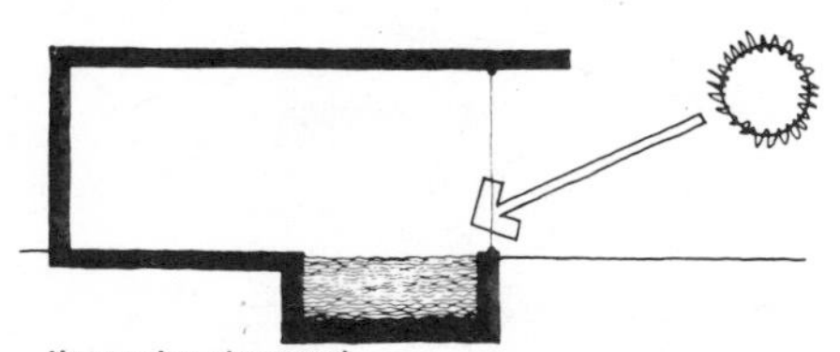

thermal water pool
with transparent cover

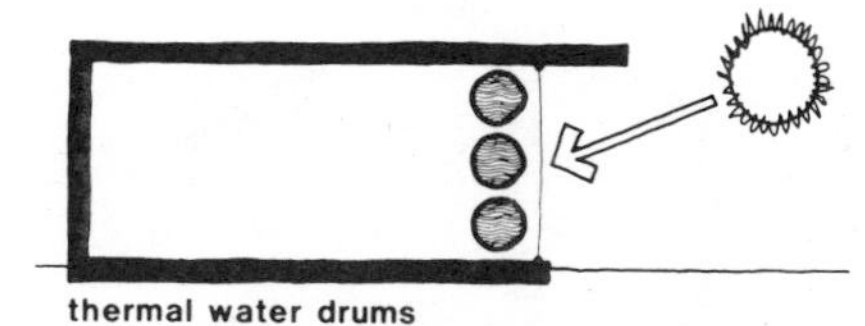

thermal water drums

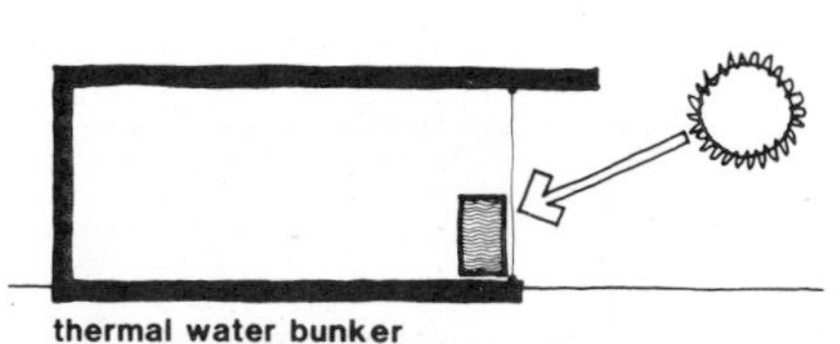

thermal water bunker
solar selective surface

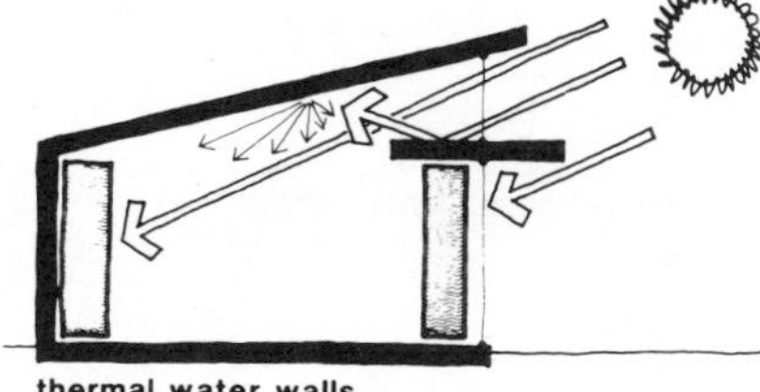

thermal water walls
solar reflective shelf

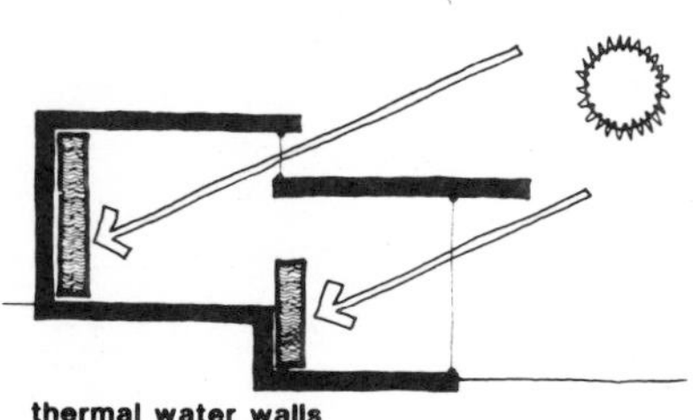

thermal water walls
solar clerestory

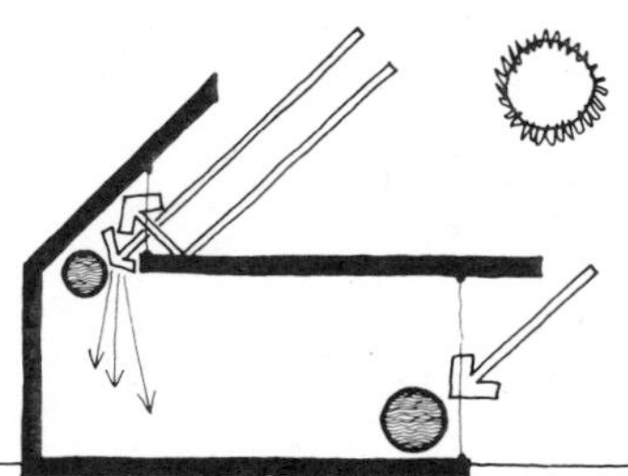

thermal water drums
with equinox position
solar reflective clerestory

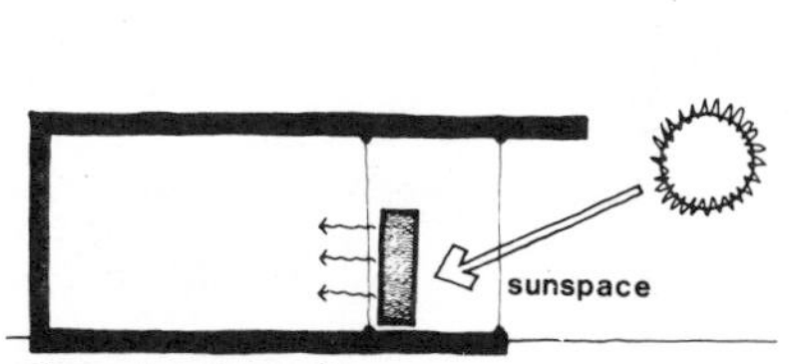

thermal water wall
thermal radiation/
secondary thermal gain

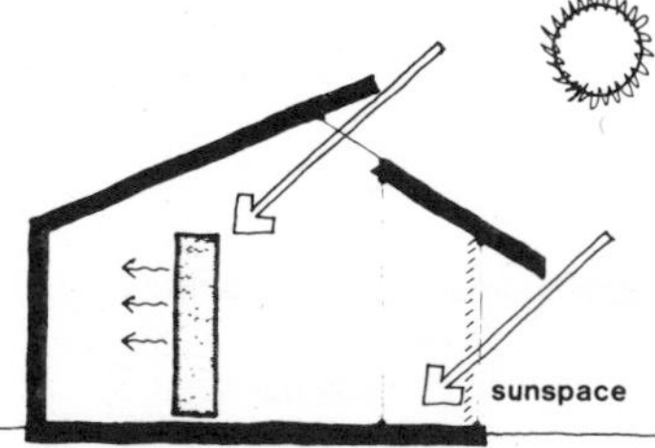

thermal water wall
reversible heat-absorbing and
reflective blinds/secondary gains

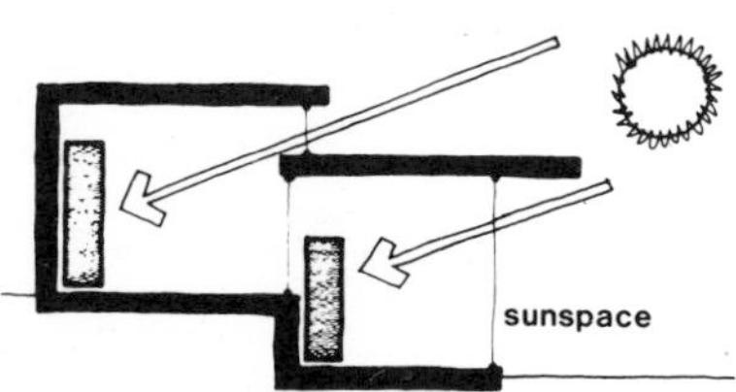

thermal water walls
secondary thermal gains

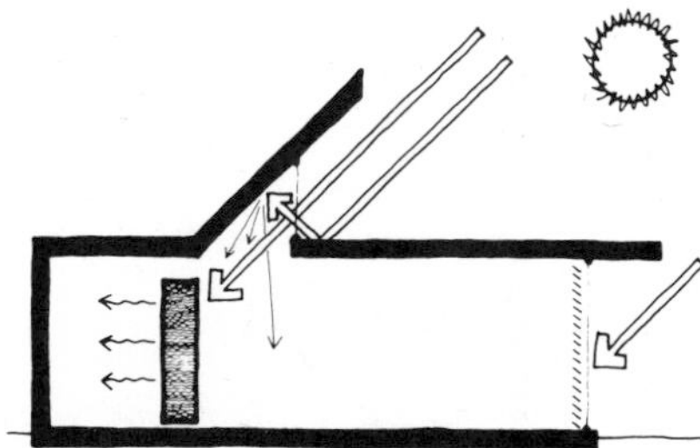

thermal water wall
thermal radiation/
direct and indirect solar gain

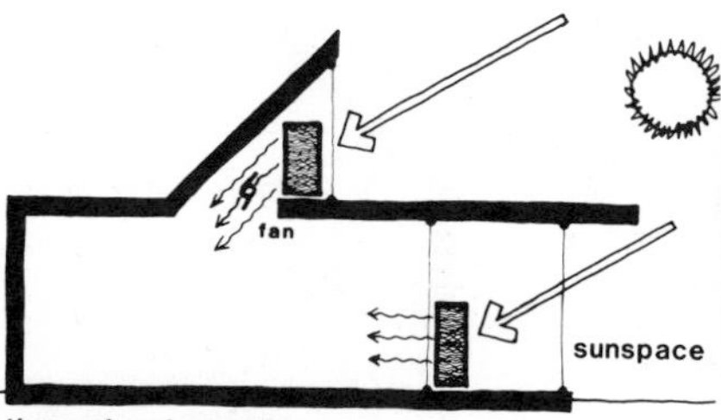

thermal water walls
secondary thermal gains

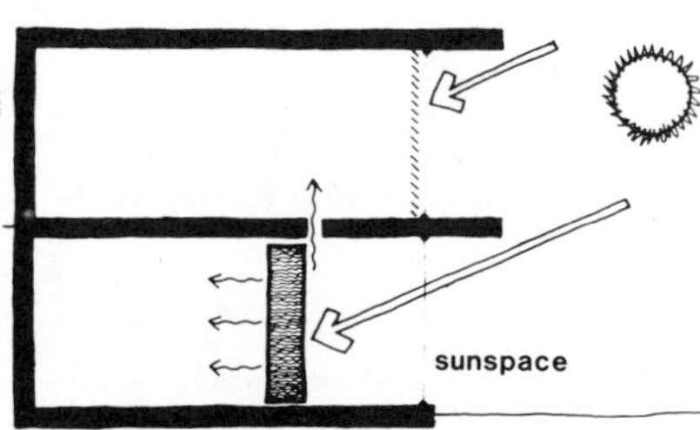

lower-level thermal water wall
thermal convection and lag
control blinds

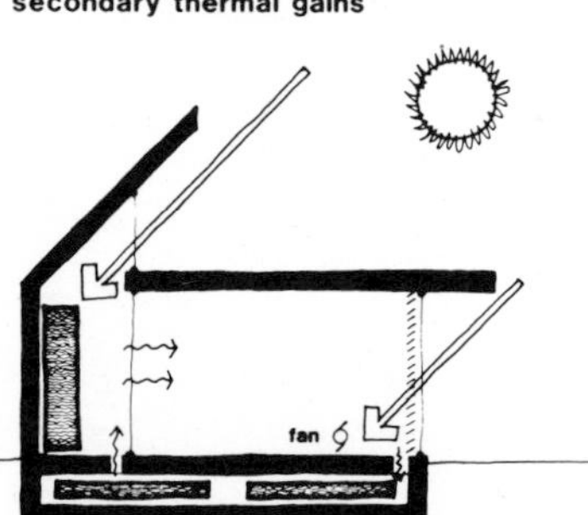

thermal water wall and reservoirs
thermal convection/heat-absorbing blinds

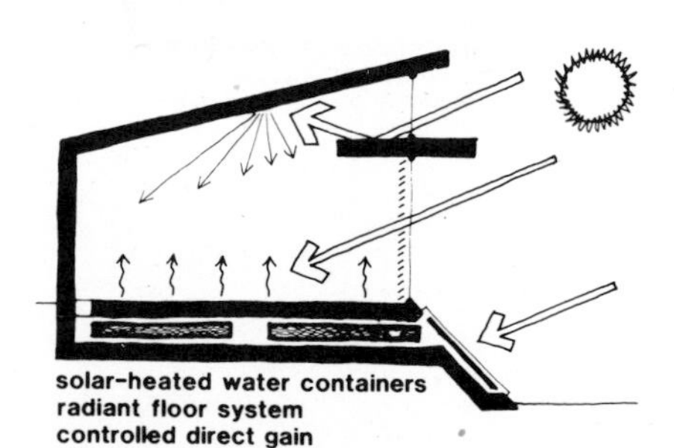

solar-heated water containers
radiant floor system
controlled direct gain

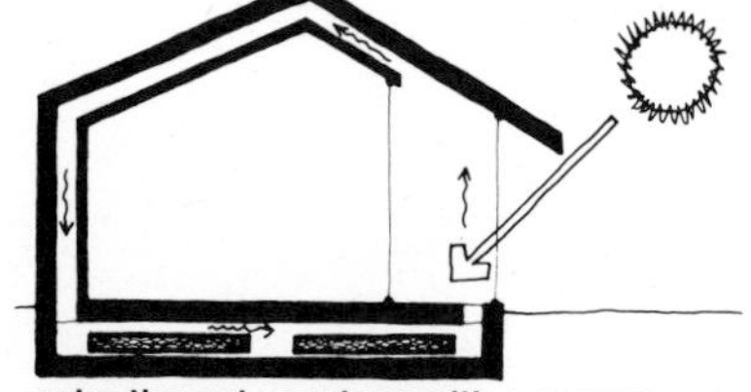

solar-thermal envelope with sunspace
underfloor water containers
secondary thermal gain

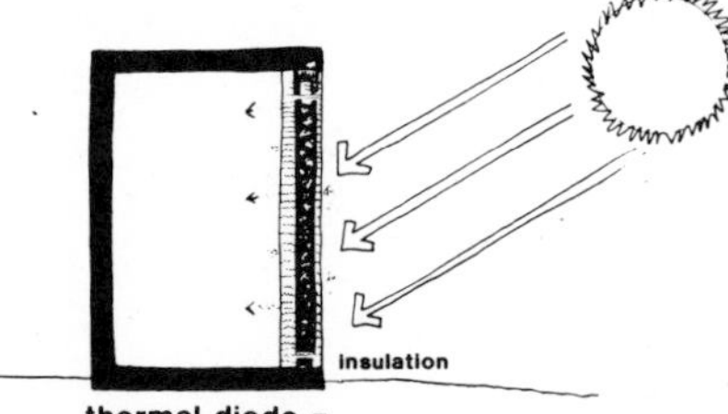

thermal diode -
solar-heated panel circulates
fluid to inner thermal
reservoir

passive solar heating / phase-change materials

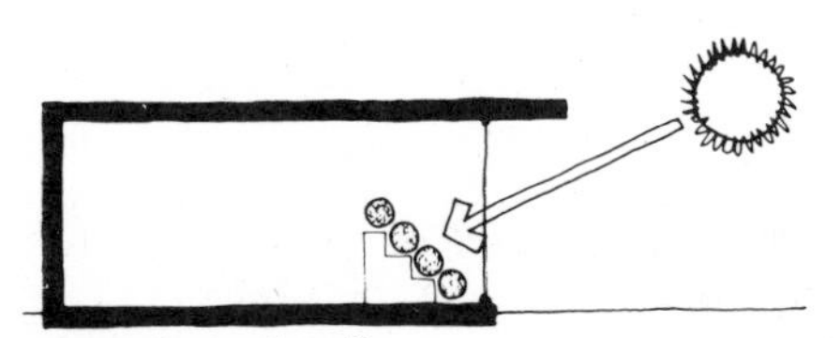
thermal eutectic-salt
containers

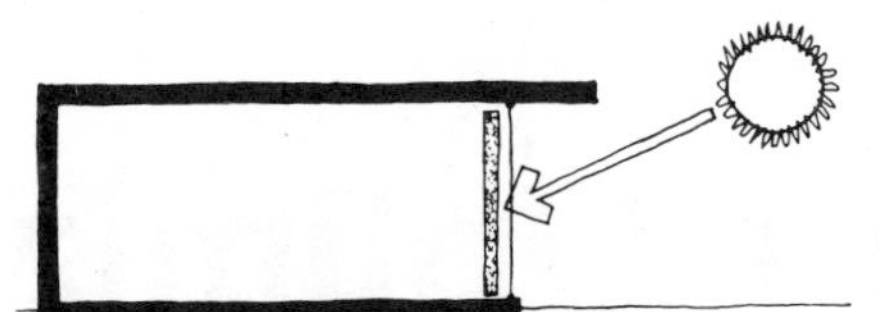
thermal eutectic-salt tubes

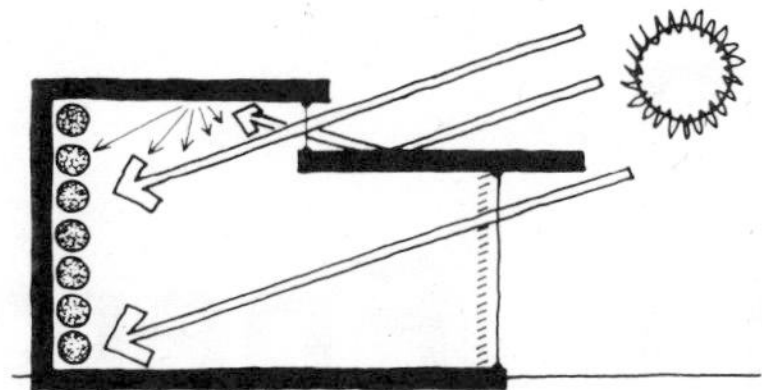
thermal eutectic-salt containers
controlled solar reflection

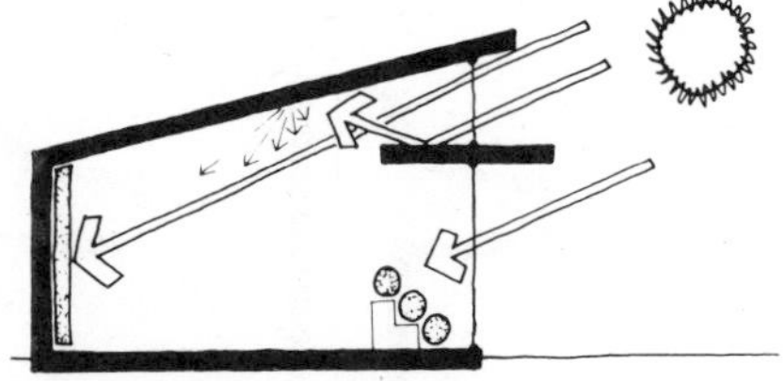
thermal eutectic-salt tubes
and containers
solar reflective shelf

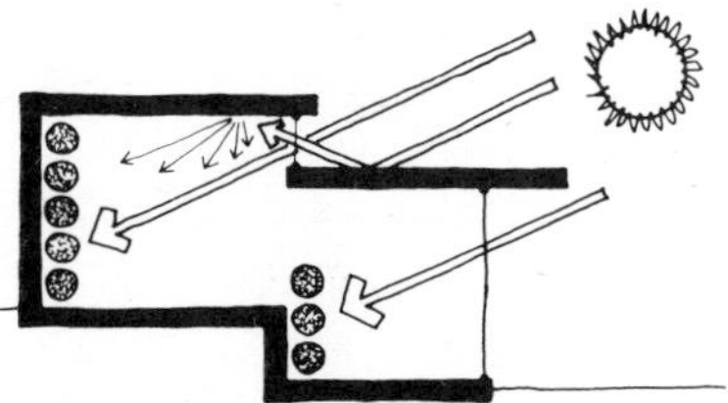
thermal eutectic-salt containers
solar clerestory

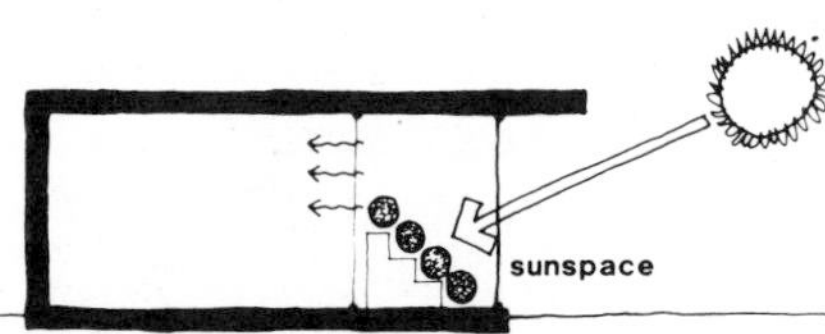

thermal eutectic-salt containers
thermal conduction

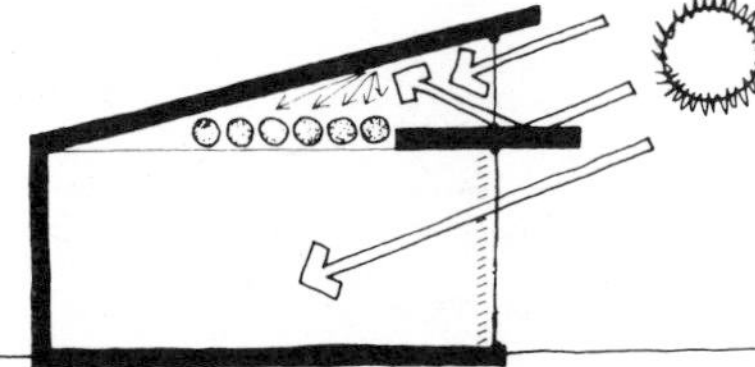
thermal eutectic-salt containers
solar radiant thermal lag

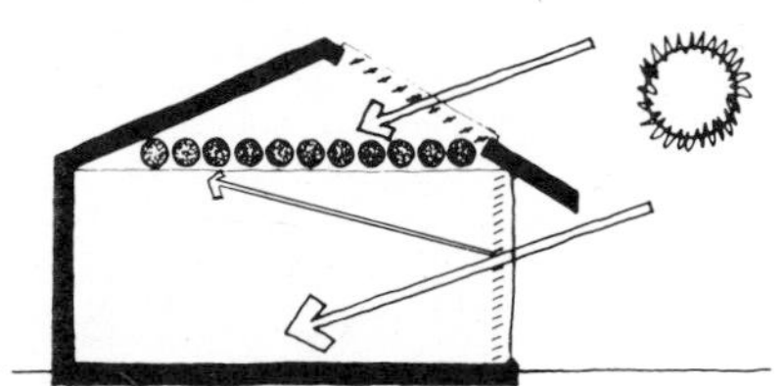
thermal eutectic-salt containers
controlled solar reflection

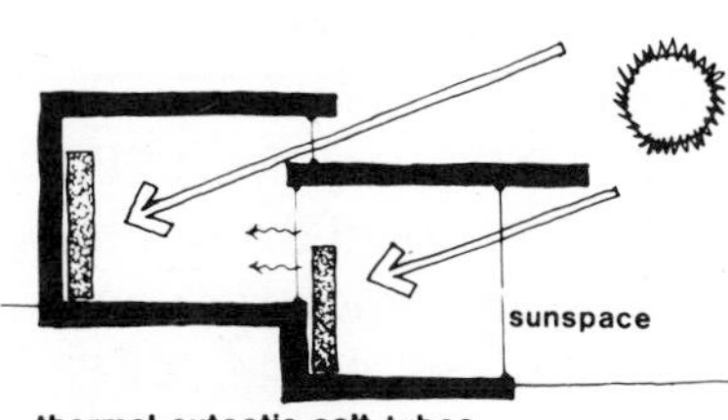

thermal eutectic-salt tubes
secondary gain

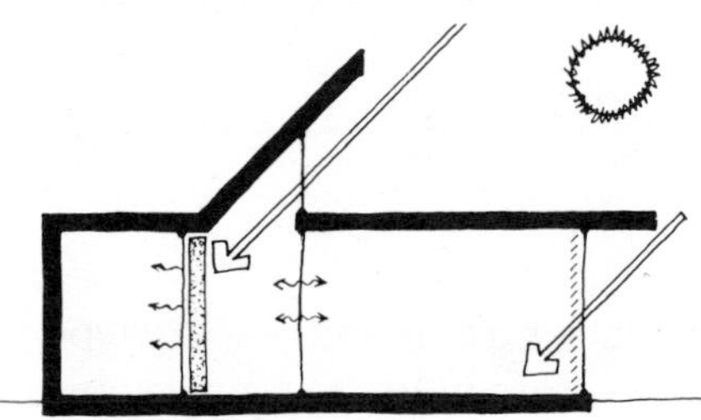
thermal eutectic-salt tubes
thermal lag/secondary gain

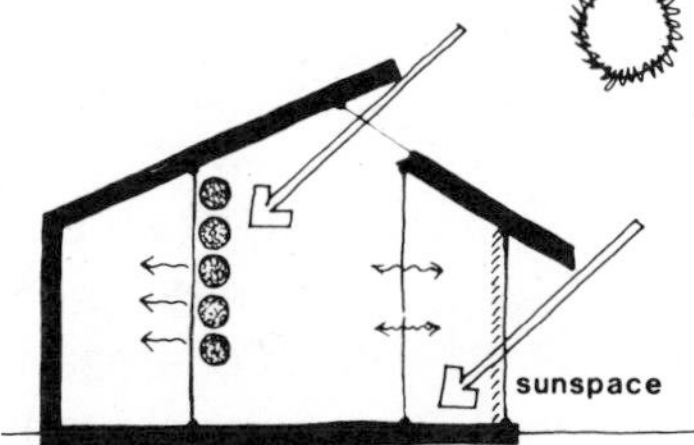

thermal eutectic-salt containers
thermal lag/secondary gain

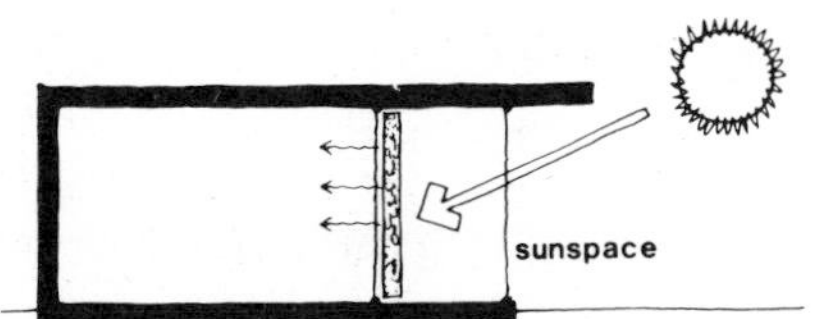

thermal eutectic-salt tubes
thermal radiation/secondary gain

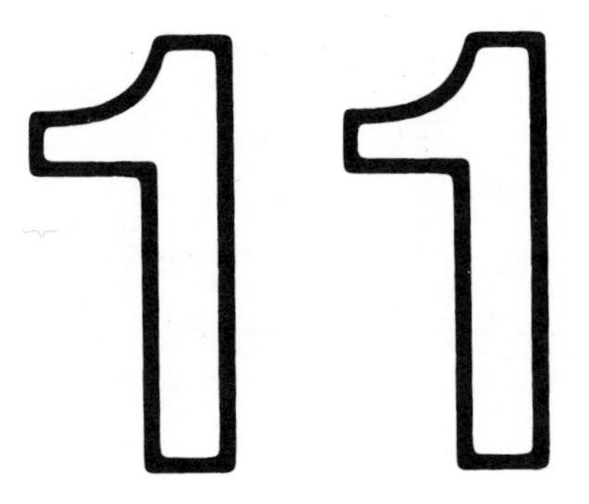

Energy-Optimized Buildings—Case Studies

In this chapter a number of case studies of energy-optimized buildings are presented. They have all been designed to optimize the conservation of energy through architectural design and site planning, to optimize the use of cost-effective natural climatic systems, and to integrate the active solar collection and thermal storage with architecture and interior design.

Principal conservation features are indicated on the drawings, and many other elements that are integrated into the architectural system are presented in the text. Each building, however, may have features that are not reported. The primary emphasis of presentation is on the individuality of systems and the important design features that make these buildings both functional and practical.

The first four houses are located in a well-established residential neighborhood, close to the Cherry Creek shopping area east of downtown Denver. Each of these buildings is unique and has certain features different from the others. This was done so that optimum designs, orientations, and configurations could be determined under actual service conditions. In addition, it clearly demonstrates that buildings employing energy conservation and solar collectors do not all have to look alike.

The first two buildings are situated on the same city lot. They are designated as "Cherry Creek Residence 419A" and "Cherry Creek Residence 419B."

The 419A building is a 186 square-meter (2,000-square-foot) residence designed on a 8.5 × 8.5 meter (28 × 28 foot) foundation. It is intended as a prototype for either on-site or modular construction for townhouses, cluster housing, or single-family use. The room arrangement is a square stacking plan with bedrooms and greenhouse on the lower level; living, dining, and kitchen on the main level; and studio at the third level. Full bathrooms, including shower, are located on the lower and main levels.

The primary entry faces east away from

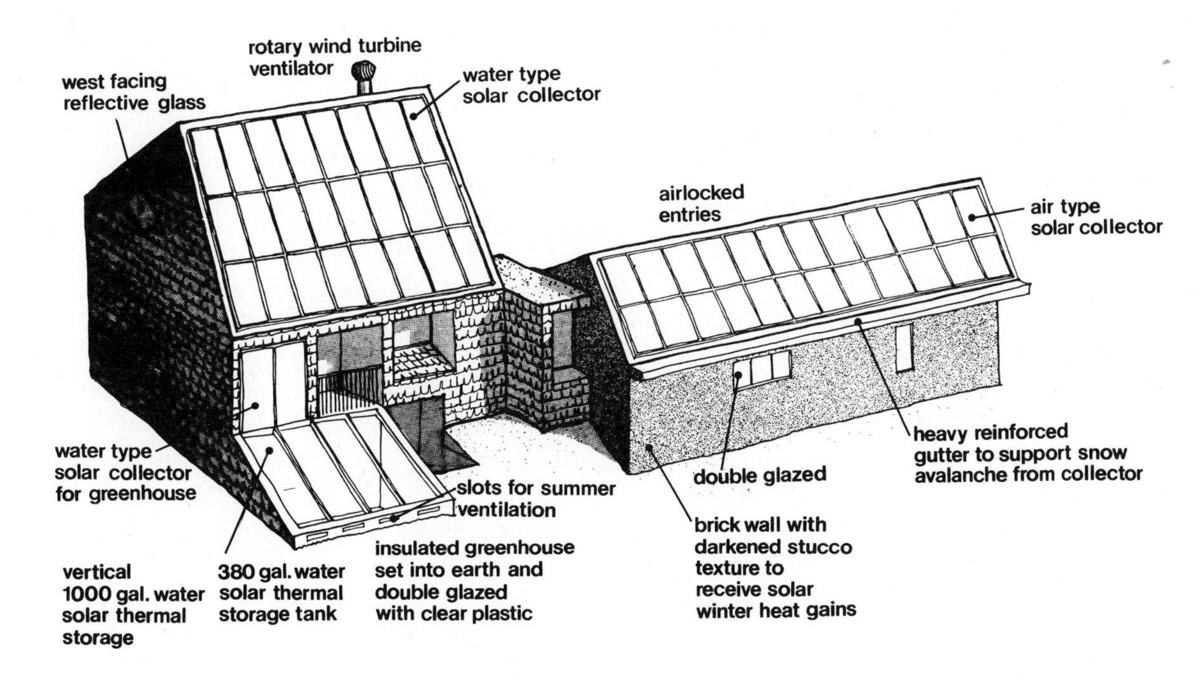

north and northwest winter winds and is an airlocked double-door unheated vestibule. Glass area is limited yet provides visual openness, attractive outdoor vistas, and natural illumination. Walls and ceilings are shaped and painted to function as light fixtures without glare. All windows are fabricated from clear float, double-glazed, insulating glass. The window position and size are calculated to receive direct winter solar gains but to exclude them in the summer.

The hood over the kitchen range recirculates the air through a charcoal filter for cleaning and deodorization. The bathrooms have recirculative fans with citrus-base filters that deodorize the air and kill bacteria. These systems prevent direct internal heat losses that occur through externally vented exhaust-fan systems.

data

CHERRY CREEK RESIDENCE "419A"

Floor area	186 square meters (2000 square feet)
Passive-solar subsystems	3
Energy centralization	yes
Solar-collector area	52 square meters (560 square feet)
Solar-collector type	liquid (water)
Solar-collector supplier	R-M Products
Heat-storage type	3,785-liter (1,000-gallon) water tank
Annual passive/active performance	95 percent
Completion date	1974
Architect	Richard L. Crowther

data

CHERRY CREEK RESIDENCE "419B"

Floor area	102 square meters (1,100 square feet)
Passive-solar subsystems	1
Energy centralization	no
Solar-collector area	35.7 square meters (384 square feet)
Solar-collector type	air
Solar-collector supplier	Solaron Corporation
Heat-storage type	27.2 metric tons (30 tons) of gravel
Annual passive/active performance	80 percent
Completion date	1974
Architect	Richard L. Crowther

During spring, summer, and fall, ventilation is provided by means of a roof exhaust turbine. Fresh air is provided in winter by the solar-heated greenhouse and from return air that passes through an electrostatic and charcoal filtration system.

Outer walls of the house are ten-inch-wood studs with 9½ inches of rockwool insulation, an outer 3/4-inch-thick insulating sheathing board, 5/8 inch of wood-ply sheathing, and wood shakes to trap air. Internal 101-micrometer (4-mil) pliofilm acts as a continuous vapor barrier on the entire interior and is covered with 12.7 mm (½ inch) interior drywall. The U-factor of this wall is less than .025, and because of its thickness it provides a higher thermal mass for retention of interior energy than conventional frame construction. The roof construction is similar to that of the walls.

The 52-square-meter (560-square-foot) solar collector, supplied by R-M Products of Denver, is a flat-plate water-type south-facing roof collector set at a 53 degree angle. Each pane of the double-glass cover plate is 1/8 inch thick. The interior coating is flat black. A reinforced gutter and top member are provided to support a movable ladder for collector access and to act as a snow trap. Heated water is piped from the rooftop solar collector to a 1,000-gallon vertical fiberglass-insulated storage tank. At night and when freezing conditions exist, the water is automatically drained from the collector into the insulated tank. This protects the collector from damage that would result from the expansion of water upon freezing.

Cooling is provided on hot days by a 153 cubic meter per minute (5,400 cubic foot per minute) evaporative cooler, combined with the roof turbine fan-exhaust system. The solar-collection system is used to heat domestic water. More than 95 percent of all hot water is provided in this way, and 96 percent of all heating requirements are satisfied by the solar-collection system in conjunction with the energy-optimized design. In actual use, 90 percent of all cooling requirements were met with the above-mentioned system.

The 419B building is the result of extensive remodeling of an older, 102-square-meter (1,100-square-foot) residence. It also has an 83.6-square-meter (900-square-foot) basement. To improve the total thermal characteristics of the house, the basement is utilized as a thermal reservoir, which moderates the heating and cooling needs of the house by means of direct-transfer grilled-floor openings. The old coal bin is reused for solar thermal storage. It is fitted with a layer of open concrete block covered with wire mesh to serve as a lower manifold and filled with 15.3 cubic meters (20 cubic yards) of large gravel. A new insulated north wall constructed the length of the house shields the north side from winter storms and winds, as well as converting the old uninsulated 20-centimeter (8-inch) brick wall to an interior thermal-inertia wall. The new north wall extends out from the old wall approximately 91 centimeters (3 feet), allowing for moderate day and cool nocturnal temperature air to be drawn in through recessed, insulated air vents. The south roof angle was altered to 53 degrees for effective use of an air-type solar flat-plate collector. Six inches of insulation was added to the roof to supplement the four inches of existing loose-fill insulation.

The main entry faces east and is an air-locked double-door unheated vestibule. A floor grille to the basement near the entry picks up cold-air intrusion. The old front entry door was replaced with a new weatherstripped door, and the old porch was converted to a solar patio. Small, old north-side windows were replaced with insulated weatherstripped wood panels that can be opened for natural ventilation. On the south side of the living room, a new Slimshade double-glazed window has been provided and is protected by a roof overhang. Two east windows with heat-absorbing gray double glazing extend to the floor and replace the old double-hung windows. The west window in the bathroom was re-

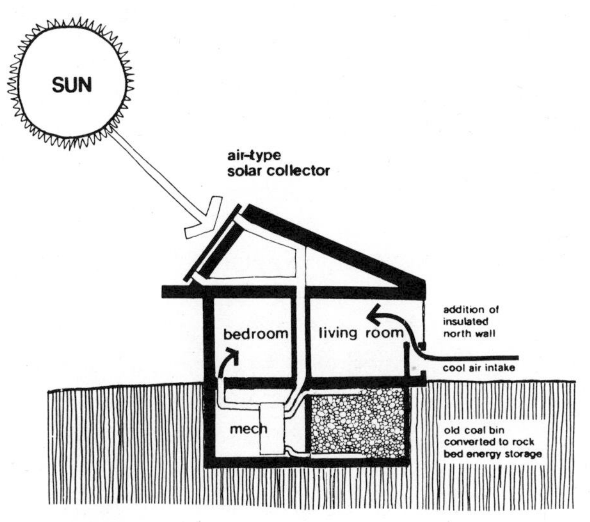

419 B

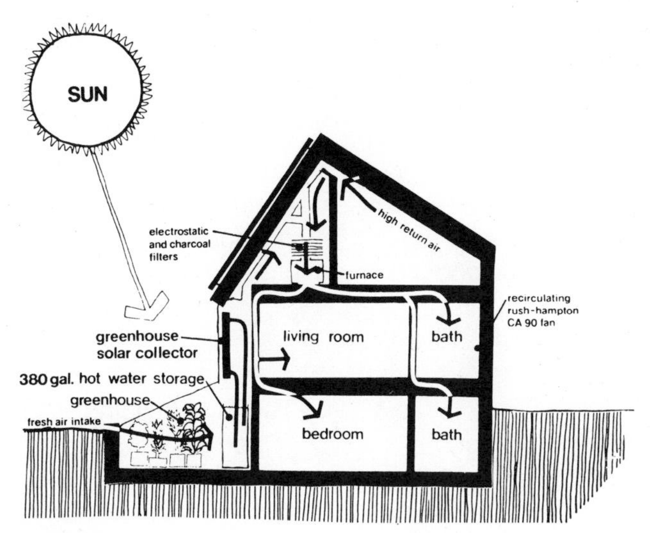

419 A

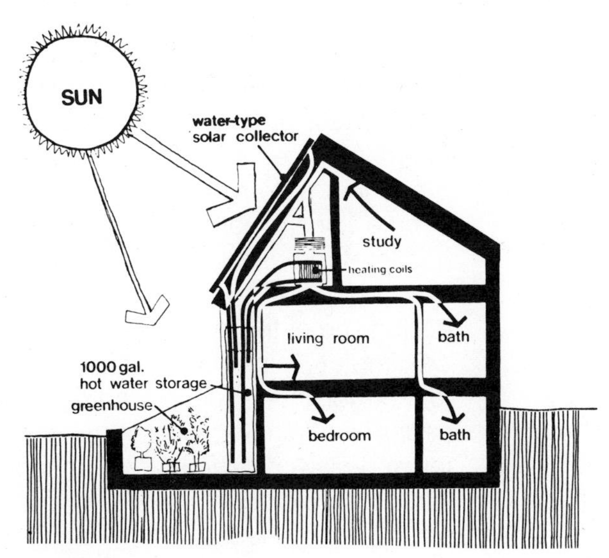

419 A

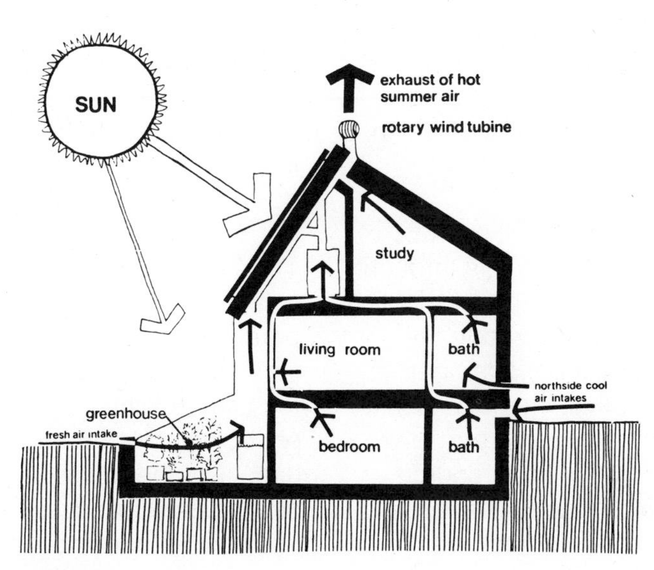

419 A

placed with fixed double-glazed reflective glass to eliminate outside air infiltration and decrease west solar radiation approximately 78 percent.

Bathroom and kitchen fans are the recirculating type to decrease heat losses. The exterior of the house was completely refinished with cedar wood shakes on the roof and sidewalls, adding extra insulation to them. Other sidewalls are finished with new stucco.

The 35.7-square-meter (384-square-foot) solar collector, supplied by Solaron Corporation of Denver, is a flat-plate air-type collector. Each pane of the double-glass cover plate is 1/8 inch thick. The interior coating is flat black. A reinforced gutter and top member are provided to support a movable ladder for collector access and to act as a snow trap. Solar-heated air is forced through a duct system to the thermal-storage gravel bin in the basement.

During daytime periods of full or diffuse sunlight, solar-heated air travels through the gravel storage bin and the existing furnace before being delivered to the house. During overcast periods or at night, the forced-air system draws solar-heated air from the gravel storage bin. Natural gas is the backup fuel source. During seasons when the house needs cooling, cool daytime and nocturnal air cools the gravel bin, and cool air is delivered as needed through the house by means of the existing forced-air system.

Eighty-five percent of all heating requirements are provided by the solar-collection system in conjunction with energy-optimized features. Cooling is provided by natural ventilation.

The third building in this group is "Cherry Creek Residence 435." This unit, on a 6 × 6-meter (20-foot × 20-foot) foundation, is designed as a prototype for single-family duplex or cluster housing. It is a two-story, low-cost, 74-square-meter (800-square-foot) home with one bedroom, small den, greenhouse, and a combination kitchen, dining, and living room. In addition it has an 18.6-square-meter (200-square foot) attic-storage area behind the solar collector, which is accessible by a pull-down ladder stairway.

The 22-square-meter (240-square-foot) solar collector, supplied by Solaron Corporation, is a flat-plate air-type south-facing roof collector, set at 50 degrees to optimize winter solar radiation. Delivery of the solar-heated air is either made directly to the house or, when internal heating is satisfied, stored in an insulated gravel bin. The storage bin is located under the house and does not take up interior space. Auxiliary heating is provided by separate radiant electric baseboard heaters located in each room. The temperature of each room, maintained by the auxiliary system, is individually controlled by remote thermostats.

A west-facing solar plenum with built-in thermal-storage capacity provides inductive ventilation. The house is also equipped with a rotary wind turbine ventilator. Individual vents are located near the lower-level floor line to allow the entry of cool nocturnal or cool daytime air. These reflective metal ventilation units have weather hoods, avoid direct wind flow, and have thick insulative doors with gaskets to prevent infiltration when closed.

The building envelope is minimized by the square configuration, low ceiling heights (2.2-meter, 7′4″) and by setting portions of the lower level below the grade line. External walls are constructed of specially treated 15-centimeter (6-inch) wood studs continuous from below grade foundations to the ceiling line of the bedrooms. Regular 15-centimeter (6-inch) wood studs continue to the roof line. Handsplit, heavy red-cedar wood shakes over water-proofed felt and plywood sheathing with taped joints to avoid air intrusion form a durable, no maintenance, energy-conserving barrier. Fifteen centimeters (6 inches) of rockwool batts in outer stud walls are used continuously from 91 centimeters (3 feet) below grade to the roof line. The roof areas have 30 centimeters (2 inches) of rockwool batts. These batts are backed with plastic vapor barriers to the interior of the house.

The total glass area to the exterior is approximately 10 percent of the total floor area. All windows are fixed insulating units set in a glazing bed to avoid infiltration and reduce thermal losses.

If produced in quantity with organized field construction methods, this home could take the place of mobile homes. This prototype was constructed in five weeks.

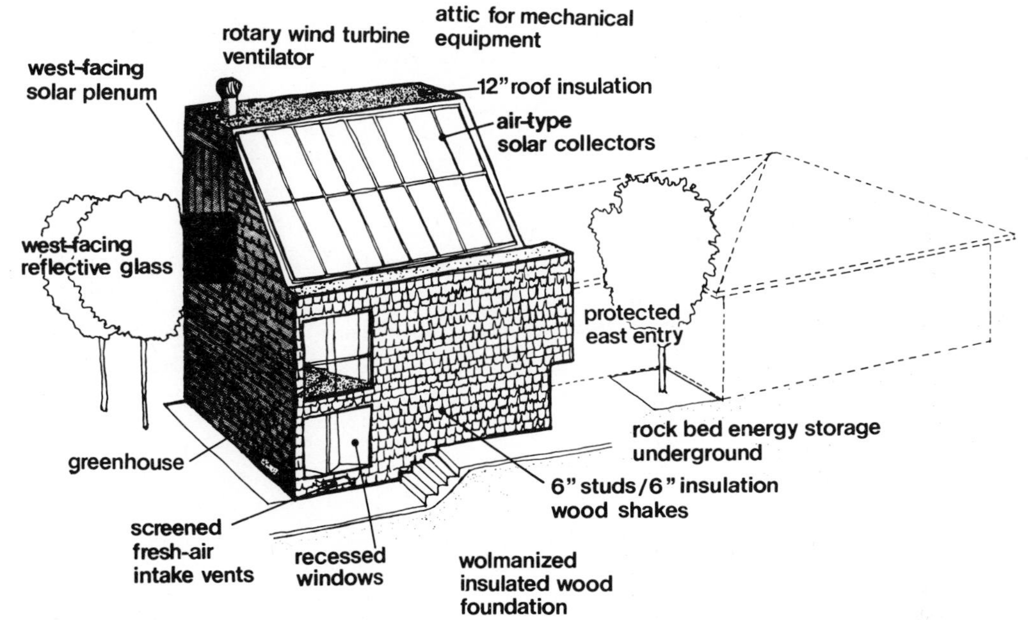

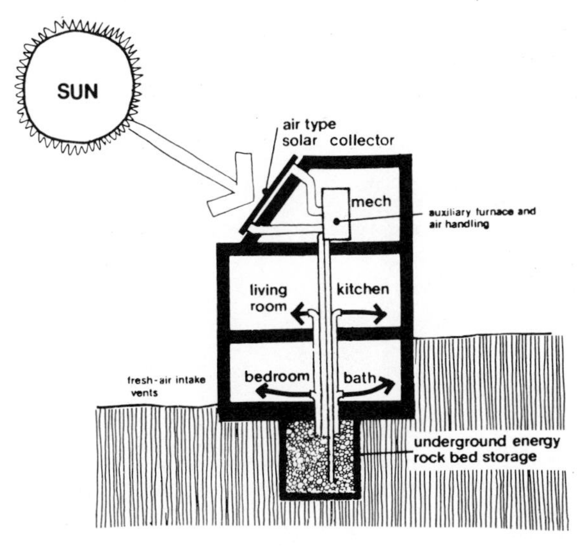

data

CHERRY CREEK RESIDENCE "435"

Floor area	74 square meters (800 square feet)
Passive-solar subsystems	2
Energy centralization	yes
Solar-collector area	22.3 square meters (240 square feet)
Solar-collector type	air
Solar-collector supplier	Solaron Corporation
Heat-storage type	10.9 metric tons (12 tons) of gravel
Annual passive/active performance	80 percent
Completion date	1975
Architect	Richard L. Crowther

Optimized energy conservation and passive systems provide 70 percent of the actual cooling of this unit, and 80 percent of the heating is provided by the solar-collection system in conjunction with the other optimization features.

The last building in this area is "Cherry Creek Residence 500." It is a 241.5-square-meter (2,600-square-foot) residence with a 116-square-meter (1,250-square-foot) apartment, designed to optimize energy conservation and utilize passive solar collection. The entire building is designed to function as a solar collector in the winter and prevent solar gains in the summer. Heat storage is provided by the thermal inertia of floors and walls.

The principal entry for each unit is recessed to protect it from prevailing winter winds. All windows are double-glazed, fixed glass. Window areas are kept to a minimum, yet their shape and location maximize interior daylight. Natural illumination is also provided by double-plastic, heat-absorbing skydomes and special Skyshafts with near-zero thermal loss. Artificial lighting levels are provided on a task basis.

Solar gain through southern exposure is maximized during winter and minimized during summer by calculated window-recess depths and outdoor deciduous vegetation combinations. The southern windows of the dining room function as two-stage winter solar collectors, with the lower unit active from September 1 until April 1 and the upper unit active from October 15 to February 15.

The exterior wall color was selected to balance radiant heat gains and losses

data

CHERRY CREEK RESIDENCE "500"

Floor area	241.5 square meters (2,600 square feet)
Passive-solar subsystems	4
Energy centralization	yes
Heat-storage type	thermal inertia of floors and walls
Annual passive performance	80 percent
Completion date	1972
Architect	Richard L. Crowther

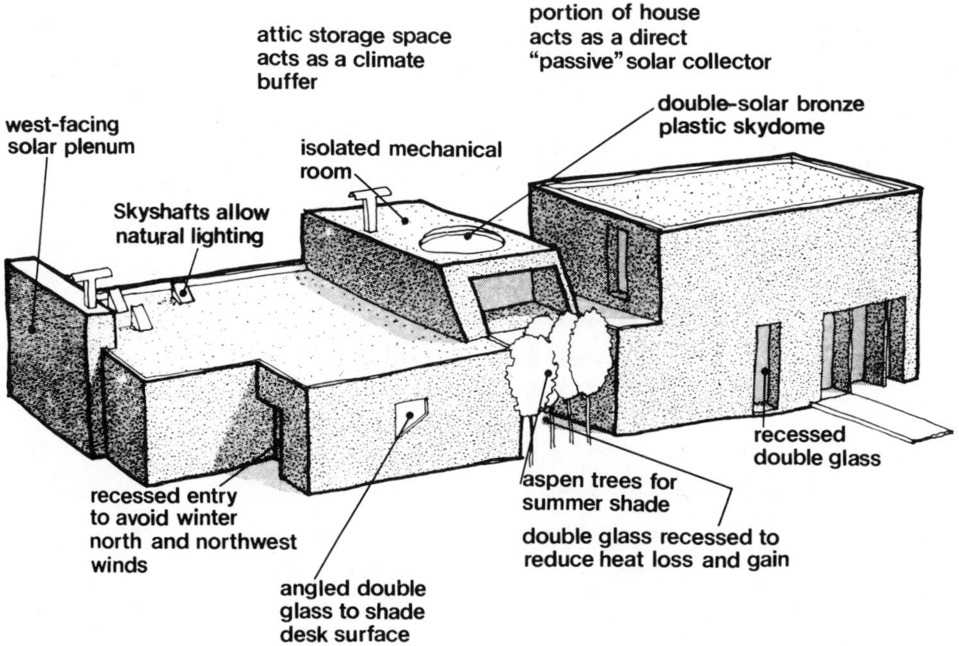

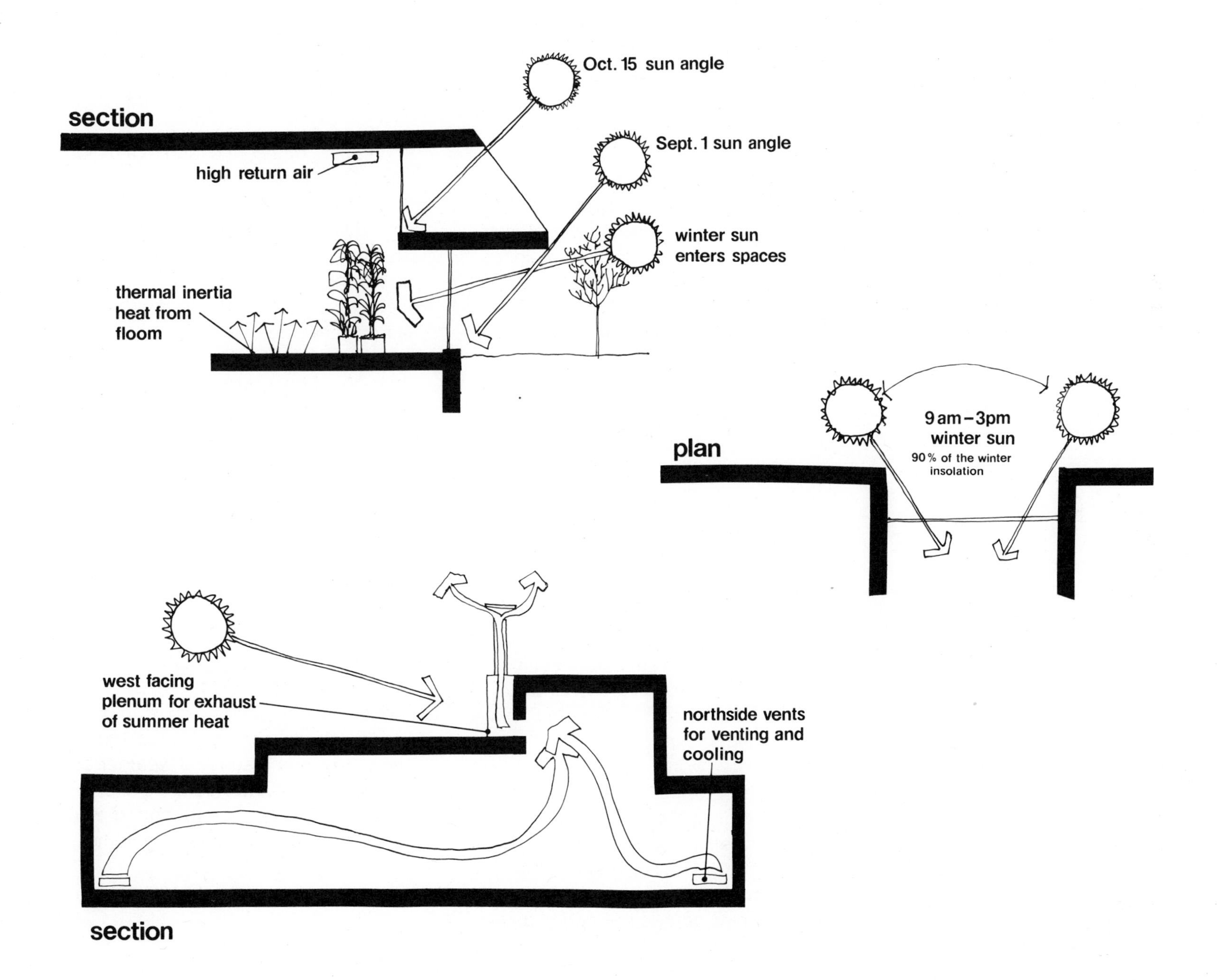
section
Oct. 15 sun angle
Sept. 1 sun angle
high return air
winter sun
enters spaces
thermal inertia
heat from
floom
plan
9 am – 3pm
winter sun
90% of the winter
insolation
west facing
plenum for exhaust
of summer heat
northside vents
for venting and
cooling
section

on a twelve-month basis. The rough exterior wall texture deepens the stationary thin air film as a thermal buffer zone. The U-factor of the walls, roof, and floor is .05.

Garage and storage-area locations provide thermal buffering of internal living spaces. Modulated ceiling heights are used to direct the hottest air to the highest point in the building, allowing interior air temperatures to be salvaged by high-level air returns in winter and low-level air returns in summer. For summer cooling roof plenums, heated by the sun, induce upward air motion, which allows outside air to enter by displacement through north vents.

The passive systems and energy-optimized features of this home provide 65 percent of the heating and 60 percent of the cooling. The house is also equipped with conventional heating and air-conditioning units. These systems contain electrostatic, charcoal, and fiber filters that allow interior air to be recirculated without the intrusion of exterior air, which may be at a different temperature, thereby salvaging thermal equilibrium.

Large interior plants provide oxygen to the air and absorb carbon dioxide. Winter humidification of interior air is produced by a water fountain, which raises the perceived temperature.

The "CSU Solar House I" was built by the Solar Energy Applications Laboratory of Colorado State University, using a grant from the National Science Foundation. It was the first residential-size building in the United States to be actively heated and cooled with solar energy. The 279-square-meter (3,000-square-foot) building was designed as a solar prototype and demonstration project.

The 71-square-meter (768-square-foot) flat-plate liquid-type solar collector is located on the southern roof area and is set at a 45 degree angle. This orientation of the collector favors the optimum collection of solar energy during the entire year.

The heat absorbed by the metal surface of the collector is transferred to the collector fluid—a 25 percent solution of ethylene glycol in water. In order to minimize the amount of ethylene glycol needed, the collector fluid does not pass directly into the storage tank but instead is piped through a heat exchanger. The heat is transferred to water, and the water is piped to storage.

The architectural energy-conservation features of this house include vertical exterior fins to reduce convective losses from window and wall areas, double-glazed wood sash windows, recessed airlock entry, garage positioned as a climatic buffer, and portions of the structure set into the earth. Overhangs provide summer shading of windows yet allow the winter sun to penetrate.

Operation, monitoring, and evaluation of the CSU Solar House I was under the direction of Dr. George O.G. Löf. Solar energy is used in this house for space heating, domestic hot-water heating, and to power a lithium-bromide absorption cooler. Heating and cooling are forced-air systems. Solar energy actually provides 83 percent of the heating and 43 percent of the cooling, as well as most of the domestic hot-water requirements. With modifications, solar energy can be expected to provide 79 percent of the cooling.

Colorado State University has constructed two additional solar laboratory buildings that are equipped with heating and cooling systems of different designs. Solar House II uses air to collect solar energy in the form of heat, while Solar House III uses a liquid. With these three houses, it has been possible to operate three different systems at the same time, under the same sun and weather conditions. With this arrangement, direct comparison of the performance of the three heating and cooling systems is possible.

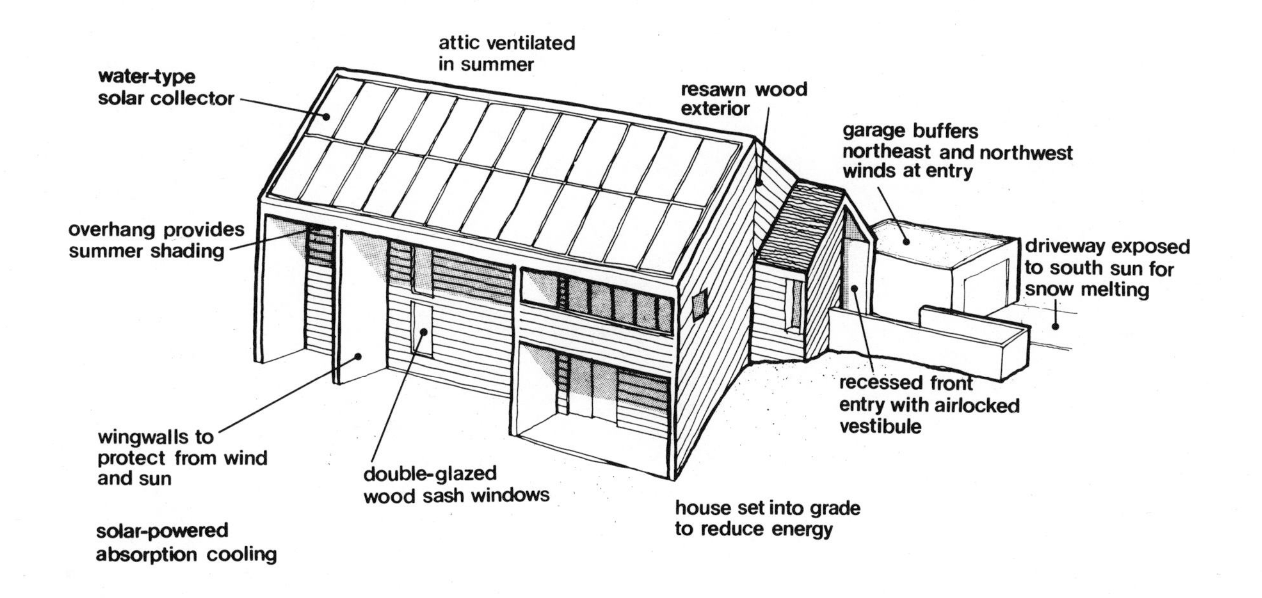

data

COLORADO STATE UNIVERSITY "SOLAR HOUSE I"

Floor area	278.7 square meters (3,000 square feet)
Passive-solar subsystems	3
Energy centralization	no
Solar-collector area	71 square meters (768 square feet)
Solar-collector type	liquid (water plus glycol)
Solar-collector supplier	Solar Applications Laboratory of Colorado State University
Heat-storage type	4,164-liter (1,100-gallon) water tank
Annual passive/active performance	85 percent
Completion date	1974
Architect	Richard L. Crowther

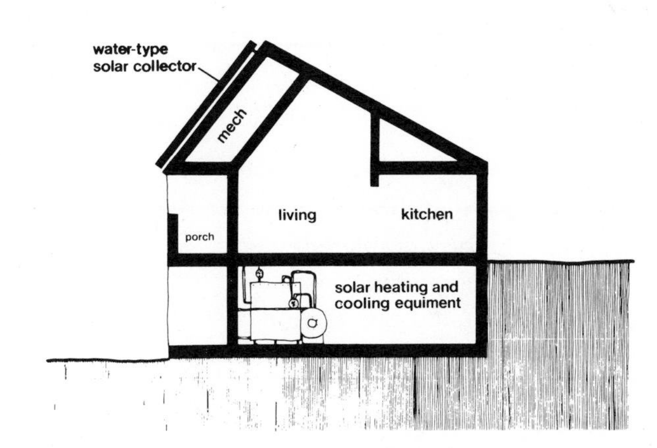

Despite its narrow urban site, this 372-square-meter (4000-square-foot), partially earth-sheltered, passive-solar home, built in 1961, optimally uses winter sunlight. Direct solar gains heat brick floors, insulated cavity sidewalls, and interior bearing walls. Concrete slabs on grade, a fallout shelter, and prestressed, second-floor slabs add to the thermal mass and stability. Double-track casement cloth and thermal drapes reduce living-room downdrafts and heat loss. The entire upper level is designed like an insulated bonnet. Exterior wood-frame walls and slightly sloping roofs are well insulated. The northside garage is a climatic buffer, and the westside utility room is an airlock. In addition to direct solar gains, earth protection, and thermal mass, a hot-water baseboard backup system with seven zones ensures discretely controlled winter comfort. The brick walls of the southern courtyard retain the sun's warmth and the use of outdoor space through cool and cold weather.

The main floor (242 square meters; 2,600 square feet) is remarkably cool in summer. Passive cooling is accomplished by evaporation and convection by air. Heavy landscaping of the front embankments and courtyard cools air by evaporation. Naturally convected outdoor air flows through open-wood grillwork in the garden. This cool, evaporative air enters through floor-line awning windows and a screened, security main door. A gravel-filled trench surrounds an east patio opening directly to the living room, studio, and bedroom. A spray of water on the gravel cools this screened, sunken patio. It provides cross-ventilation through the house to another patio on the north. Above the open stairway in the center of the house, a powered exhaust fan provides additional ventilation as desired. The upper-level recreation and guest rooms are comfortable in warm weather, but a small air conditioner can be used for social gatherings in hot weather. An upper, outdoor deck to the north offers a cool, shaded place to relax.

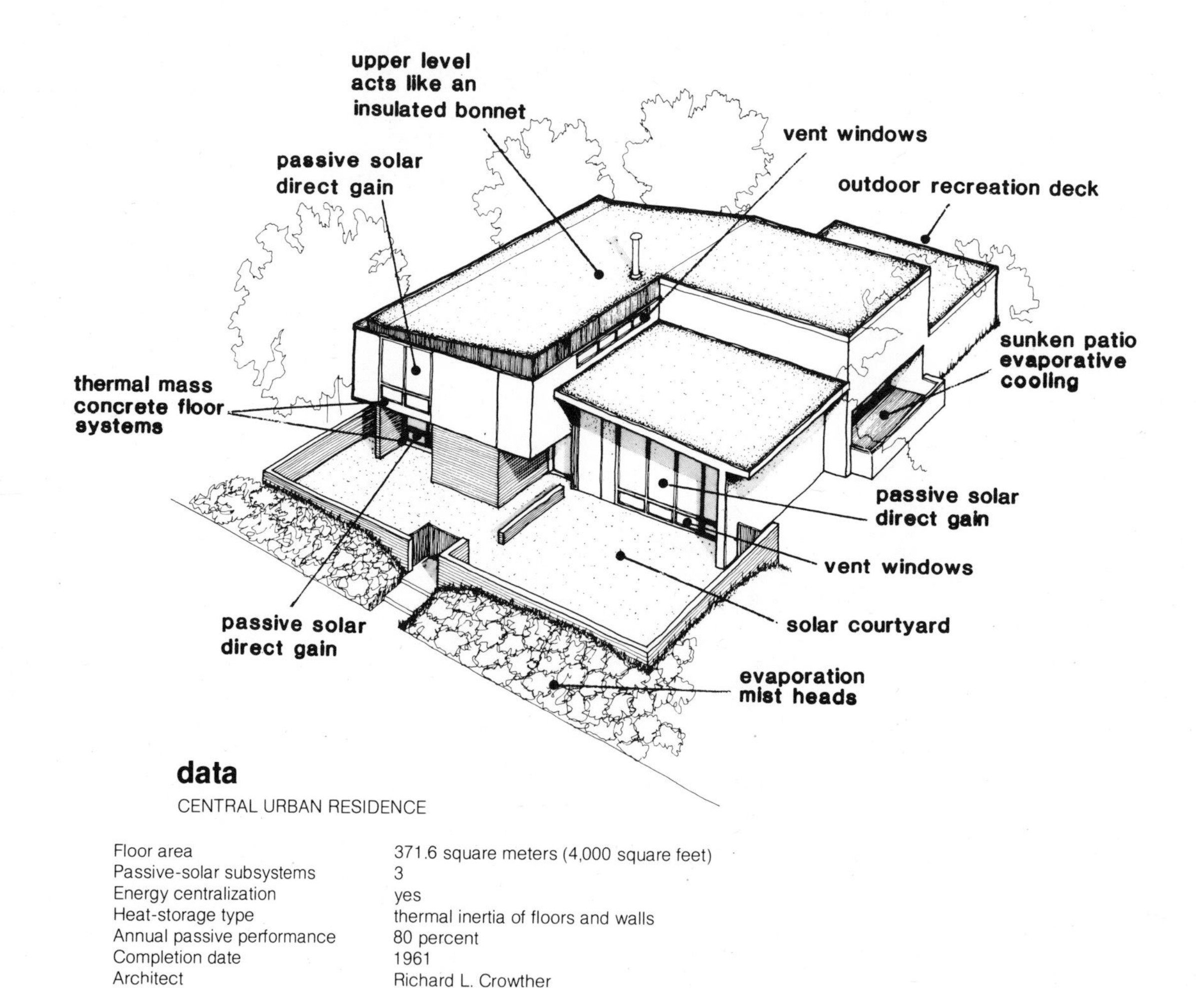

data

CENTRAL URBAN RESIDENCE

Floor area	371.6 square meters (4,000 square feet)
Passive-solar subsystems	3
Energy centralization	yes
Heat-storage type	thermal inertia of floors and walls
Annual passive performance	80 percent
Completion date	1961
Architect	Richard L. Crowther

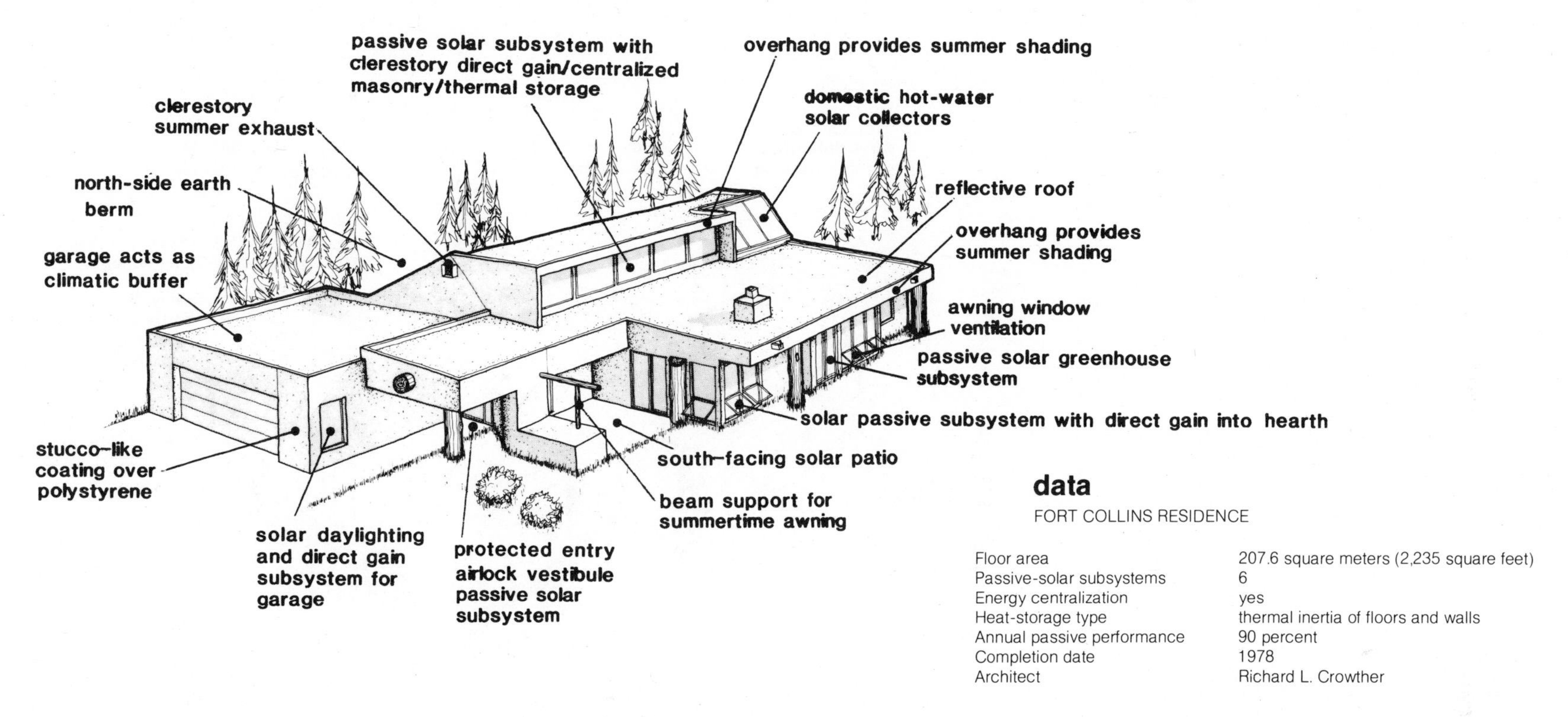

data

FORT COLLINS RESIDENCE

Floor area	207.6 square meters (2,235 square feet)
Passive-solar subsystems	6
Energy centralization	yes
Heat-storage type	thermal inertia of floors and walls
Annual passive performance	90 percent
Completion date	1978
Architect	Richard L. Crowther

This moderate-cost 208 square-meter (2,235-square-foot) home, located in northern Colorado, depends on several passive solar subsystems and active solar domestic-water heating. The Southwest architectural idiom uses entire sections of dried trees for columns and beams and an external, earth-tone, Settef coating over polystyrene. The home is situated on a spacious site in an open terrain.

The principal system uses the effectiveness of energy centralization. Direct solar radiation strikes an interior, thermal mass masonry wall through a row of south-facing, clerestory windows. Airborne heat from passive solar subsystems flows across living-space ceilings to the high, central gallery.

The passive solar subsystems include an airlocked, front entry with a tile-covered, thermal mass floor; the central, solar gallery with a thermal mass wall and tile-covered floor; a thermal mass masonry wall and hearth for a wood stove; an attached greenhouse with eutectic thermal rods; and a protected outdoor patio—all providing agreeable daylighting and seasonally appropriate thermal gains. A northside earth embankment against the house has retaining walls of railroad ties to expose two small windows. The northside garage, airlocked entry, southside greenhouse, and outside patio act as thermal buffers.

Ventilation and nighttime and daytime cooling use the height of the central gallery to inductively exhaust hot indoor air displaced by cooler outdoor air. This home, built in 1978, performs with gratifying thermal efficiency and remarkably low amounts of supplemental energy. No mechanical cooling is needed for summer comfort.

This three-level residence is located on a densely forested site. A very small clearing barely accommodates its boundaries. To benefit from the sun, this dwelling reaches from the dense-forest matrix toward the sky.

The possibility of using south-wall, passive-solar gains in wintertime was nil. A bank of air-type solar collectors was located above the garage in the only solar exposure and was directly connected by circulated ductwork to a rock-storage bin within the house.

The house virtually becomes part of the natural forest. The architecture of the forest moderates the extremes of climate by acting as a buffer in winter and an evaporative and reflective cooler in summer. The house responds harmoniously to this naturally conditioned environment.

Capturing magnificent mountain views to the northeast was complicated by the most unfavorable winter exposure. The design resolution was to strategically locate and discretely minimize windows at the top-most level to frame the most impressive angle of the mountains. The cold northwest winds of this treetop vista can be interrupted on the interior by insulative blankets.

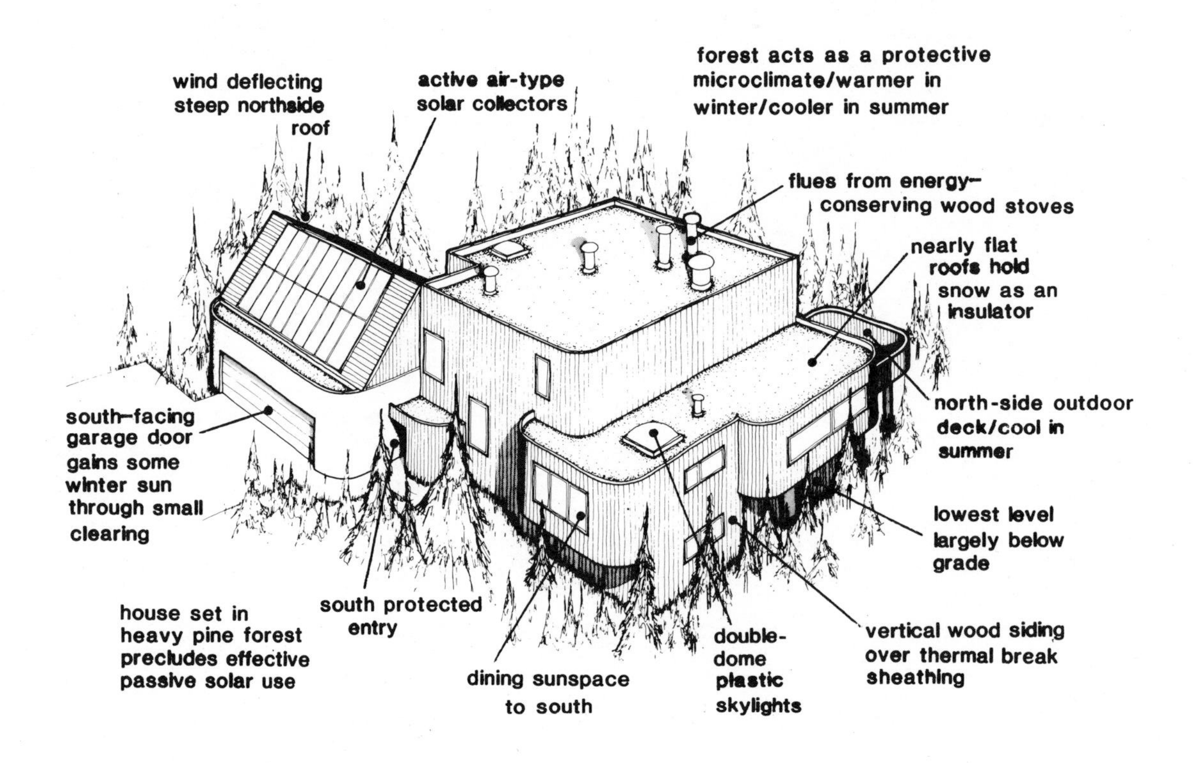

data

MOUNTAIN FOREST RESIDENCE

Floor area	244 square meters (2,630 square feet)
Passive-solar subsystems	0
Energy centralization	yes
Solar-collector area	39.9 square meters (429 square feet)
Solar-collector type	air
Solar-collector supplier	Solaron Corporation
Heat-storage type	9.98 metric tons (11 tons) of gravel
Annual active performance	75 percent
Completion date	1978
Architect	Richard L. Crowther

This executive, solar, ranch-style guest house hovers, with an unparalleled view of the mountains, above a ravine looking down on a swiftly flowing river. It consists of four studio bedroom suites, a caretaker's apartment, a large living space, numerous other amenities, and a loft-type room that overlooks the entire mountain panorama. The form of the house was dictated by the surrounding magnificent vistas and the constraint of optimizing passive- and active-solar energies.

The split-level plan has balconies projecting over the river flowing to the north side. The house surrounds a mountain-viewing patio, which is kept private from a major highway by means of berms and planting. The rear of the house follows the steep line of the river embankment, and the lower level cuts deeply into it, enhancing contact with the river and its sounds. The house is superinsulated to match the climatic extremes of the 2,438-meter (8,000-foot) elevation. Extensive south-facing glass with a shading canopy allows direct solar gains to a dark charcoal, tile floor over a thick, concrete-insulated base.

The south-facing, airlocked entry acts as a direct gain passive solar collector. The garage is located to the southwest to receive and maintain the warmth of the winter sun in this cold climate and to give easier egress during periods of very heavy snow. A long, clerestory window

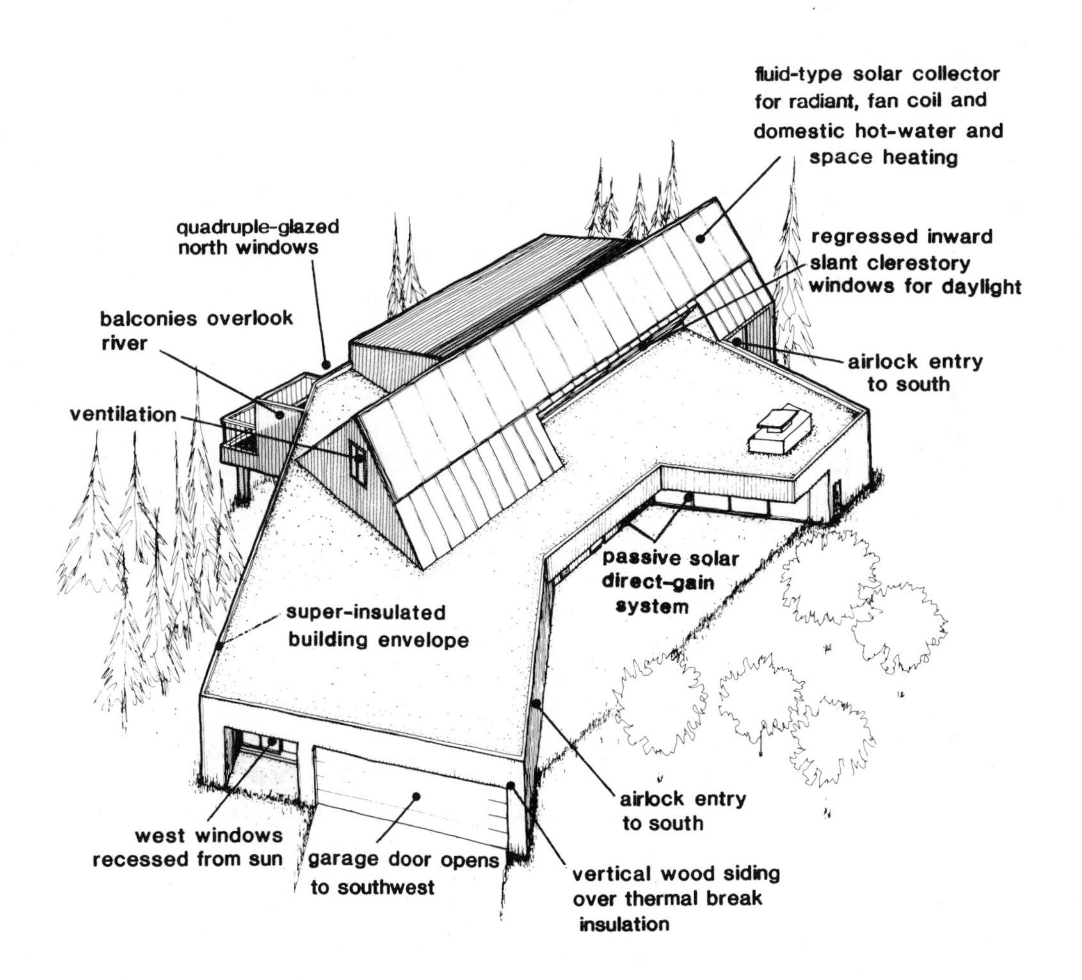

data

MOUNTAIN GUEST RANCH

Floor area	390 square meters (4,200 square feet)
Passive-solar subsystems	4
Energy centralization	yes
Solar-collector area	42.7 square meters (460 square feet)
Solar-collector type	liquid (water plus glycol)
Solar-collector supplier	Solar Shelter (Rollbound)
Heat-storage type	3,785-liter (1,000-gallon) water tank
Annual passive/active performance	85 percent
Completion date	1981
Architect	Richard L. Crowther

slopes inward so that from the vernal equinox through the summer it is in shadow and receives reflected light from the relatively flat roof deck. In the winter months, the sunlight penetrates the upper gallery and multipurpose loft room that the owner added in the building process.

The liquid-type solar collectors heat a 3,785-liter (1,000-gallon), concrete tank, which in turn serves a radiant floor system and a portion of the main living area not directly in the passive solar path. The four bedroom suites and the caretaker's apartment are equipped with individually controlled, radiant fan-coil units that deliver solar heat to those areas. The solar thermal tank also heats domestic water and water for a large hot tub by means of heat exchangers.

This guest facility is designed for optimal year-round living and recreation. A northside earth berm protects the caretaker's apartment. The guest rooms have quadruple glazing plus exterior, rolling shutters operated from the inside. A pressurized, circulating, heat-exchanger fireplace provides heat with high efficiency during the winter and can be effectively used for nocturnal cooling during the summer.

Temperature swings are rapid, both in winter and between the summer day and night in this high-altitude climate. Full-sized, screened, louvered doors on the north and west ground level and louvered ventilators on the other elevations allow nighttime and cooler daytime air to enter inductively and be exhausted either naturally or mechanically from the ridge line. In the winter, warm, stratified air near the ridge point is brought down to the lowest levels and recirculated to make use of the hot-air strata in the high, central, clerestory area. The living-area glass, with interior, thermal shutters that roll down, extends from the floor to the cornice line for optimal views and winter passive solar gains. The high earth berms that surround the site are well landscaped by the owner and provide a protective winter microclimate and a cool surround during the warm and hot months.

The exterior and portions of the interior of the house have stained, vertical, resawn-cedar siding. Other portions vary from reflective light colors to white to give the interior a daylit brilliance. The panoramic mountain and river views are spectacular from the south patio and northside balcony decks. Visual enhancement and accommodation are provided by the fully glazed areas, the overhead canopy, and the internal luminosity, which reduces visual contrast to the scenic views.

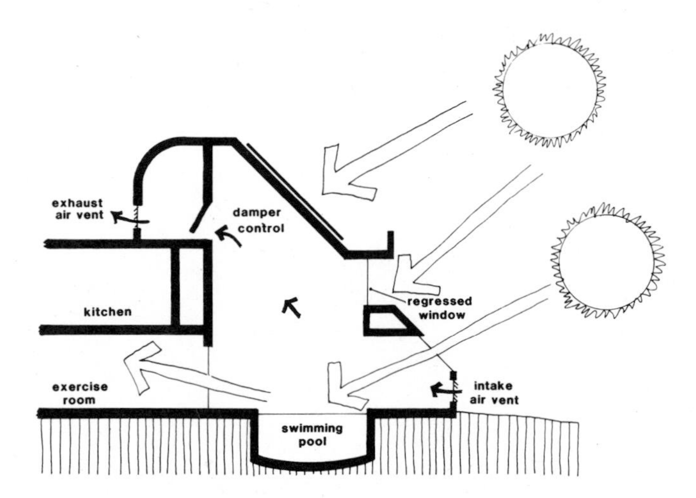

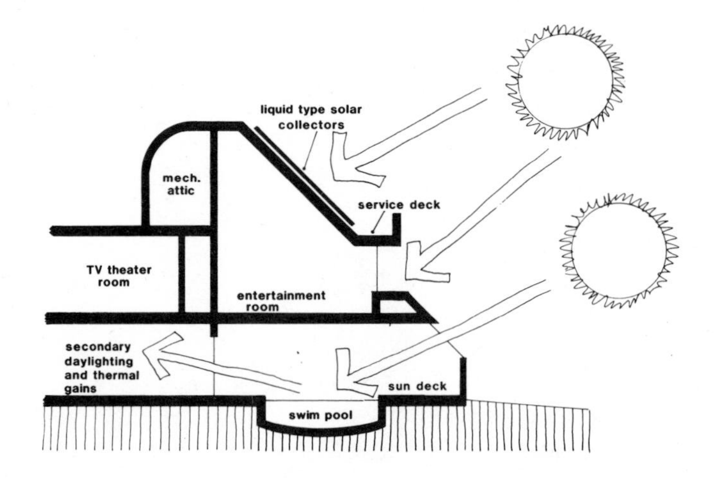

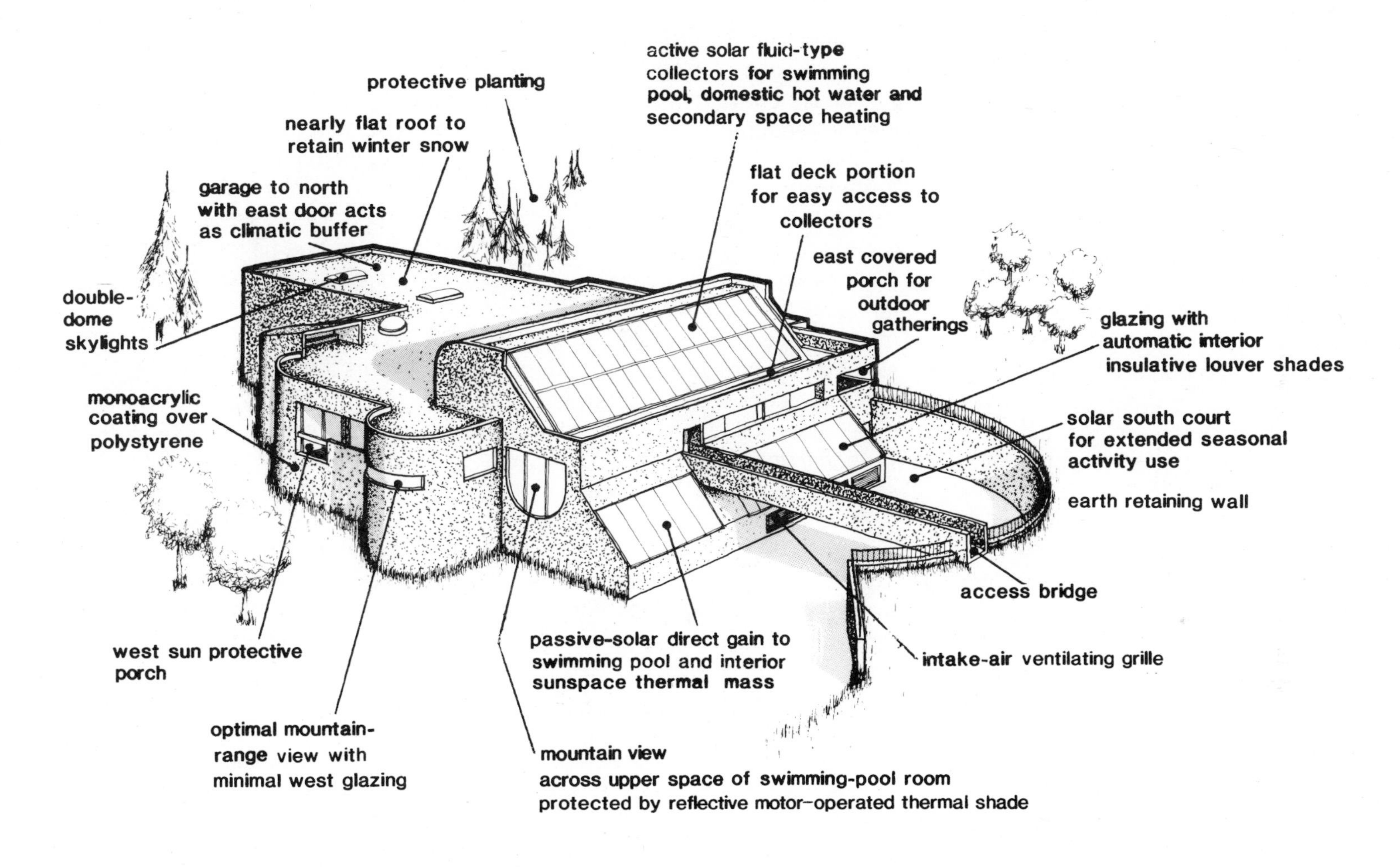

A prime energy question in the design of this solar residence was how to pack a 15-meter (50-foot) swimming pool, a gymnasium, a racquet-ball court, a home workshop, a television theater room, a recreation room with a pool table, an entertainment room, and space for the kitchen, dining room, and bedrooms into an interior volume with a min-

data

SPORTS-ORIENTED SUBURBAN RESIDENCE

Floor area	920 square meters (9,900 square feet)
Passive-solar subsystems	2
Energy centralization	no
Solar-collector area	59 square meters (640 square feet)
Solar-collector type	liquid (water plus glycol)
Solar-collector supplier	Sunworks
Heat-storage type	7,571-liter (2,000-gallon) water tank
Annual passive/active performance	85 percent
Completion date	1981
Architect	Richard L. Crowther

imal exterior surface. The minimal energy form is enclosed in a total of 929 square meters (10,000 square feet). The two-acre-plus site has a tennis court, a small playing field, and berms and planting for microclimatic modification.

An outdoor patio, accessible from the pool, has a cylindrical courtyard spanned by a bridge to the upper level. This lower court faces south to extend seasonal use and augment direct solar gains to the indoor swimming pool. The pool is also heated by an array of roof-mounted, fluid-type solar collectors. A 7,571-liter (2,000-gallon) tank stores heat from the 59 square meters (640 square feet) of collectors. Heat exchangers convert the solar gains to heat domestic water.

The earth-coupled lower level has concrete walls and floors. The main floor and ceiling system are of prestressed-concrete core slabs. Insulative, lower-level ceilings act as a partial barrier to the direct transfer of lower-level heat to upper-level space. The swimming pool acts as a heat sink for reverse-cycle, water-to-air heat pumps zoned for interior space heating.

Secondary daylighting and solar gains reach the lower-level gymnasium through a fully glazed, interior wall. Artificial lighting is minimized. Daylighting is optimized with minimal thermal-energy loss. Dark-bronze vertical blinds, transparent for living areas and opaque for bedrooms, allow daylight control. Curved wall spaces have flat-panel, bypassing, thermal, decorative draperies.

A garage acts as a northside thermal buffer. The upper level has no north windows. Southwest windows are few and carefully treated as narrow, horizontal bands to capture the panoramic mountain view while limiting solar intrusion and optimizing visual accommodation. All interior space is arranged for the most effective balance among functions, seasonal energy flows, earth coupling, solar gains, and changing atmospheric temperatures.

The sculptural forms of the internal and external architecture and the modified landforms evolved from the building's interface with and response to purpose and environment. The color of the external, synthetic coating matches the earth of the site. Interior colors were carefully selected for optimized reflectivity or modulation to accommodate vision. Interior design items and treatments are coordinated with the energy and bio-functional needs of the building and its occupants.

A southwestern architectural flavor is expressed in the exterior and interior of this large, two-level, solar residence, constructed on a large, flat site west of Denver. Rough-sawn timbers preserve the southwestern idiom. Extensive, landscaped earth berms surround the house to moderate climatic extremes. A resulting high front embankment has a wood bridge spanning to the house. Heavy beams shade the front deck used for gatherings and family outdoor living.

A high-ceilinged, clerestory gallery daylights the upper level and centralizes solar and internal energies. A stuccoed masonry mass wall and a dark tile floor over a concrete slab store direct and indirect solar thermal gains. A horizontal duct along the gallery ceiling destratifies the air with a thermocontrolled fan. Roof-mounted, liquid-type solar collectors heat domestic water. Deeply framed walls, stuccoed on the exterior, shade the regressed glass in summer but not in winter. Limited glazing areas in the bedrooms reduce solar impacts. Screened, insulated vents, operable windows, and clerestory exhaust provide nocturnal and moderate-temperature daytime cooling and ventilation.

The lower level is earth-sheltered on the north, east, and west. This earth-coupling of the lower level and fireplace provides a summer-cooling and winter-stabilizing thermal mass. A northside, connected garage has east-facing doors, visually shielded from the approach view of the house by a landscaped them.

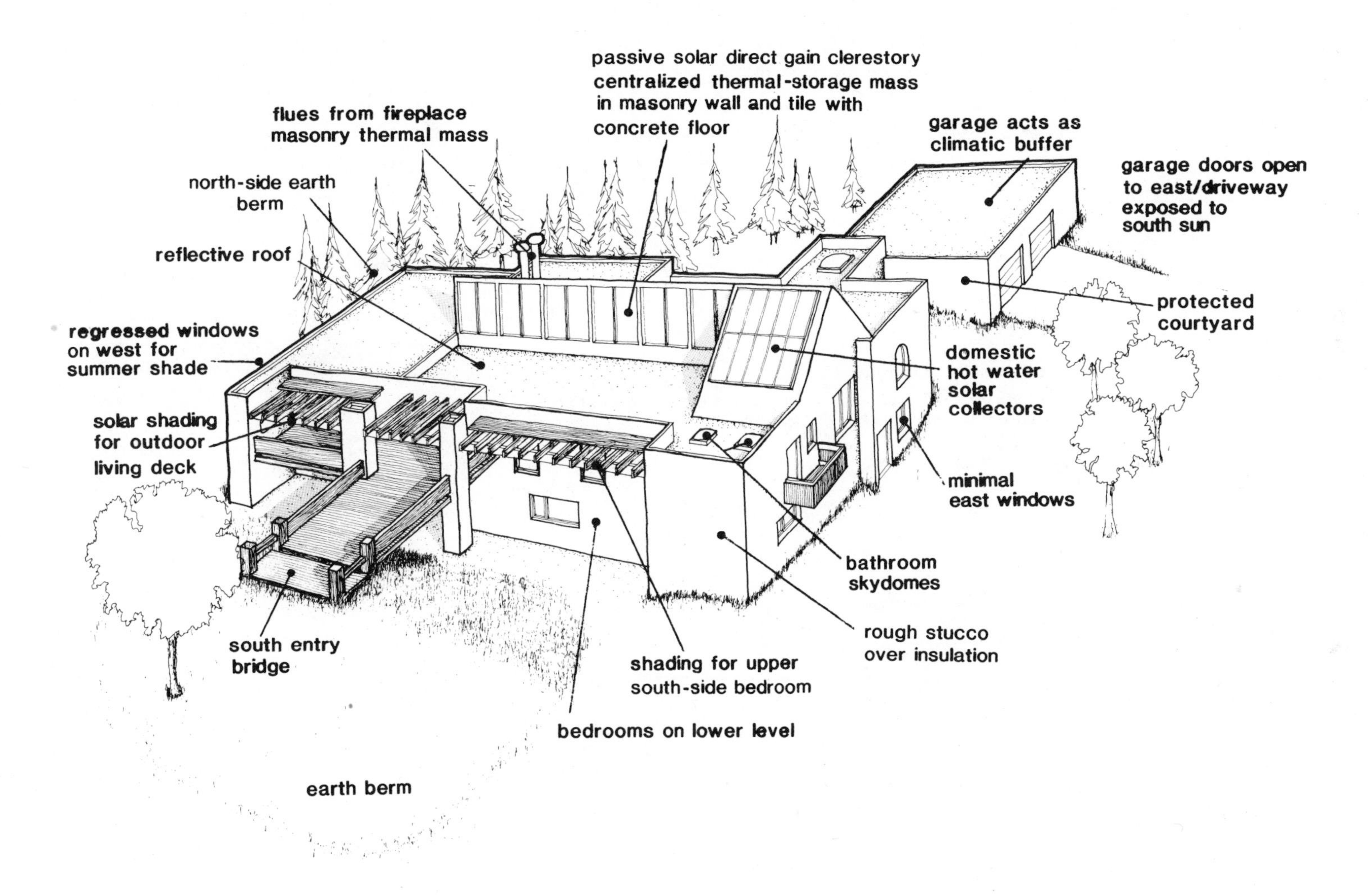

data

RESIDENCE NORTHWEST OF DENVER

Floor area	390 square meters (4,200 square feet)
Passive-solar subsystems	2
Energy centralization	yes
Heat-storage type	thermal inertia of floors and walls
Annual passive/hybrid performance	75 percent
Completion date	1978
Architect	Richard L. Crowther

This low-cost, 156-square-meter (1,680-square-foot), one-story house is notable for its simplicity and practical use of passive solar energy. The peripheral, concrete-foundation retaining wall to the north, east, and west supports earth berms and acts, in cold weather, as a frostline wall. It is insulated from the earth with Styrofoam and finished on the interior with dark tile. The floor is a thickened, insulated, tile-covered, concrete slab on grade. These surfaces and the thermal mass receive direct solar gains.

South-facing, door-height, double glazing is set into the floor slab to optimize winter solar gains. Living, dining, and greenhouse spaces occupy the entire width of the house. A raised roof section accommodates liquid-type solar collectors for domestic water heating.

Fan-assisted destratification is designed to equalize floor- and ceiling-level temperatures. A landscaped earth berm in front ensures a reasonable degree of privacy.

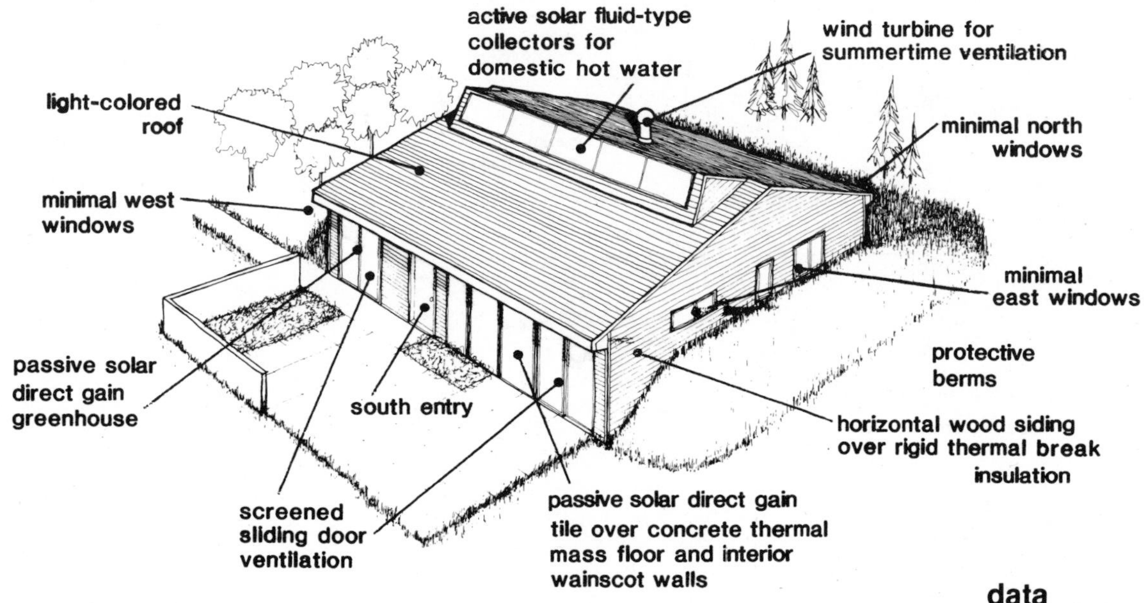

data

URBAN LOW-COST RESIDENCE

Floor area	156 square meters (1,680 square feet)
Passive-solar subsystems	2
Energy centralization	yes
Heat-storage type	thermal inertia of floors and walls
Annual passive performance	75 percent
Completion date	1980
Architect	Richard L. Crowther

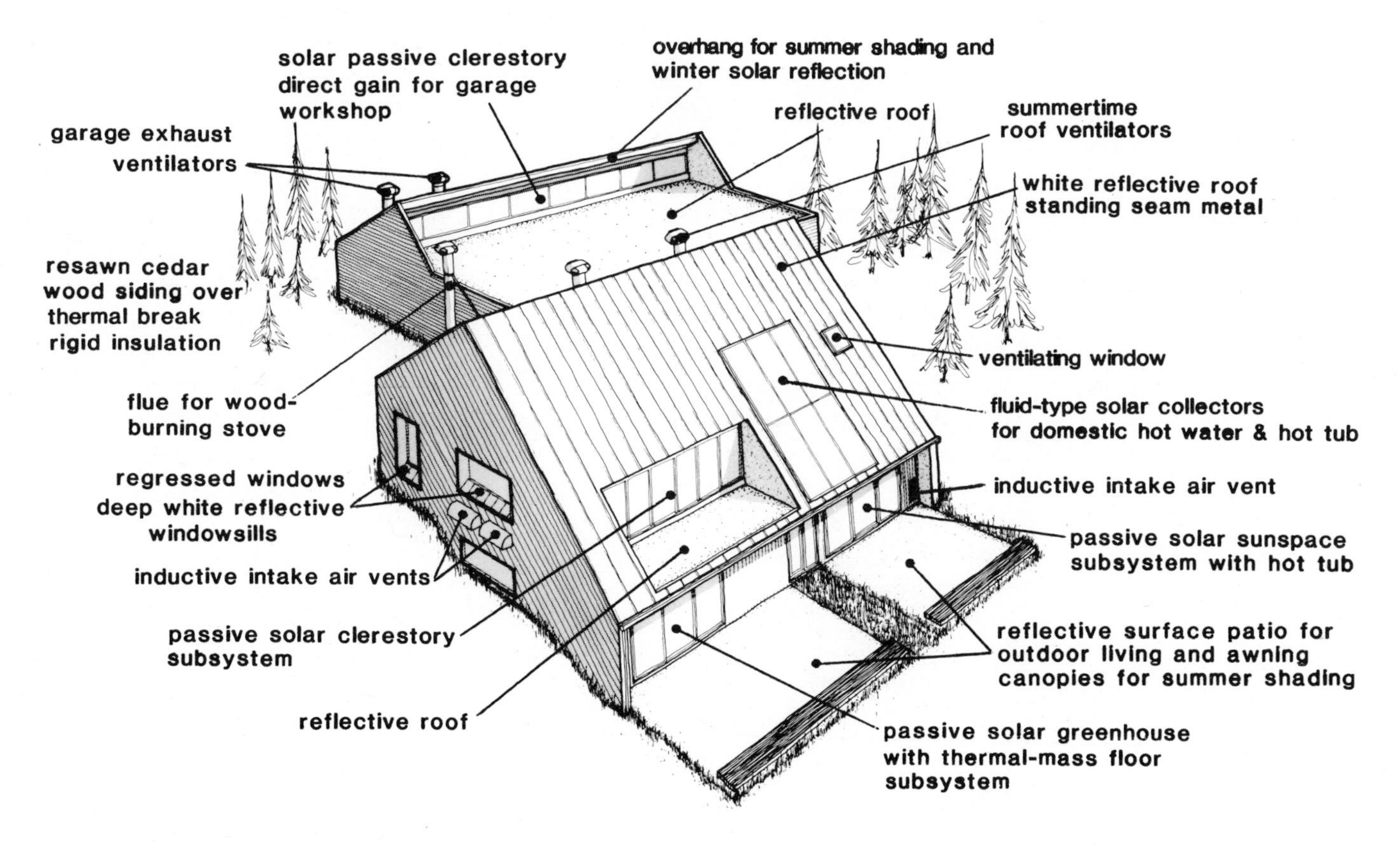

Located in an urban, wooded setting near Denver, this solar home realizes a high level of energy efficiency at moderate cost. The lower level is earth sheltered. A two-story-high, northside, double-envelope, masonry, thermal-mass wall is insulated on the exterior. The lowest inner space of the double envelope is filled with graded gravel. Solar gains from the southside solar subsystems rise by natural convection into an attic space where a ducted fan

data

ISOLATED WOODED-LOCALE RESIDENCE

Floor area	288 square meters (3,100 square feet)
Passive-solar subsystems	4
Energy centralization	yes
Heat-storage type	thermal inertia of floors and walls
Annual passive/hybrid performance	85 percent (estimated)
Completion date	1982
Architect	Richard L. Crowther

forces this heated air into the masonry cavity. Interior relief vents through the masonry wall distribute residual heated air to the lowest level.

A lower-level sunspace the continuous width of the house acts as a passive solar collector and climatic buffer. A line of southside double glazing with sliding, glass access doors is matched by a parallel inner line of glazing that defines the sunspace. Vertical clerestory windows, set in the south roof, add passive solar gains, with the primary and secondary gains from the sunspace and attic, to the northside thermal mass. Heat is also captured in the thermal-mass walls. Concrete walls, insulated on the exterior, and a concrete slab-on-grade floor add to the lower-level thermal mass. A wood-burning stove, mounted on a thick, semicircular masonry hearth immediately in front of the north envelope wall, provides supplemental heat as needed.

The airlock connects a three-car garage and a workshop heated and daylighted with a clerestory. The house is well daylighted by the sunspace, clerestory, open-loft study, and recessed windows. Secondary solar gains and daylighting are an important design element. This house has a low exterior-surface-to-interior-volume ratio. The roof is white, ribbed metal to reflect solar radiation back to the sky. Exterior sidewalls are resawn cedar.

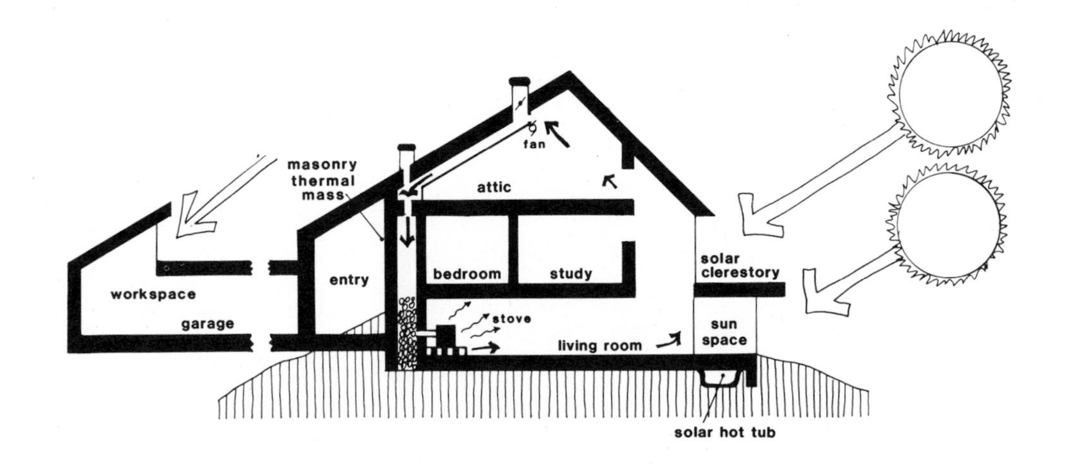

This spacious Kansas home, located on a very large site, is oriented due south. Its plan has angular portions to the west and southeast, decreasing glass reflectivity in the morning and afternoon and thereby increasing direct gains to the thermal mass of dark-tiled, concrete floors. Sunspaces are geometrically created by the plan and include the indoor spa, recreation, and living areas. These solar subsystems act as thermal buffers with inner, sliding glass doors that accept secondary thermal gains to the interior and provide a barrier to control convective solar thermal gains. Solar fluid-type collectors heat domestic water.

The series of passive solar subsystems, each specifically tailored to an activity area, contribute thermal energy to a central corridor space. This central corridor has direct gain, lateral clerestory windows that concentrate reflective roof and overhead canopy energy upon an interior, parabolic surface that reflects to a lateral, black air duct. A photovoltaic-powered, motor-driven fan distributes the accumulated solar-heated air through ductwork manifolded to the open cores of the prestressed-concrete floor system. Discharge is through the core slabs acting as a radiant surface to the basement, with concrete-mass, perimeter-insulated foundation walls, and as air distributors for upper-level living spaces.

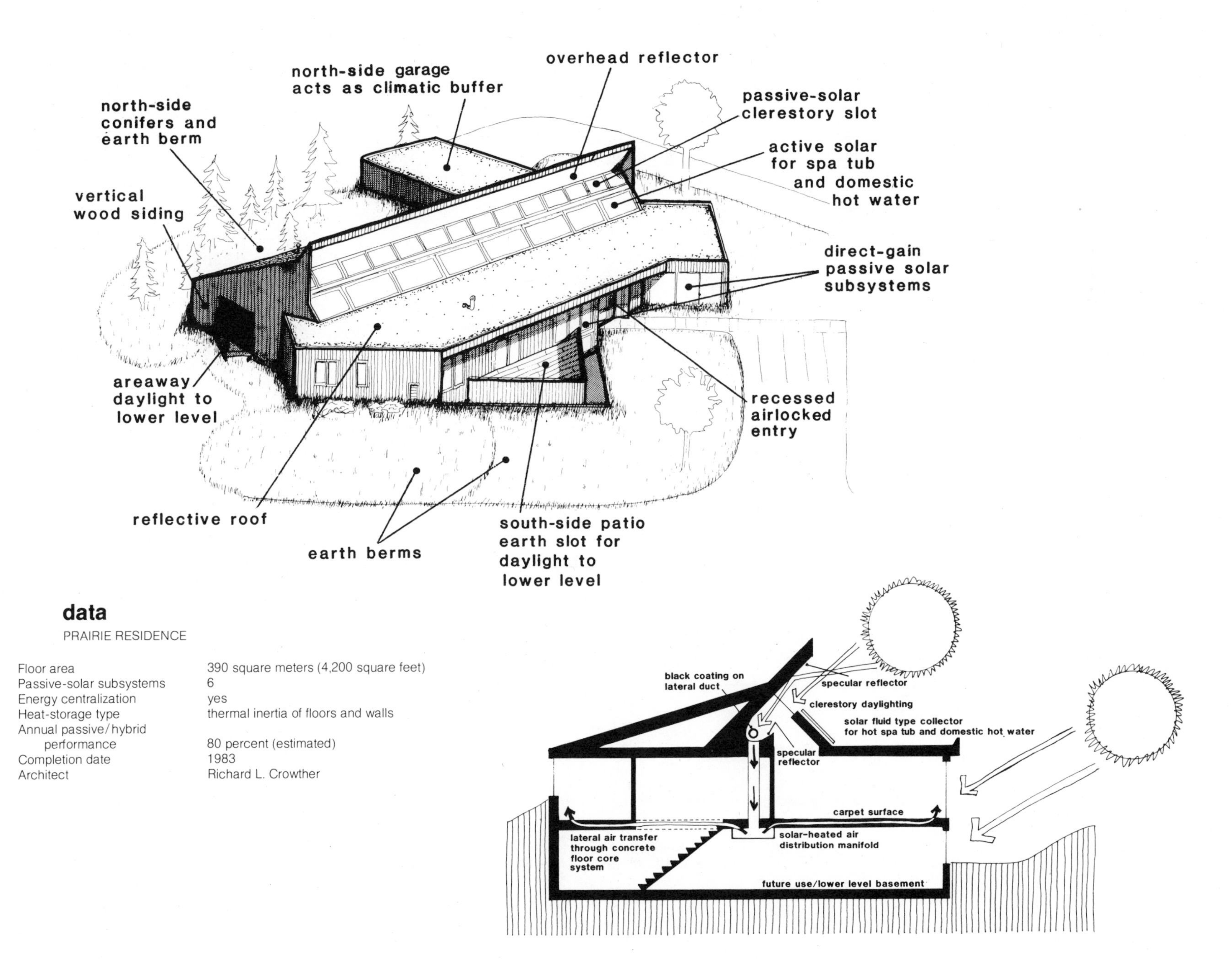

data

PRAIRIE RESIDENCE

Floor area	390 square meters (4,200 square feet)
Passive-solar subsystems	6
Energy centralization	yes
Heat-storage type	thermal inertia of floors and walls
Annual passive/hybrid performance	80 percent (estimated)
Completion date	1983
Architect	Richard L. Crowther

All of the solar subsystems provide effective seasonal and diurnal daylighting. A moat-type light well projects as a triangular form following the line of a grade-level patio. A west-facing, triangular light well adds to the daylighting of the large basement. Secondary daylighting is designed to illuminate the interior space of the den and a hall to the bedroom area. The orientation of interior space is benefited by a breakfast area that receives early-morning light. Living spaces face south, and recreation, spa, and outdoor-patio areas face east to avoid hot, late-in-the-day summer sun.

The attached, northside garage acts as a wintertime climatic buffer. East-facing garage doors avoid the north exposure. Earth berms and landscaping are judiciously located to attenuate cold north and northwest winter winds.

The mountainous region near Gunnison, Colorado, forms a beautiful setting for this 232-square-meter (2,500-square-foot) residence. The house is supported on wooden foundations that were chemically treated with chromated arsenate for permanence. Wall and foundation construction of 15-centimeter (6-inch) wood studs provides space for 15 centimeters (6 inches) of fiberglass batt insulation (R-19) to be continuous from the wooden foundation plate straight to the roof line.

The 55.7-square-meter (600-square-foot) air-type solar collector, supplied by Solaron Corporation of Denver, provides space heating and domestic water preheating. Heat is stored for distribution in a well-insulated bin filled with 13.6 metric tons (15 tons) of gravel. A double-dampered fireplace and an auxiliary system of radiant electric baseboard heaters supply the remainder of the heating needs. The fireplace is equipped with a system of low-level cool air-intake vents, ducts running next to the firebox in which the air is heated, and a hot-air outlet that returns the air to the room. This system salvages some of the heat that would normally be lost up the flue.

The solar-collection system in conjunction with optimized-energy conservation features supplies this house with 85 percent of its annual space heating and a large percentage of its hot water. Cooling is accomplished by natural ventilation.

data

GUNNISON RESIDENCE

Floor area	232 square meters (2,500 square feet)
Passive-solar subsystems	2
Energy centralization	yes
Solar-collector area	55.7 square meters (600 square feet)
Solar-collector type	air
Solar-collector supplier	Solaron Corporation
Heat-storage type	13.6 metric tons (15 tons) of gravel
Annual passive/active performance	85 percent
Completion date	1976
Architect	Richard L. Crowther

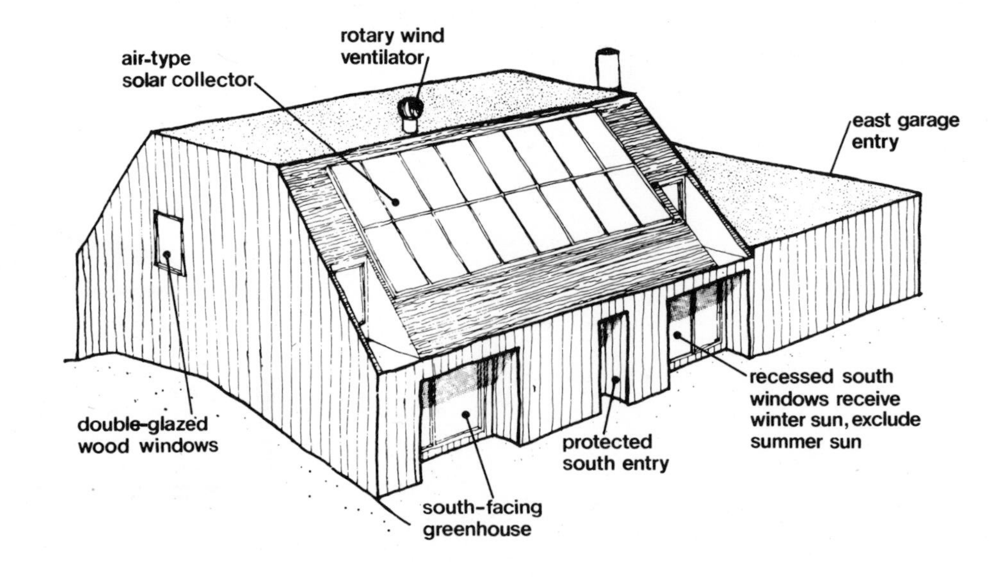

This mountain park residence has a commanding view of Denver to the east. It is designed with a daylighted, two-level gallery for the display of East Indian artifacts. The important views require extensive glass to the east. External, rolling shutters reduce cold-weather heat losses. Reflective blinds control early morning sunlight.

The forested area and the high rise in terrain to the south preclude adequate wintertime solar exposure. The roof, in consequence, is designed with exterior, reflective bays that heat containers of eutectic salts with direct solar radiation. The bays also provide daylighting to the upper-level living areas where nighttime illumination is largely indirect, by argon, mercury-vapor fluorescent tubes. A lateral heat-pipe and direct water-line solar collector are located within the bays.

A hybrid destratification system, thermostatically controlled, distributes direct and stored solar heat to the lower level of the house. The lower-level recreation room has a passive solar water wall. The garage acts as a northside climatic buffer and protects the airlocked main entry to the west. This vestibule receives direct solar radiation. The house is superinsulated and finished with a sprayed-on monoacrylic coating over polystyrene.

A subsurface atomic fallout shelter is served by batteries powered by photovoltaic cells. These cells also power a thermal destratification fan that terminates its distribution with a filter in the fallout shelter.

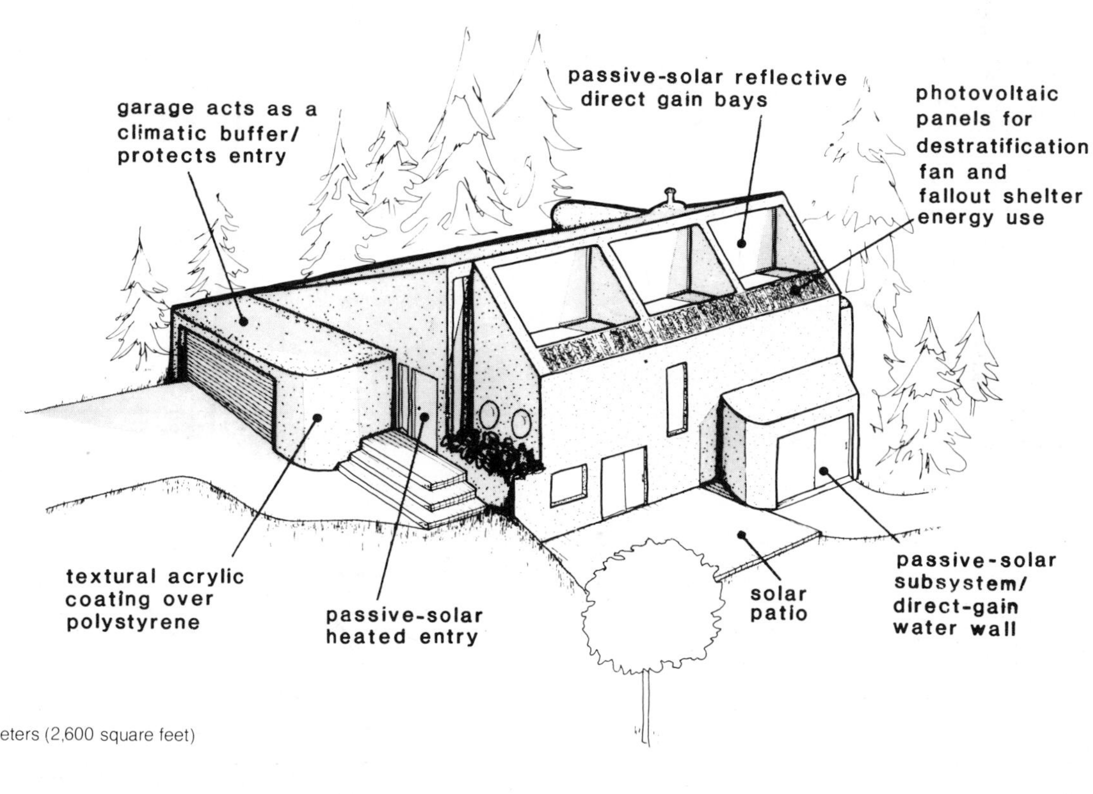

data

MOUNTAIN PARK RESIDENCE

Floor area	241.5 square meters (2,600 square feet)
Passive-solar subsystems	3
Energy centralization	yes
Solar photovoltaic cells	(future)
Heat-storage type	thermal inertia of floors and walls
Annual passive/hybrid performance	70 percent (estimated)
Completion date	1983
Architect	Richard L. Crowther

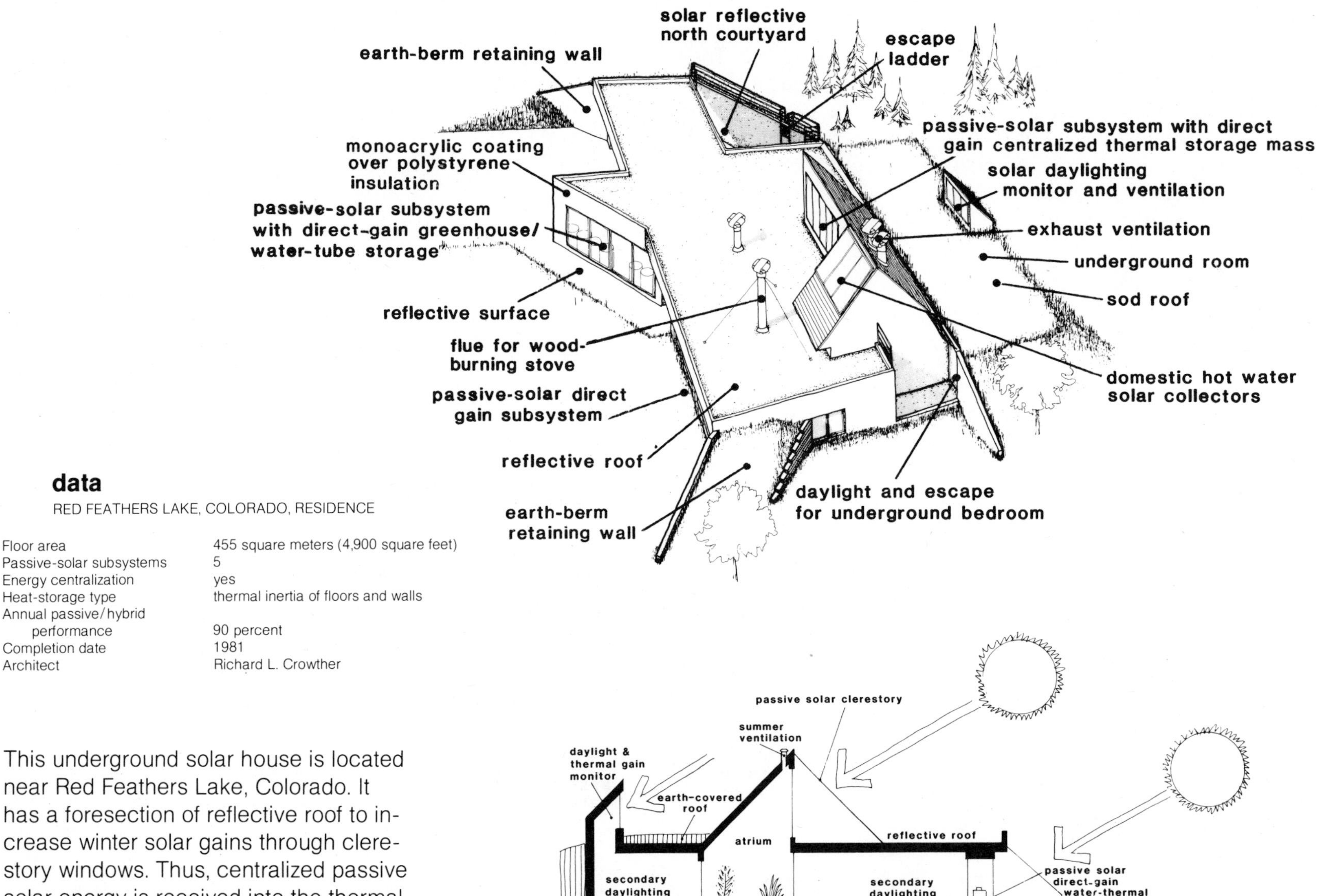

data

RED FEATHERS LAKE, COLORADO, RESIDENCE

Floor area	455 square meters (4,900 square feet)
Passive-solar subsystems	5
Energy centralization	yes
Heat-storage type	thermal inertia of floors and walls
Annual passive/hybrid performance	90 percent
Completion date	1981
Architect	Richard L. Crowther

This underground solar house is located near Red Feathers Lake, Colorado. It has a foresection of reflective roof to increase winter solar gains through clerestory windows. Thus, centralized passive solar energy is received into the thermal mass of the house. A triangular greenhouse projects to the south. Open-to-the sky courtyard atriums provide light, air, and an escape route from the bedrooms.

Exterior walls and interior bearing walls are of poured concrete, as are the slab-on-grade floors. The rear portion of the roof is earth covered. All exposed, concrete walls are covered with polystyrene insulation, nylon netting, and Settef—a synthetic weather coating.

The large thermal mass of the house, insulated with Styrofoam, is in earth contact at the level of the concrete footings and floor. A wood stove has been installed for cool-weather heating. Direct passive solar gains are stored in 208-liter (55-gallon) drums filled with water. This thermal storage is hidden from interior view by a cabinet partition with air slots that allow convective air from the drums to slowly pervade the interior space.

This 232-square-meter (2,500-square-foot) residence with a mountain setting is located in a pine-forest clearing northwest of Golden, Colorado. It is designed to have optimal views in the mountain forest, yet it has a total glass area of less than 10 percent of the total floor area. All glass is double-pane insulating glass set in wooden frames.

The exterior siding is natural, exposed wood. The house is set partially into the mountainside to reduce energy losses. The four bedrooms are located on the lower level, with the primary living areas on the upper level. This allows for sleeping areas to be cooler than living areas. The plan arrangement provides interior buffering elements on the north side to protect from prevailing winter winds. Ceiling heights, room volumes, and openings are designed to accumulate and combine internal energy with solar energy.

The air-type solar collector was supplied

data

MOUNTAIN RESIDENCE NEAR GOLDEN, COLORADO

Floor area	232 square meters (2,500 square feet)
Passive-solar subsystems	2
Energy centralization	yes
Solar-collector area	34.4 square meters (370 square feet)
Solar-collector type	air
Solar-collector supplier	Solaron Corporation
Heat-storage type	6.35 metric tons (7 tons) of gravel
Annual passive/active performance	95 percent
Completion date	1975
Architect	Richard L. Crowther

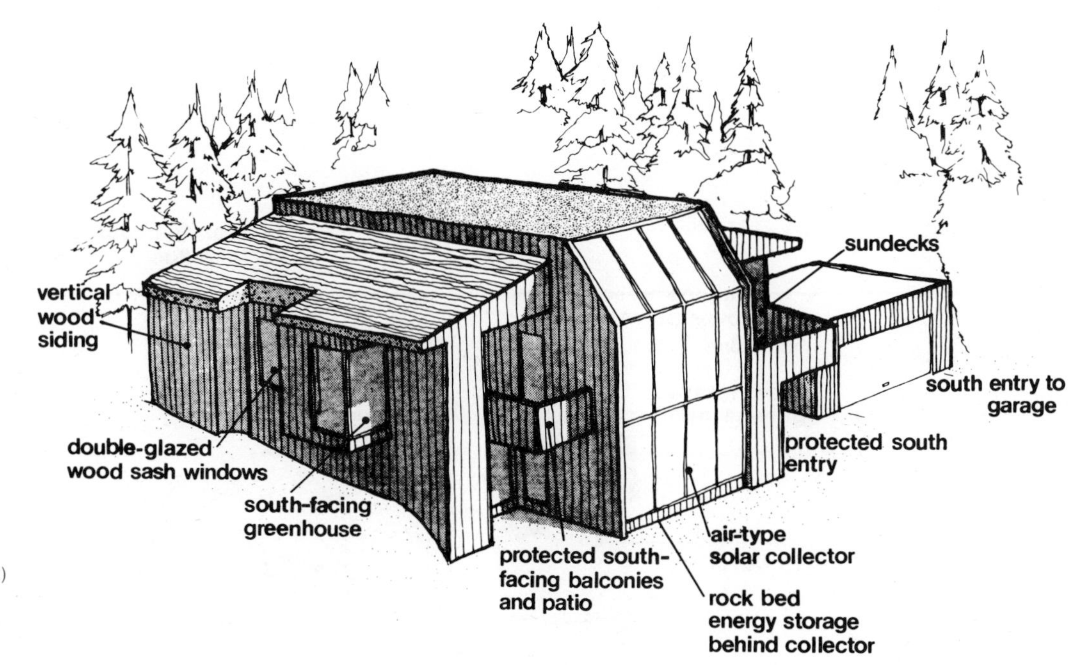

by Solaron Corporation of Denver. It has 34 square meters (370 square feet) of area and supplies heat to seven tons of gravel storage. Part of the collector is tilted and part is vertical. The vertical collector provides about 10 percent less energy during the heating season than one tilted to the optimum heating orientation. The solar heating system is also used to heat domestic hot water. Cooling is provided by natural ventilation.

This two-story, ranch-type residence is located southwest of Denver in the fringes of a pine forest and has a magnificent view of the entire Rocky Mountain front range. The house is oriented due south but still provides a full scope of vision to the natural environment and the mountains. The concrete-tile roof slopes sharply to the north side, and the house is largely buried into a deep berm that protects the entry and garage and acts as an air foil for the entire building envelope.

The garage doors face east, which leaves the driveway open to the south sun, giving easy access and avoiding cold, north winds and snow accumulation. There is one small, triple-glazed window on the north side. The entry is unheated and airlocked.

The south side resolves into a deeper, excavated, circular courtyard that permits direct solar gains to a large, vertically glazed, lower greenhouse. The greenhouse-heated air travels up through the envelope space of outer double glazing and dark, inner heat-absorbing glass, which adds to solar entrapment. The heated air rises upward and across the living-room ceiling to a clerestory area and then circulates down through a large masonry fireplace filled in the lower levels with gravel. Thus, the peripheral solar energies and those received directly through the clerestory (augmented by an overhead reflector and reflective roof deck) are literally captured at the lowest level at the center of the house and thereby can thermally radiate to all spaces. The

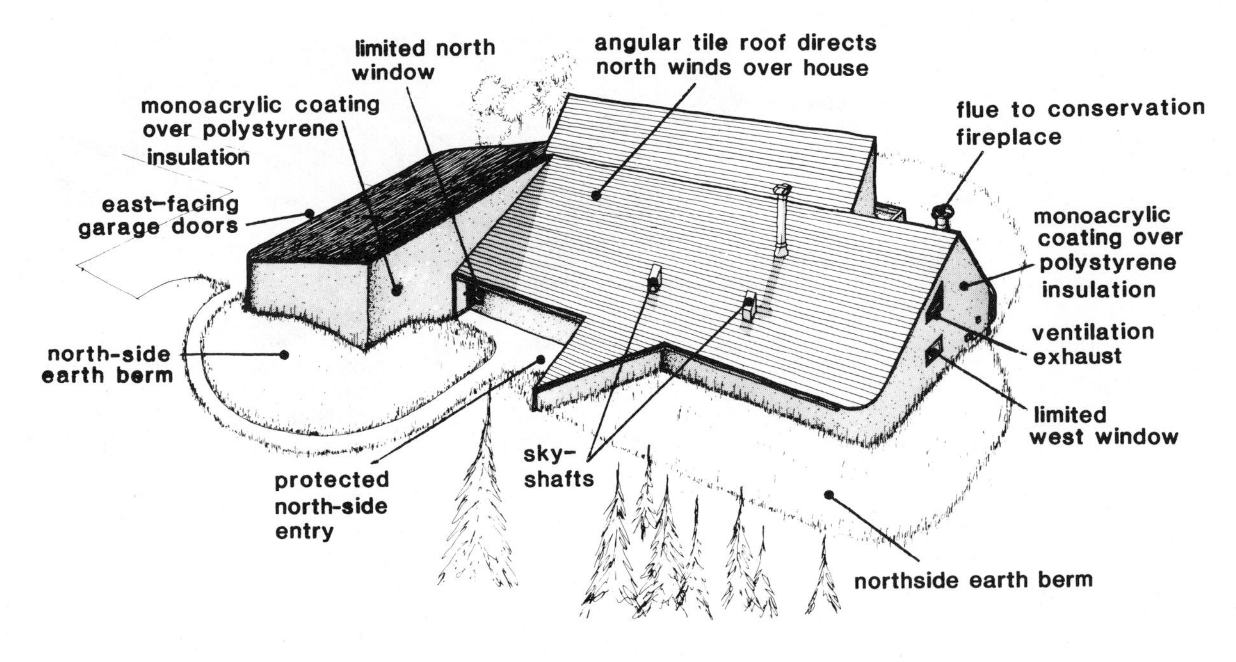

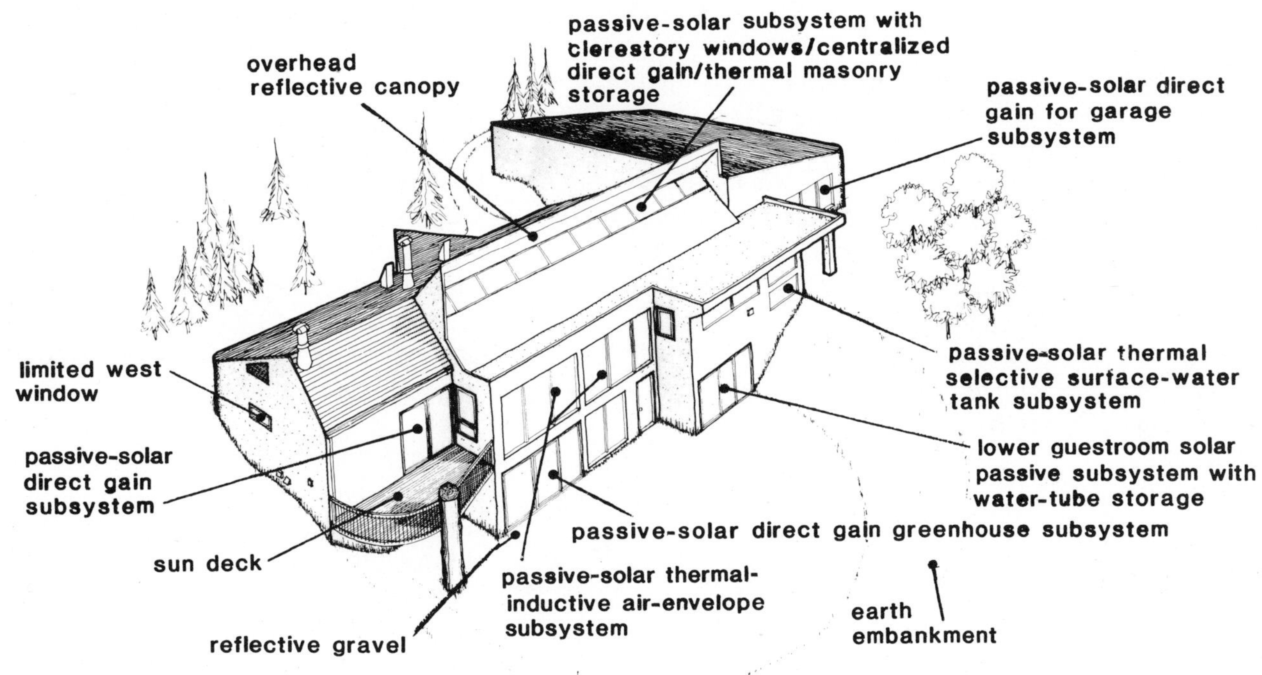

data

HIGH COUNTRY RESIDENCE

Floor area	372 square meters (4,000 square feet)
Passive-solar subsystems	6
Energy centralization	yes
Heat-storage type	thermal inertia of floors and walls
Annual passive/hybrid performance	85 percent (estimated)
Completion date	1982
Architect	Richard L. Crowther

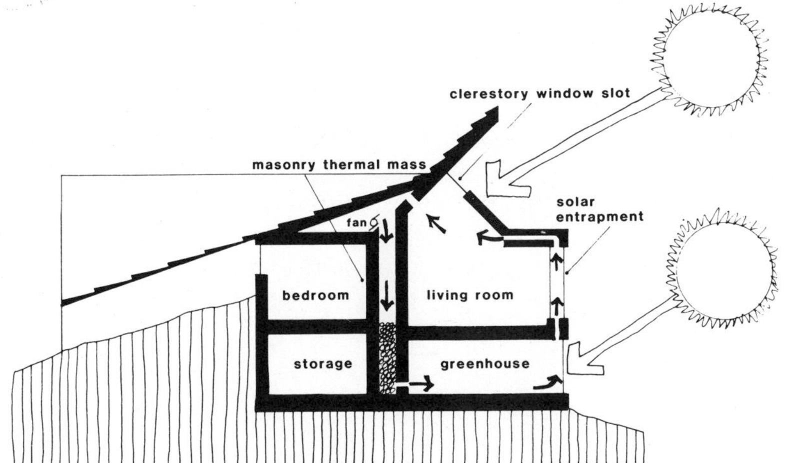

energy-conserving central fireplace draws in outside air for combustion, which it adds to the existent heat of the central mass.

The bedroom wing enjoys an open sun deck and direct passive gains to a thermal tile floor. A lower, guest bedroom also has a direct gain system, which stores solar gains in water-filled tubes. The garage also has a water-drum, direct gain subsystem to provide heat. Thus, the entire south side consists of a series of passive solar subsystems. All of the subsystems are designed for the internal space use. A kitchen overhang provides summer shade but allows the sun's energy to penetrate the selective-surface, thermal water bunker, which looks like a cabinet, in the breakfast-nook area.

This house has a monoacrylic finish over a wraparound of polystyrene insulation that extends into the earth. The greenhouse placed at the lower level also corresponds with the thermodynamics of the heat it produces. During the winter months, its radiant energy rises to the living spaces; in summer, the height (to the ridge line) of the stack action accelerates ventilation. The texture of the house will tend to preserve a thin air film and reduce heat losses to cold winter air. The need for movable insulation is largely avoided by the combination of the lower-level greenhouse and the upper-level envelope acting as a buffer. The band of clerestory windows is narrow to concentrate entering solar energy and avoid the nighttime or daytime cold-weather losses to the sky.The destratification system, which uses thermal-mass storage, prevents the normally higher temperatures at the ridge line, thus reducing reradiant and convective losses to the external climate.

This underground house consists largely of a matrix of concrete perimeter and interior divisional walls. Clerestory windows with a south-facing reflective roof deck light the interior during the day and provide solar gains to a concrete wall with an abstract, reverse-cast design created by the author to maximize the surface absorption of direct solar radiation. The large thermal mass of the building structure stabilizes indoor temperatures year round.

An indoor atrium court providing daylight to interior room spaces can be opened for ventilation and cooling. Living, recreation, and hot-tub spaces receive direct solar gains. Plastic water tubes act as a passive storage system. Domestic water is heated by roof-top, fluid-type solar collectors. A large, northeast garage/workshop is connected to the house and acts as a climatic buffer. A glazed roof monitor provides daylight to the workspace area.

Exposed concrete walls are covered with polystyrene, nylon mesh, and Settef in a finish color closely matched to that of the earth.

The airlocked entry and earth-covered garage face east and are protected from the wind by the surrounding pine forest. The entry is further shielded by the garage and daylighted by a quadruple-glazed skydome. A south-facing deck for warm-season use is suspended above a large, excavated, triangular areaway, with stairway access from under the deck.

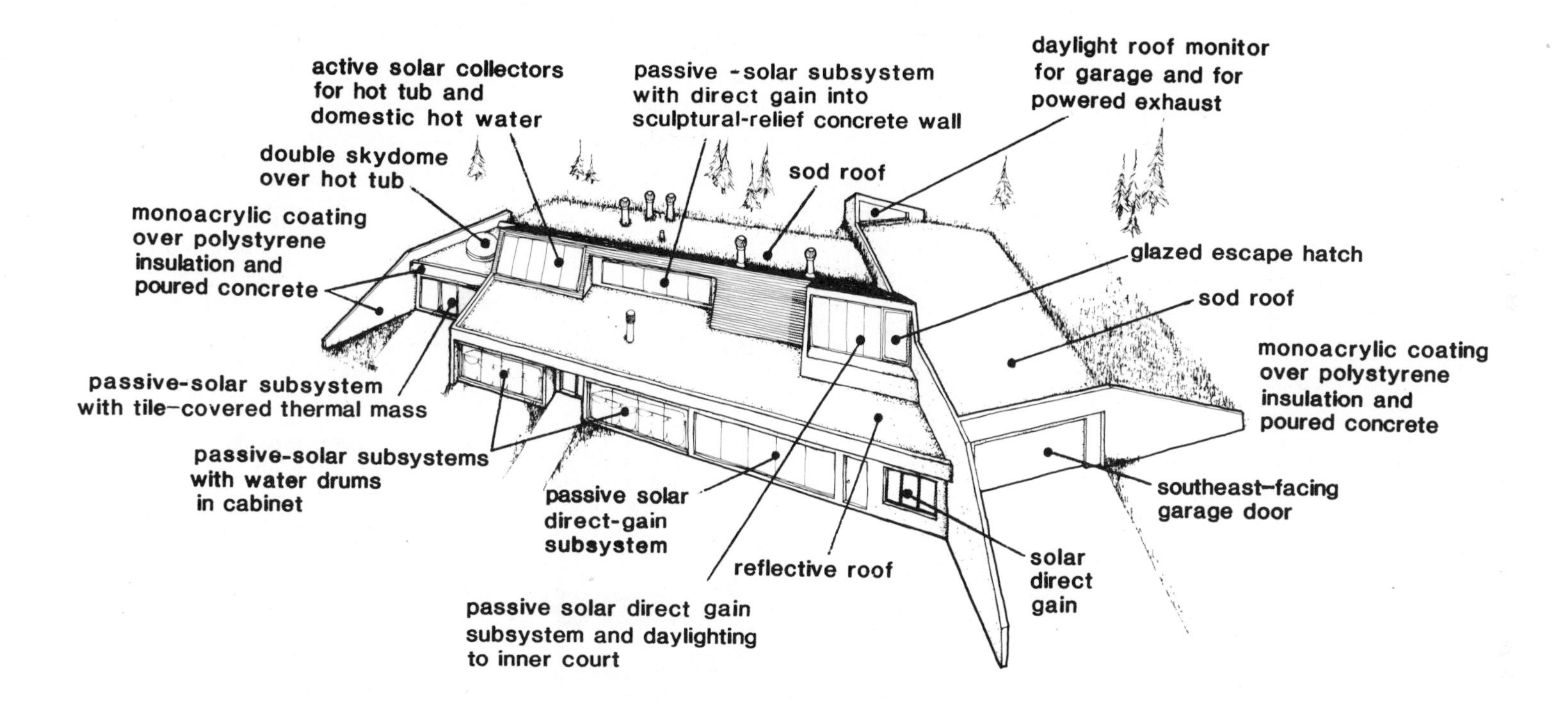

data

KANSAS UNDERGROUND RESIDENCE

Floor area	260 square meters (2,800 square feet)
Passive-solar subsystems	6
Energy centralization	yes
Heat-storage type	thermal inertia of floors and walls
Annual passive/hybrid performance	85 percent
Completion date	1981
Architect	Richard L. Crowther

This triangular, two-level mountain retreat is designed to minimize the surface-to-air ratio and divert winter winds from south-facing windows. The lower-level main floors and north, east, and west foundation and retaining walls are of exterior-insulated concrete. The roof is a superinsulated thermal cover.

A solar-entrapment system allows for a comfortable, uninterrupted view of the complete, spectacular mountain range to the west and south. Solar radiation passing through clear double glazing is absorbed by dark transparent plastic with protective inner glazing. The heat-absorbing plastic does not break as glass may under the thermal stress of shading and direct solar exposure.

The entrapment system, wide enough for internal cleaning, acts as a solar envelope. Heated air rises in the chamber and is displaced by cooier lower-level air. A fan powered by photovoltaic cells captures the heated air at the roof line and returns it to the thermal mass of the lower northside level.

A companion solar system supplies internal illumination and direct wintertime thermal gains. It concentrates sunlight through a canted clerestory window, into a black, large-diameter, lateral duct and then to storage in the gravel plenum of a central masonry mass, which can be heated by a catalytic wood and coal stove. Circular, spiral-type solar collectors on the south reflective roof heat domestic hot water.

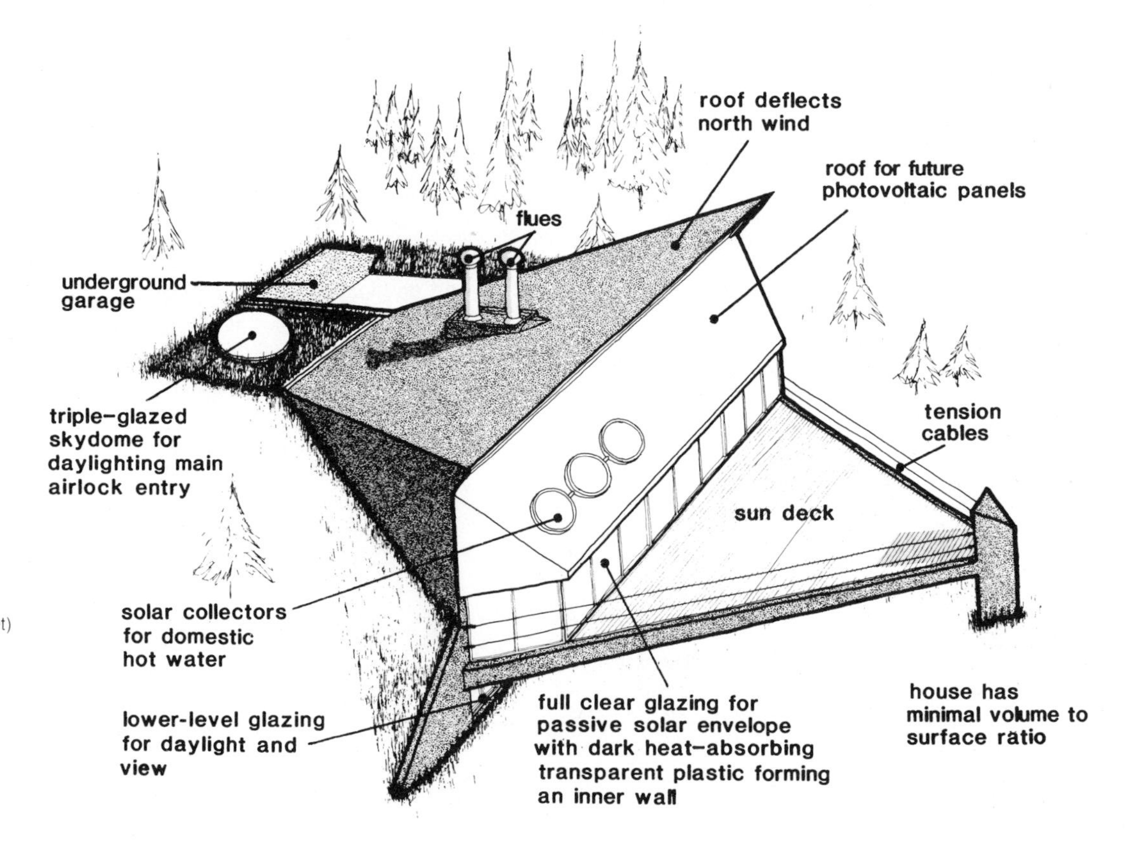

data

MOUNTAIN RETREAT RESIDENCE

Floor area	260 square meters (2,800 square feet)
Passive-solar subsystems	2
Energy centralization	yes
Solar-entrapment envelope	yes
Heat-storage type	13.6 metric tons (15 tons) of gravel
Annual passive/hybrid performance	75 percent (estimated)
Completion date	not yet constructed
Architect	Richard L. Crowther

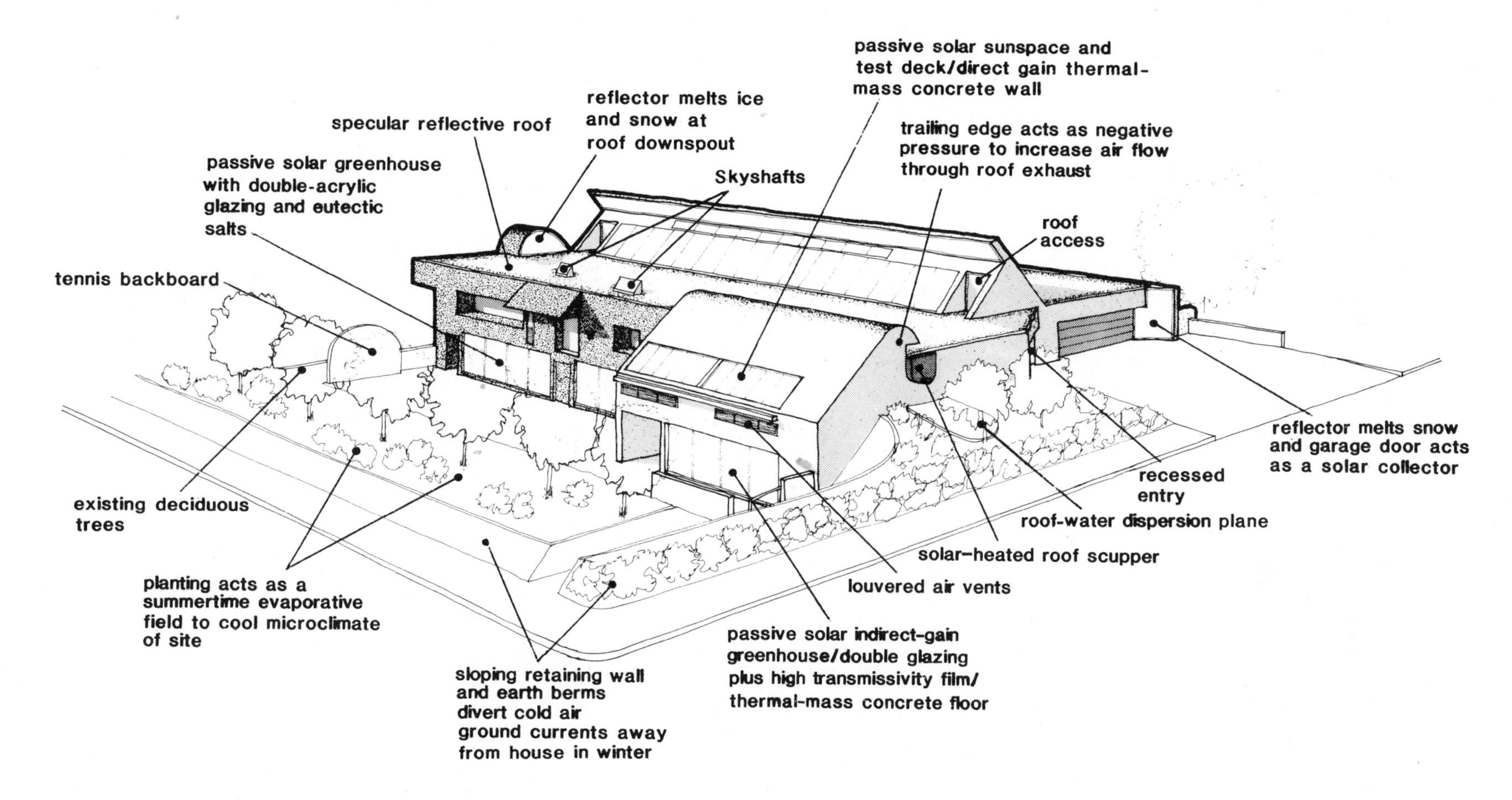

This research-test facility for active- and passive-solar and other natural-energy systems is translated into energy-responsive topographic and architectural form. The microclimate of the site is modified with earth berms, a sunken courtyard, landscaping, and retaining walls to reduce summer and winter climatic extremes. External landforms and

data

ENERGY RESEARCH FACILITY

Floor area	650 square meters (7,000 square feet)
Passive-solar subsystems	10
Energy centralization	yes
Solar-collector area	33.4 square meters (360 square feet)
Solar-collector type	liquid (water plus hydrocarbon fluid)
Solar-collector supplier	Lordan Solar Energy Systems
Heat-storage type	3,785-liter (1,000-gallon) water tank 41,639-liter (11,000-gallon) swimming pool
Annual passive/active/hybrid performance	90 percent
Completion date	1980
Architect	Richard L. Crowther

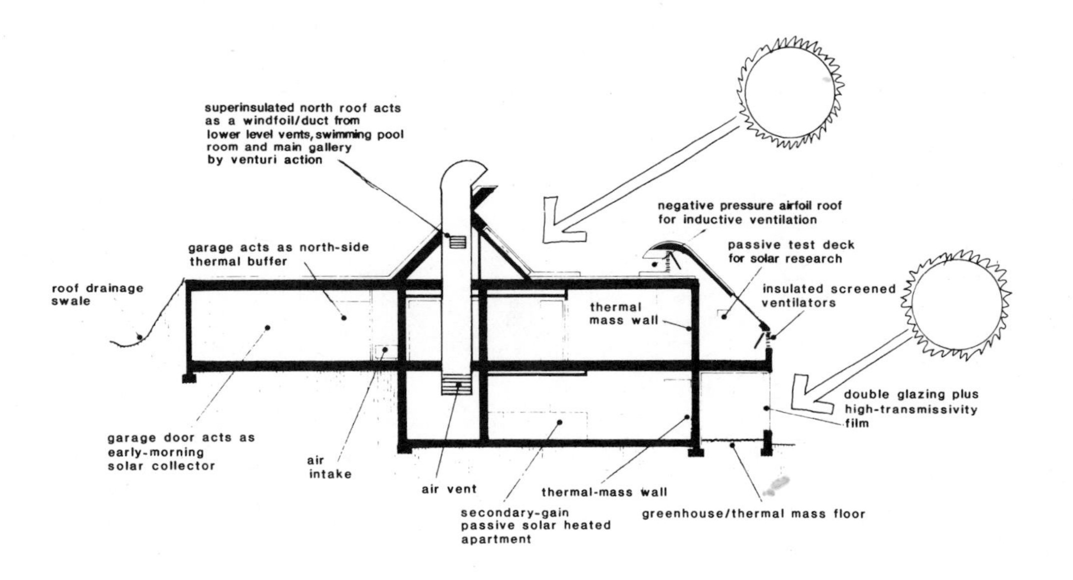

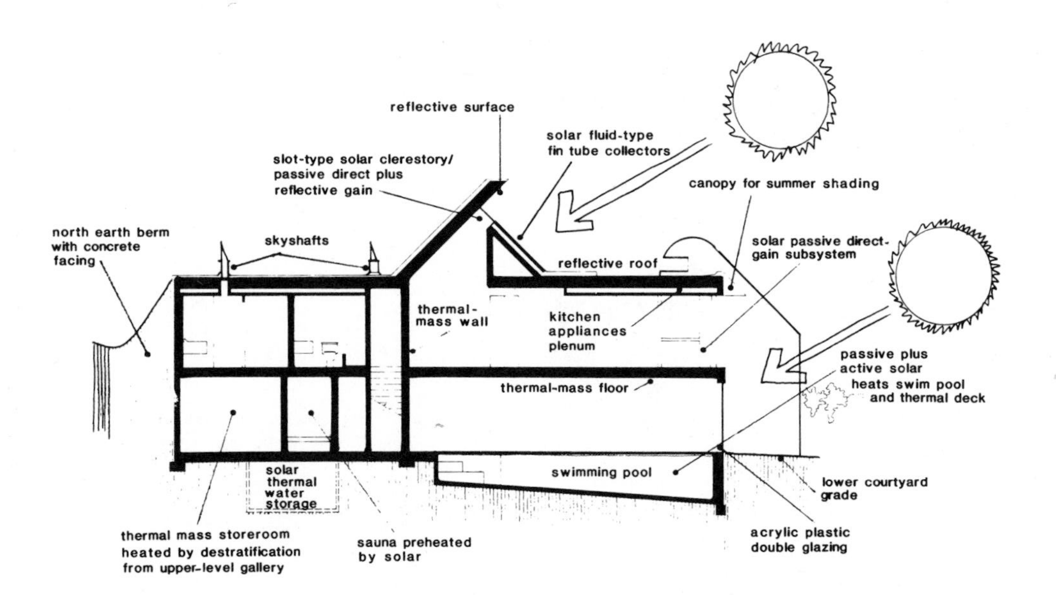

landscape textures direct cold air flows away from the building structure during winter. Deciduous trees to the south allow solar penetration to reflective surfaces in the courtyard, direct and indirect sunlight into two greenhouses, and passive solar gains to the indoor swimming pool.

East, southeast, and west exterior surfaces moderately reflect the sun's energy, while the south exterior building envelope is a very dark color to absorb winter sun. Dark-colored, recessed garage doors receive low-angled winter sun and transmit heat to the interior. A specular (reflective) surface melts snow in front of the garage. A large, east-facing scupper drains the roof and heats up with the winter sun, usually forming a large ice column. A roof-top reflector melts snow and ice through a large, west downspout.

A 30-square-meter (320-square-foot) bank of fluid-type fin-type collectors receives direct sunlight and reflected radiation from the roof and upper angular canopy. These reflective surfaces also concentrate solar radiation through a horizontal slot of clerestory windows. Thus, during the winter, thermal energy is centralized in the interior, high-ceilinged gallery. A dark, solid-concrete wall receives direct passive gains, residual energy, and gains from eight other passive solar subsystems, plus kitchen and utility-room heat. This centralized energy is relocated into thermal mass storage at the lower northside level by means of insulated ductwork and a fan.

The active solar system heats a 3,785-liter (1,000-gallon) insulated water tank for domestic hot water and an 41,639-liter (11,000-gallon) swimming pool through a series of heat exchangers.

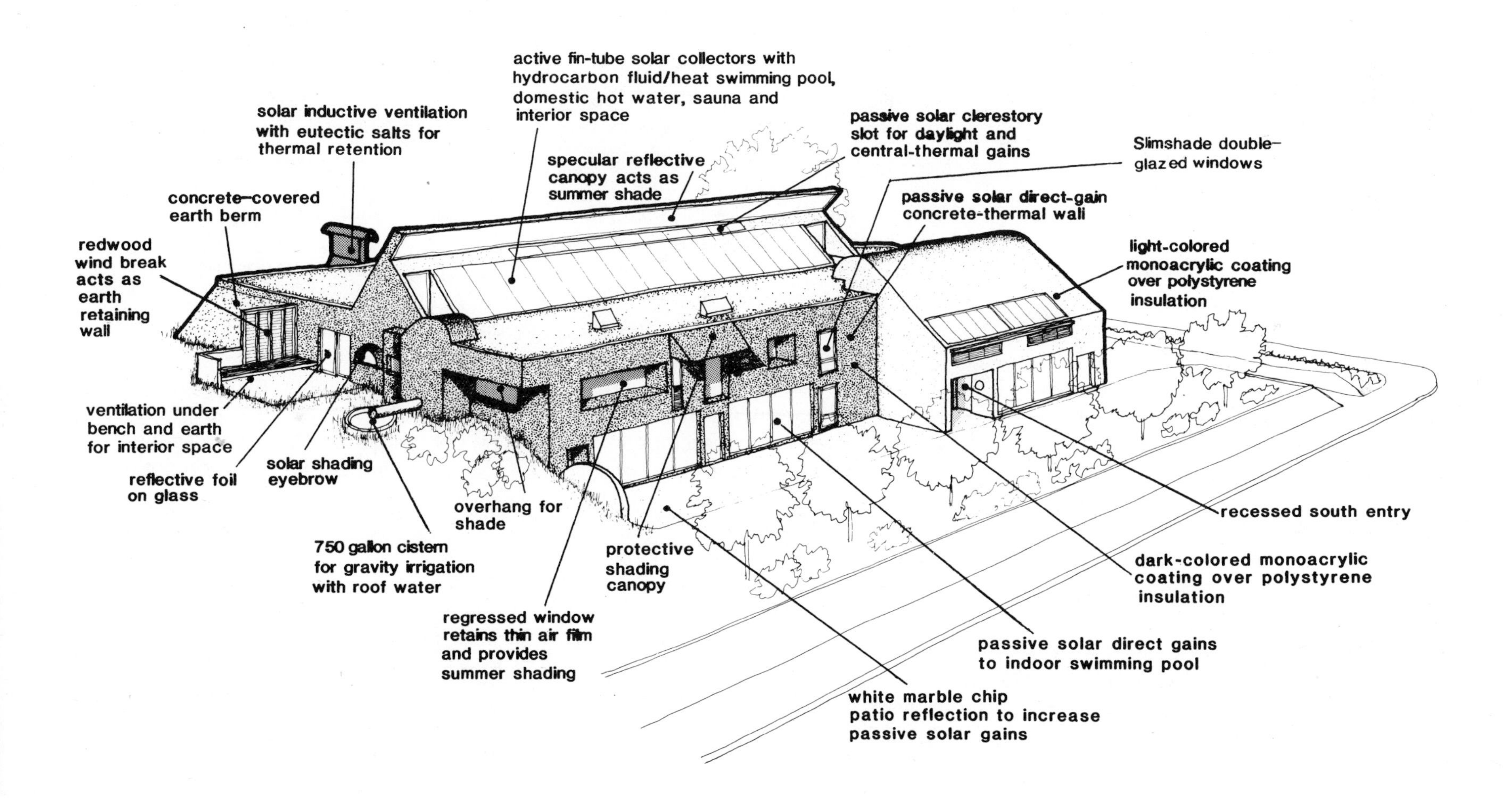

The solar-thermal storage tank and its residual heat preheat the sauna and the shower room above the sauna.

The passive solar-heating subsystems are:

- lower-level greenhouse with double, cellular, clear-acrylic plastic glazing; eutectic salts
- lower-level swimming pool with double, cellular, clear-acrylic plastic glazing
- lower-level greenhouse with double glazing and intrasuspended, high-transmissivity film; dark, concrete floor; eutectic salts
- upper-level, angular, double-glazed, solar test deck
- upper-level, direct gain, dark, solid-concrete thermal wall
- upper-level, direct gain, dark tile-and-concrete floor system; clear acrylic plastic glazing
- upper-level direct gain with horizontal, double glazing; eutectic-salt bin; dark tile and concrete slab
- central, solar-clerestory slot; direct gain, dark, concrete, thermal wall; hybrid subsystem
- upper-level west sunroom with double glazing and reflective, removable foil

The passive cooling and ventilating subsystems are:

- lower-level greenhouse: screened, louvered panel with insulated, magnetic-weatherstripped door and vent-to-roof grille
- lower-level swimming pool: steel-grilled, screened, security door; inner, insulated, magnetic-weatherstripped door for intake air; cross-ventilation intake grille; large, cylindrical exhaust to a hooded roof exhaust that acts as a venturi and trailing edge, negative-pressure ventilator
- upper-level test deck acts as a negative-pressure air foil: large, louvered openings at top and bottom induce optimal ventilation; suspended, specular mesh reflects the hot summer sun while allowing soft secondary daylighting to interior room space
- upper-level awning window with Slimshade louvers in the glazing assemblage adjacent to thermal wall provides both cooling and ventilation
- upper-level, screened ventilation louvers with interior weatherstripped, insulated, shutter panel and canopy
- upper-level southwest room: horizontal, recessed windows extend daytime interior illumination; their dark exterior surround favors visual accommodation (skyshaft acts as a reading light)
- upper-level west room: a deep architectural eyebrow for a semicircular window, double glazed with inner, dark-tinted glass, attenuates the intensity and thermal gains of west sunlight
- upper-level sunroom/climatic-buffer space with reflective glazing and fully glazed inner doors provides secondary daylighting to adjoining rooms
- upper-level northwest room has no windows and borrows light from sunroom
- upper-level bath and utility rooms illuminated by skyshafts

A pressurized, fan-powered, filtered air intake provides nocturnal cooling; roof and other air-intake vents can act as exhaust vents, thereby selectively cooling interior spaces. This system can be regulated to use the lower-level southwest greenhouse as a solar-collection, air-tempering, humidification, and botanic air filtration system.

Daylight pervades interior space with a daily and seasonal balance. Diffuse softer daylight prevails during summer and direct, intense daylight during winter. Since light is heat, thermal gains are minimized in summer and optimized during cold winter months. During spring and fall an agreeable transition between effective daylighting and thermal gains is realized.

Peripheral windows are recessed and provided with exterior canopies, and a limited number have vertical translucent blinds. Outside the lower level, reflective gravel augments low-angled, secondary interior daylighting during winter. The selective reflectivity of interior surfaces is designed for visual accommodation and comfort at the optimal depth of interior penetration.

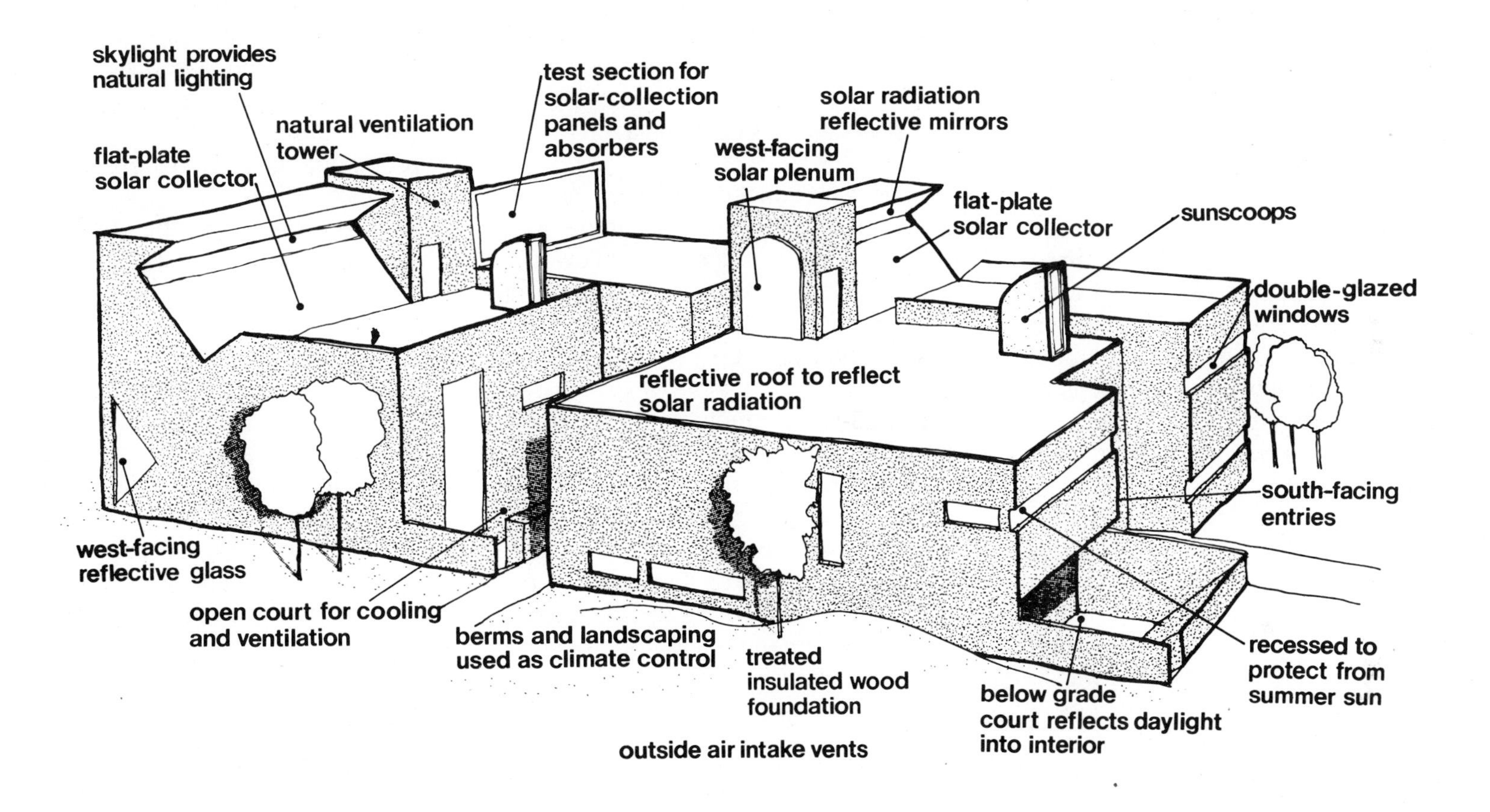

These energy-optimized solar office buildings of unified design are located in the Cherry Creek area of Denver. Each one has 437 square meters (4,700 square feet) of floor area on two levels. The levels are divided by a central foyer to produce space that is suitable for use by one to four tenants.

data

CHERRY CREEK OFFICE BUILDINGS

Floor area	436.6 square meters (4,700 square feet) each
Passive-solar subsystems	6
Energy centralization	yes
Solar-collector area	16.3 square meters (175 square feet), south building 27.9 square meters (300 square feet), north building
Solar-collector type	air
Solar-collector supplier	Solaron Corporation
Heat-storage type	7.26 metric tons (8 tons) of gravel
Annual active/hybrid performance	80 percent
Architect	Richard L. Crowther

Architectural energy-conservation features are similar for both buildings and are described below.

The primary building entry faces south for winter sun exposure and for protection from cold winter winds. The total glass area is limited to less than 10 percent of the total floor area, yet the size, location, and shape of windows provide visual openness and natural illumination. Some windows are recessed so that they receive direct winter solar gains but are excluded from the hot summer sun. West-facing windows are covered with a reflective surface and resist the penetration of 78 percent of incident summer radiation. All windows are double glazed with insulating glass to minimize thermal transfer.

Wooden foundations are treated with chromated arsenate for permanence. Wall-construction type is continuous from the wooden foundation plate to the top of the roof parapet. This system permits uninterrupted insulation to be placed in the walls from below-floor slabs to above the roof line. Heat loss is greatly reduced using this system. Outer walls are 15-centimeter (6-inch) wood studs, with 15 centimeters (6 inches) of fiberglass batt insulation, an outer 12.7-millimeter (½-inch) thick plywood layer, 38 millimeters (1½ inches) of Styrofoam insulation, and Tri-lite exterior finish. Internal 152-micrometer (6-mil) pliofilm provides a continuous vapor barrier and is covered with 12.7 millimeters (½ inch) of interior drywall. The roof structure has 40.6 centimeters (16 inches) of high-density fiberglass batt insulation.

Heat loss is reduced by having the lower level of the building set into grade. Earth berms are used to direct wind flow away from building surfaces, reducing the infiltration of automobile exhaust and dust from a nearby street.

The 16.3-square-meter (175-square-foot) flat-plate air-type solar collector, supplied by Solaron Corporation, is south facing and is tilted at a 45 degree angle to the horizon. A skylight is located on the same plane as the collector, at the collector's upper edge. An upper mirrored overhang reflects winter sun directly into the skylight for interior illumination and interior heat gains to the north side of the building.

Direct light enters the skylight through the winter months, but not during the summer months when the sun is at a high angle.

The roof of the building is surfaced with white marble chips. During the winter when the sun angle is low, diffuse radiation is reflected into the collector, increasing the amount of heat at the absorber. In the summer when the sun angle is high, a large portion of the diffuse radiation is reflected back to the sky. This scheme increases heat gains in the winter and decreases them in the summer. Exterior walls are colored darker than the roof to balance year-round heat gains and losses.

The sun scoop over the entry stairway foyer has a south-facing aperture glazed with double-insulating glass. Inside the scoop behind the glass is a curved surface covered with reflective aluminum. Sunlight enters the glass, reflects from the aluminum, and is directed downward through an opening into the interior space for light and heating of the entryway. A duct with a fan on one end, reaching up inside the sun scoop, draws hot air down and delivers it to a storage bin filled with gravel located beneath the opening in the ceiling. Heat from the bin is radiated and conveyed to the entryway, thermally tempering this area.

Natural cooling is provided by inductive ventilation. Once the warm air reaches the highest point in the building, it enters ducts and is either vented by a wind turbine or is injected into the bottom of the solar collector and ventilates the collector before being exhausted by wind-powered roof turbines. Cool nocturnal or late-afternoon air is drawn into the building through ground-level vents. These vents have weather hoods, prevent direct wind flow, and have insulated panels with gaskets to block infiltration when closed. Heat pumps are provided to remove heat from the interior when the outside temperatures are too high to be used for natural cooling.

Auxiliary heating is also provided by the rotary high-performance heat pumps. One of their primary advantages is that they can be reversed. To cool the building they reject heat through roof-mounted condensing units; to heat the building the roof condensers function as evaporators that extract heat from cold outside air and supply it to the building through interior condensing units.

The building is divided into two heating and cooling zones, the temperatures of which can be controlled independently. One heat pump serves each zone, resulting in efficient energy usage. During heating periods, warm air that rises to the highest point in the building is captured and transported down to the lowest level. Before being reintroduced into the heating system, it is cleaned by fiber, charcoal, and electrostatic filters. This system allows internal building heat to be conserved and recycled.

In addition to the above features, the north building has a greenhouse on the lower level and a negative-ion water fountain. The plants in the greenhouse absorb carbon dioxide and produce humidity.

The unique feature of the south building is that it has a west-facing heat-box plenum. The plenum is warmed by the sun in the afternoon during spring, summer, and fall, and, combined with the building's internal heat loads, it increases inductive ventilation by producing stack action that is used to ventilate the building.

The initial cost of the solar systems has paid out in three and a half years. The two buildings combined presently save more than $6,000 per annum in utility costs.

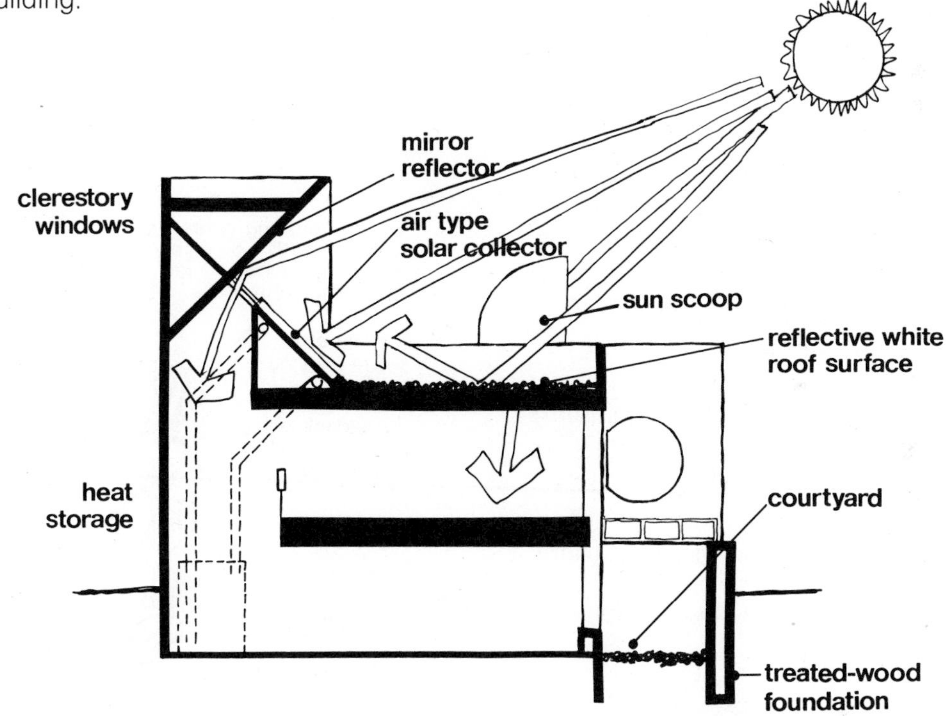

A new solar-daylighted office building is being planned for a small, 15 × 38-meter (50 × 125-foot) site. The boxlike configuration of the three-level building will be relieved by the northside, sculptural architectural form designed to receive sunlight.

An airlocked entry to the east will give access to stairs, to all levels. The lower-level, east courtyard will gather morning sun and provide outdoor access to the lowest level. The bottom of the angular, upper stairway on the west will reflect daylight and thermal radiation to a narrow areaway and lower-level interior.

The northside, sculptural architectural form will roll over into a south-facing clerestory designed to concentrate solar reflection from the roof deck in winter for daylighting and air tempering while providing shade and soft, indirect illumination to the interior in summer. On the southside roof, a double-glazed skylight will also be curved. Both the northside clerestory and southside skylight will ventilate.

The building will visually terminate the city block, with a contemporary earth form, at its northernmost corner. At the street level, it will be textured concrete. Two 2.13-meter (7-foot) wide skydomes will top the sculptural form and provide daylight to the lowest level. The site is limited to daylighting, air tempering, and ventilation; solar collectors are not planned due to the probable interference of future buildings to the south.

The building's east and west elevations will be highly reflective. The north and south sides will be constructed of a heat-absorbing dark finish. Projecting downspouts will catch the winter sun to free them of ice.

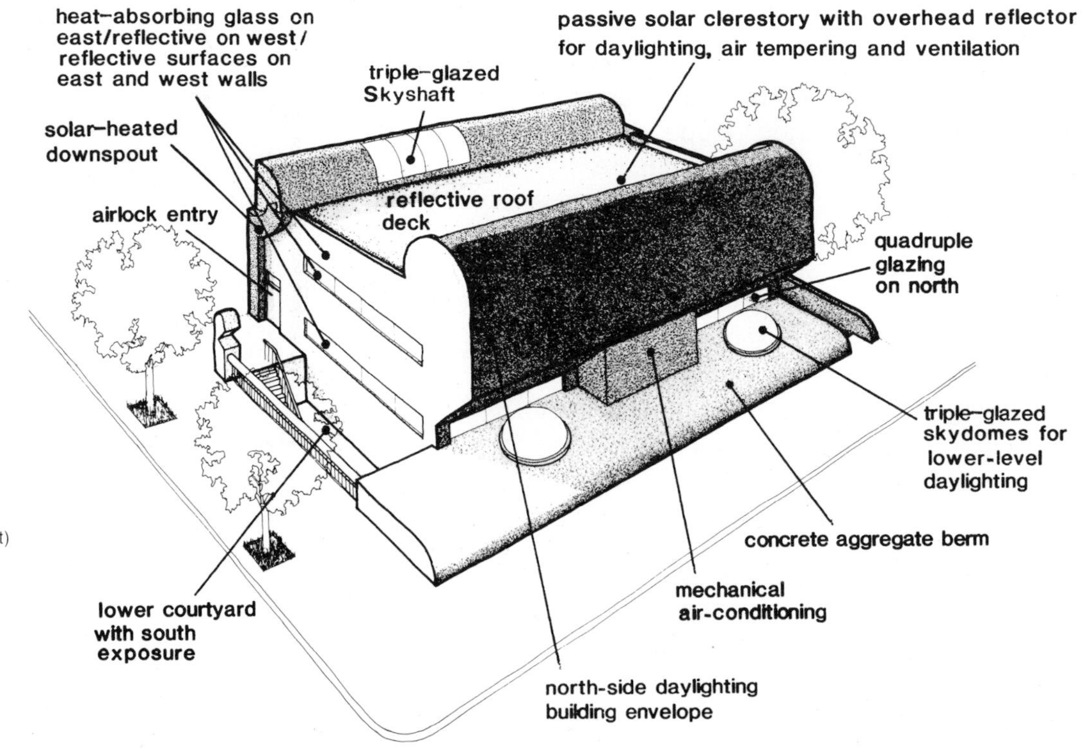

data

DENVER COMPACT OFFICE

Floor area	557 square meters (6,000 square feet)
Passive-solar subsystems	4
Solar photovoltaic cells	(future)
Heat-storage type	thermal inertia of floors and walls
Annual hybrid performance	70 percent (estimated)
Completion date	1983
Architect	Richard L. Crowther

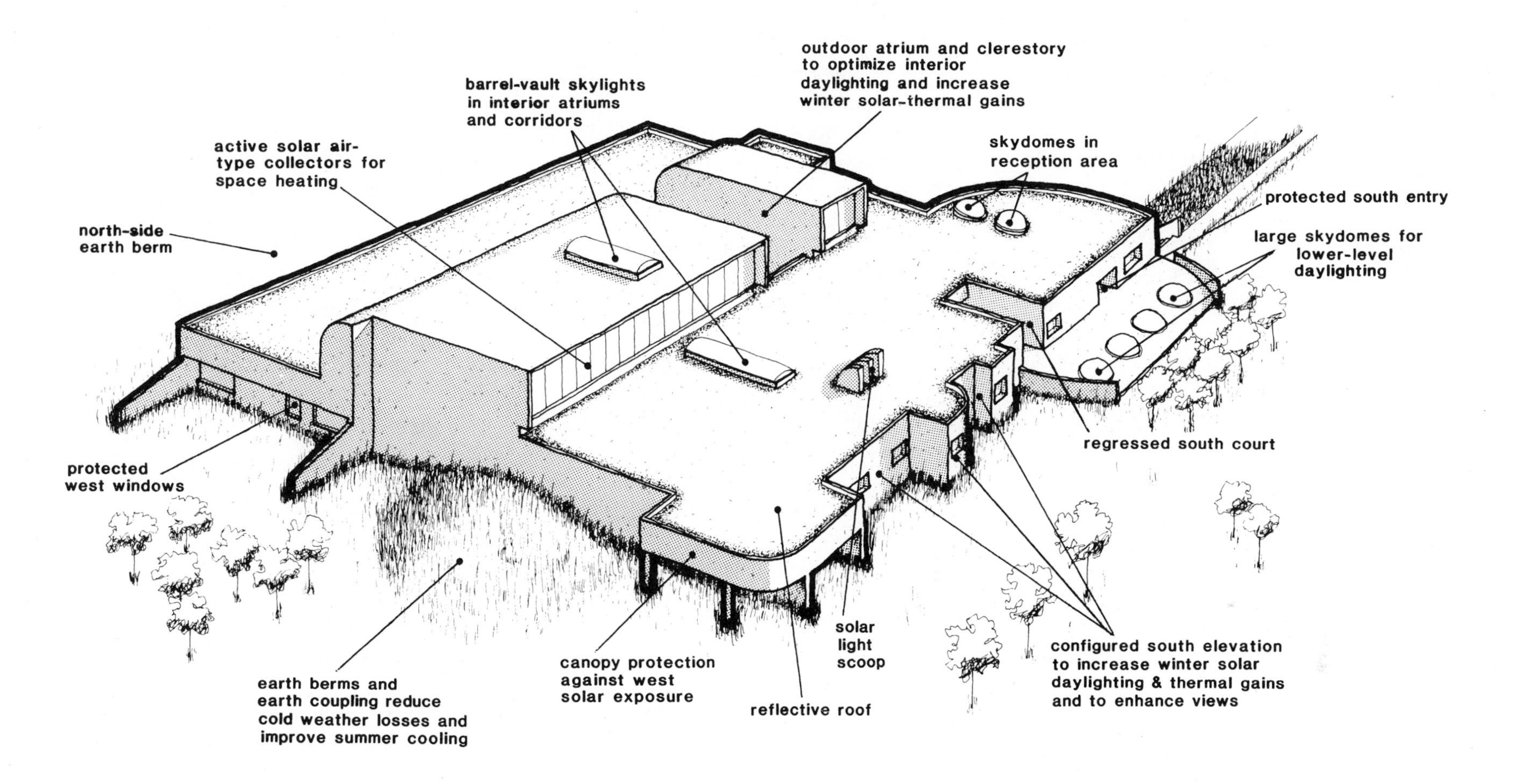

This holistic, functional, energy-conserving solar office building was designed for executive, sales, educational-training, accounting, and computer functions. The exterior landscape, architecture, and interior design are energy-integrated entities.

Controlled landscaping attenuates climatic extremes. Strategically placed earth berms buffer the microclimate of

data

CORPORATE HEADQUARTERS

Floor area	2,787 square meters (30,000 square feet)
Passive-solar subsystems	6
Energy centralization	no; variable-volume energy balance
Solar-collector area	44.6 square meters (480 square feet)
Solar-collector type	air
Solar-collector supplier	Solaron Corporation
Heat-storage type	27.2 metric tons (30 tons) of gravel
Annual passive/active performance	80 percent
Completion date	1981
Architect	Richard L. Crowther

the building. Native grasses, shrubs, and trees demanding little moisture reduce irrigation needs. Outdoor parking areas are screened from view but are located where direct sun exposure rapidly melts winter snow.

Outer insulation wraps the thermal mass of the building. Careful construction largely eliminates thermal bridging, infiltration, and exfiltration. The exterior building surface is textured to preserve a thin surface film of air. The building is oriented due south, and its dark south-face color favorably increases winter solar gains. Irregular exterior walls further increase the surface area of the south face for even greater energy gain. North and west exterior wall areas are minimized to conserve energy relative to the use of interior space and air volume. The east exterior wall deflects cold winter winds, and regressed windows reduce energy losses. The principal entries are weather protected.

The primary source of energy for north-side interior areas is air-type solar collectors. Rock storage sustains heat through night and day hours and for early-morning start-up. On the penthouse roof a small canopy shades the collectors to prevent summer overheating.

On the east side of the building, a lower level accommodates training, demonstration, lunchroom, and storage facilities. The lower level is principally below grade and has poured concrete foundations. An upper floor system of prestressed-concrete panels and an extensive posttensioned system enclose the lower-level demonstration area. This provides an energy-stabilizing thermal mass that includes prestressed-concrete roof slabs.

The variable-volume air-conditioning system uses outdoor air when outdoor temperatures are moderate. It provides refrigerated air as required for indoor comfort.

Solar shading, daylighting, and control is accomplished with regressed east windows, west canopies, an open-to-the-sky atrium, skylights over nonwork areas, interior borrowed light, and reflective blinds. For energy conservation, few windows face north and most face south. The strategic location, size, and proportion of exterior and interior windows, atriums, skydomes, and skyvaults assure that all interior offices and space are accommodated by daylight.

Natural daylight penetrates the lower-level demonstration area through four impressive hemispheric skydomes, each 2.1 meters (7 feet) in diameter. A similar skydome introduces a cylindrical shaft of daylight into the main reception lobby. Other barrel-vaulted roof penetrations, a smaller skydome, and a clerestory over the main stairway bring daylight into corridors.

An open-to-the-sky atrium brings daylight to upper and lower levels and provides a garden for the lower-level lunchroom. From the south, daylight enters the central reception area and private offices. The careful location and size of daylight and interior borrowed-light openings provide visual accommodation but allocate privacy for a discreet view of indoor activities.

Transparent, reflective, dark-bronze vertical blinds control daylighting and thermal gains in the interior and reflect sunlight in summer or as needed. They considerably reduce direct solar radiation into the interior yet afford an attenuated outdoor view.

The Rocky Mountains can be seen from many offices, and most others allow outdoor views. The aspen and pine trees of the central atrium give an outdoor presence to interior office spaces. The classroom has a view across the corridor of the atrium yet retains a subdued interior for back-lighted audiovisual projection.

The interior lighting adds an agreeable dimension and serves visual tasks. Artificial lighting for tasks and task-lighting levels can be controlled in all areas. Total lighting for the building is less than 6.83 Btu/hr (2 watts) per square foot. Heat from lighting and people nevertheless contributes substantially to wintertime heating needs. The interior was designed with tones and colors that respond to seasonal daylight to accommodate vision. Balanced lighting gives the entire interior a luminosity.

The building interior emphasizes task areas. Ergometric work areas were planned for effective interior communication and circulation. Circulation areas—the reception foyer, display area, corridors, and stairways—were designed by the architect with stimulating, intensive color; natural daylighting; abstract, graphic-design wall plaques; and green plants. Carpet colors and textures are unobtrusive. Large planters add greenery and reflectivity. Work areas were designed to be visually relaxing, with low-key colors and understated furnishings.

This solar condominium-office building in metropolitan Denver is divided into eight three-level condominium spaces that can be flexibly subleased. Reflective overhead canopies and side fins introduce direct and indirect solar light through the clerestory into a central, landscaped atrium that acts as a solar luminaire, internally daylighting the surrounding offices. The building optimizes

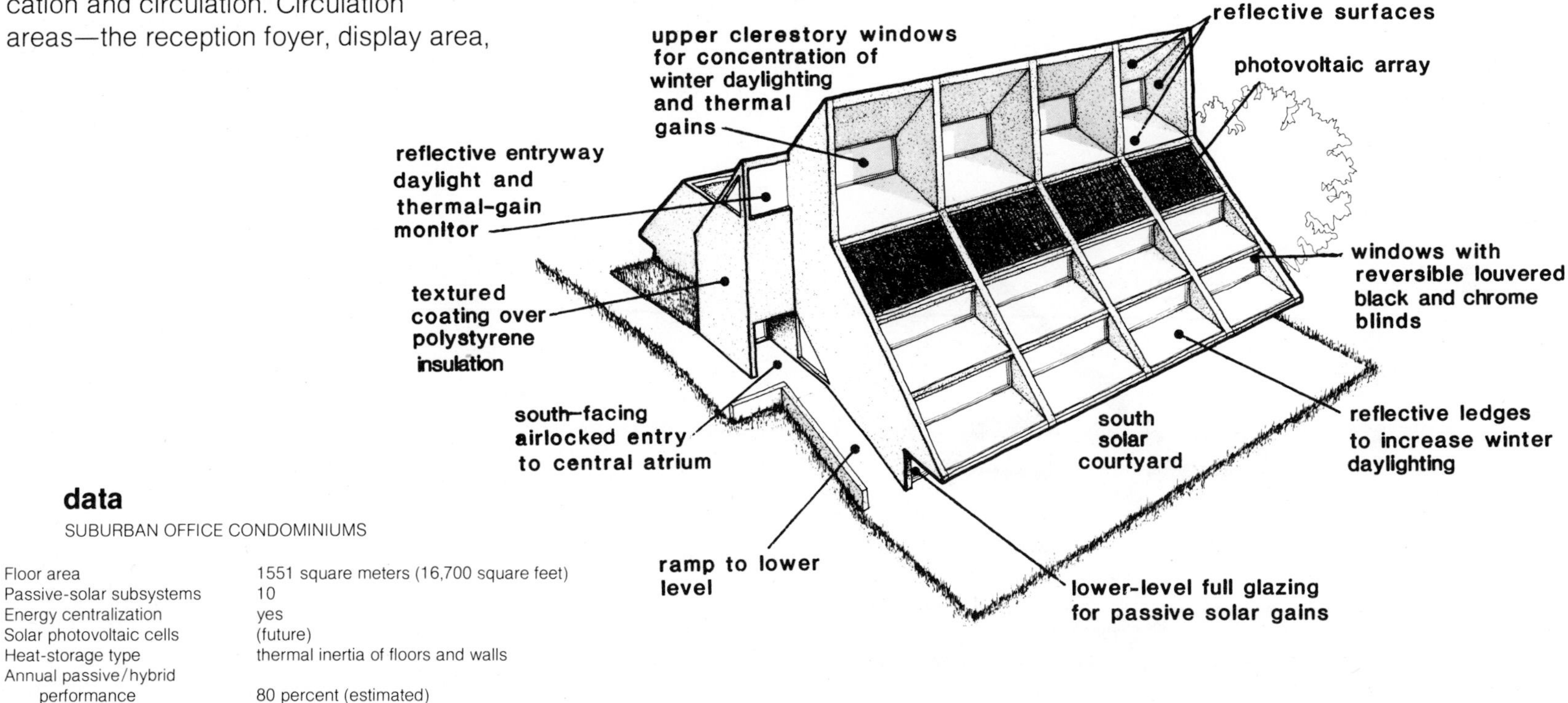

data

SUBURBAN OFFICE CONDOMINIUMS

Floor area	1551 square meters (16,700 square feet)
Passive-solar subsystems	10
Energy centralization	yes
Solar photovoltaic cells	(future)
Heat-storage type	thermal inertia of floors and walls
Annual passive/hybrid performance	80 percent (estimated)
Completion date	1983
Architect	Richard L. Crowther

both southern sunlight and northern skyvault daylighting. Reflective windowsills maximize ceiling-reflective illumination. All exterior and interior surfaces reflect or absorb daylight and thermal gains as appropriate to interior spaces. The north and south fenestration is slightly regressed for some summer shading, with horizontal, reflective blinds to direct light to ceilings from the reflective windowsills.

An airlocked street entrance opens into the central atrium—essentially a pressurized, air-tempering chamber. The atrium supplies filtered air to the offices and joins the individual condominiums. Winter solar energy circulates from a high point in the atrium through a lateral, low-temperature, gravel, radiant plenum immediately under the atrium floor. Supplemental mechanical equipment is located in the atrium attic.

The building's pyramiding form has a minimal surface-to-interior-volume ratio yet a maximum surface area to enhance interior daylighting and skyvault reflection. Photovoltaic cells in the south-facing roof supply common-area electricity. High-efficiency battery arrays will be part of this system.

A sunken southern courtyard admits maximum daylight to the lowest building level and provides separate entries. The drop in grade to the north permits lower-level walk-out. Earth berms and topographic landscaping are climatic, visual, and noise buffers to the north and south. Landscaping hides parking areas and heightens seasonal microclimatic control. A large underground reservoir for surface water compensates for limited storm-sewer capability and supplies water for toilets and irrigation.

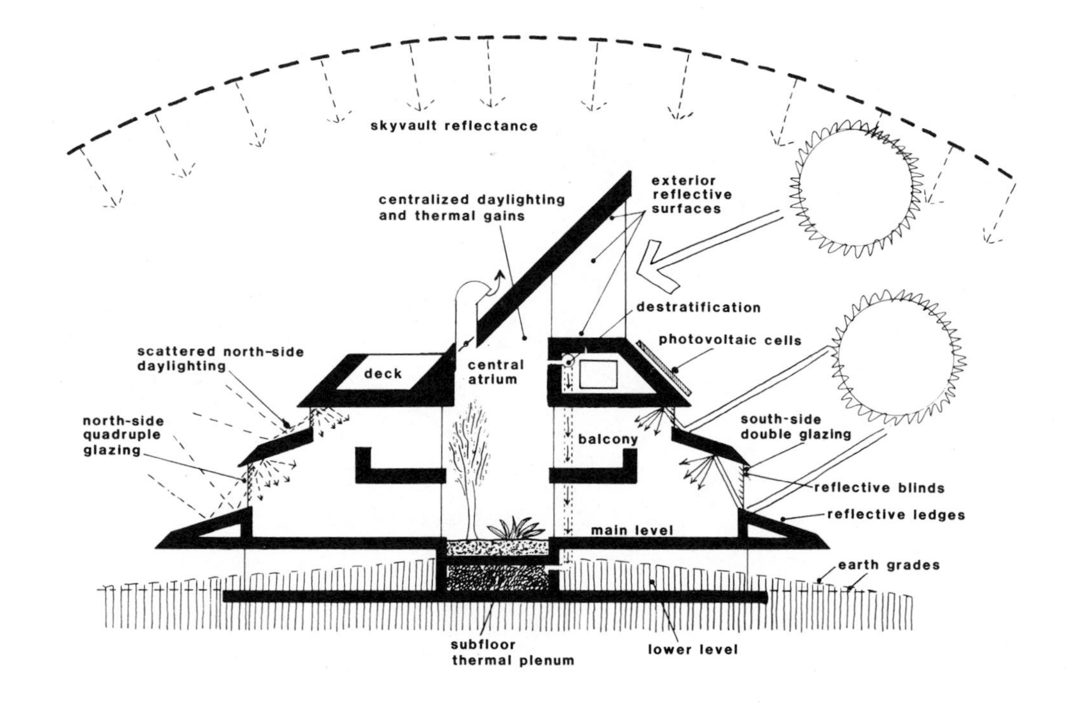

The Sunplace building of shops in Denver is designed to optimize an economic balance between initial, long-term operational, and maintenance expenses; flexible occupancy; and a responsive work environment. The three levels of this 1,115-square-meter (12,000-square-foot) solar building can have combined or separated access.

The main and lower levels, designed as showrooms, require display lighting, and the entire building is fitted with tracks for incandescent, fluorescent, or halogen lighting. While daylighting, unfortunately, can play but a subordinate role, artificial lighting is minimized by careful planning. Light wells introduce direct and diffuse

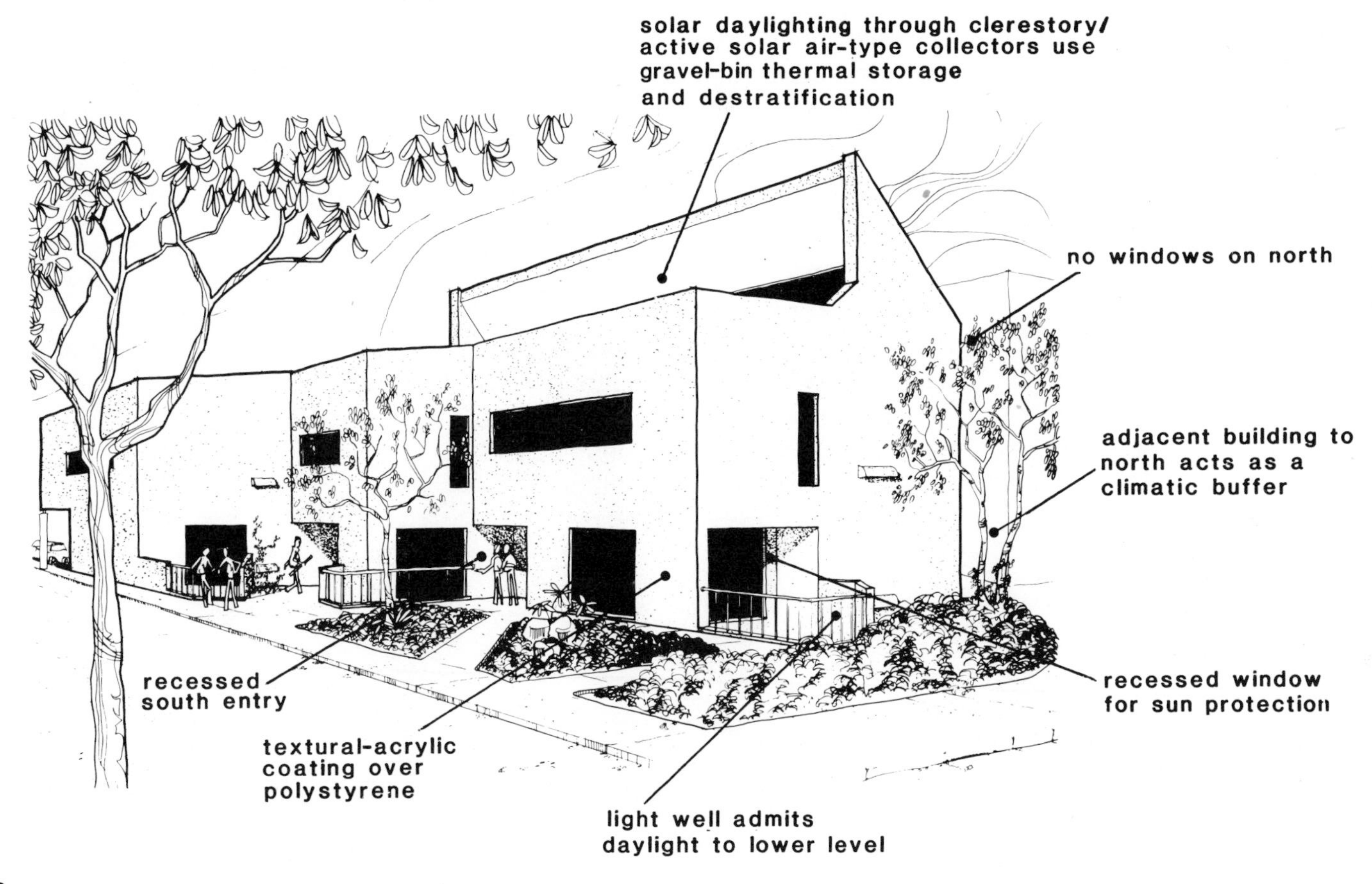

data

SUNPLACE SHOPS

Floor area	1,115 square meters (12,000 square feet)
Passive-solar subsystems	0
Energy centralization	yes
Heat-storage type	thermal inertia of floors and walls
Annual active/hybrid performance	70 percent (estimated)
Completion date	1983
Architect	Richard L. Crowther

daylight into the lower level. Main-level windows principally face south, a few east, and none faces north or west. To the northside main-level interior, a clerestory adds light as well as thermal gains, which are destratified from the apex of the clerestory by the building's air-handling system. The hottest air, which could have been lost from the building's most energy-vulnerable place, is instead recovered for use throughout the main and lower levels.

The building has air-type solar collectors that deliver the sun's thermal energy to the centralized rock-bin storage at the lowest level, principally for nights and weekends, when the building is not occupied, and for early-morning start-up.

The total architectural-energy concept has three main features: first, reusing the internal heat of lights, people, and equipment; second, optimizing the direct use of sun, earth-coupling, and air energies; and third, zoning and controlling the energy for the most effective seasonal use. Economizer, reverse-cycle, air-to-air heat pumps deliver moderate-temperature outside air, cool it (the principal need during the occupant period), or heat it, either by thermal extraction or by means of electric resistance coils during excessively cold weather.

12 Holistic Energy Design Process

The Holistic Viewpoint

The world's energies are interrelated in a whole manner. All forms of existence are in a constant state of energy exchange and transformation. As human beings at home on the earth, we are part of this holistic energy matrix.

Holistic energy architecture reflects the total interactions and transactions of society. The holistic (total) energy design equation includes clientele, climate, conservation, function, light, site, and structure. Each is a symbiotic part of the holistic design process: the clientele, by its choices; the climate, with its seasonal vagaries; energy conservation, as a design practicality; the dynamics of human energy; optimized daylighting; functional networks; and the energy specifics of the land and architecture. Successful holistic energy design involves carefully weighing these elements for a given project and putting them to the test of economic practicality.

Each project requires separate analysis. Every project, with its specific range of uses, specific location, specific microclimate, specific earth relationship, and specific neighboring environment, should be viewed as having its own specific character and specific economic profile.

Architectural Vocabulary and Economics

In holistic design terms, our rich vocabulary of contemporary architectural form offers a freedom of expression, material, and methods. Passive-energy systems and conservation can reduce annual heating, cooling, and ventilation needs 40 to 80 percent in northern latitudes and 50 to 100 percent in southern latitudes at a less than 5 to 7 percent increase in initial construction cost. Given acceptable design and construction trade-offs, most projects can realize substantial utility savings at no initial-cost increase. Each project, however, should be analyzed in a holistic manner through all seasons and parameters of energy use. Design decisions commit architecture to long-term energy excess or frugality.

Holistic Energy Design Process of Architecture

Comprehensive design directs and balances energy flows through all seasons. Architectural energy is invested in the initial materials, structural components, and methods, the ongoing expenditures of time and people, the total energy networks, and the useful life of the project. Holistic architecture employs renewable and appropriate energy techniques in the most economic life-cycle context.

Baseline Criteria

Criteria stimulate or impede effective energy design, and those criteria that are predisposed to waste energy or that exceed the budget should be modified. "Baseline Criteria" is a principal heading of the Holistic Energy Design Process chart, in this book. The category "Occupancy Functions" relates space planning to energy and time. "Policy Functions" deals with the design limitations, restrictions, and opportunities of urban codes and planning. "Objectives" includes present and future planning, self-image, and the sociocultural and occupational images conveyed to others.

"Needs" concerns the volumes, forms, light, and communications required to fulfill hurnan desires.

The major concern of most persons is bottom-line economics, including the initial and life-cycle investment of talent, energy, materials, and finance.

Humanization

People are energy, too. Human patterns of motion, metabolism, and gregariousness are disciplined by the environment. Planning for the use of human energy is fundamental to the energy-design process.

Under the principal heading of "Humanization," the prevailing subdivision "Outdoor Environs" offers opportunities as well as constraints. The challenge here is to design a humane environment that stimulates as well as relaxes our psychoneural responses and vitalizes our learning. Indoor environs are largely what we make them. We are particularly affected by our perceptive encounters with the focal elements, surfaces, and boundaries of our immediate environment.

All of our observations, activities, and transactions are bound by time. Our responses to time directly condition our sensory stimulation, perception, and memory.

Site

The specific microclimate of the site should be moderated to respond to seasonal use and architectural energy. The design should recognize on-site energy attributes that relate to day, night, and seasonal change. As we alter the form of any site by excavation or berming, we displace earth with equal volumes of atmosphere. This changes the natural energy transactions between the earth and the atmosphere.

The soil is teeming with life-supportive organisms and microorganisms (see chart). Harming one organism can imperil others, including ourselves. Influences on and from the neighborhood should not be neglected, nor should the symbolic strength of our design as a cultural force.

Architecture

Architecture is the most immutable element of the holistic energy design equation, as it inherently controls passive-energy flows by establishing thermal and private zones. Secondary energy gains flow from the primary spaces heated by solar or other energies.

Architecture, although a fixed element in the dynamic climate and active human environment, can create many horizontal planes of occupancy, commonly expressed as floor levels. The result is verticality, movement upward or into the earth. "Ergometrics" defines a planned environment in which human movement is more efficient in holistic-energy terms. "Energy transfers" radiate, conduct, convey, absorb, or transmit without materially altering the energy form. "Transformations" are exchanges that alter the energy form of the substance. "Thermonics" deals with the phenomena of heating and cooling, and "servomechanics" with supplementary or

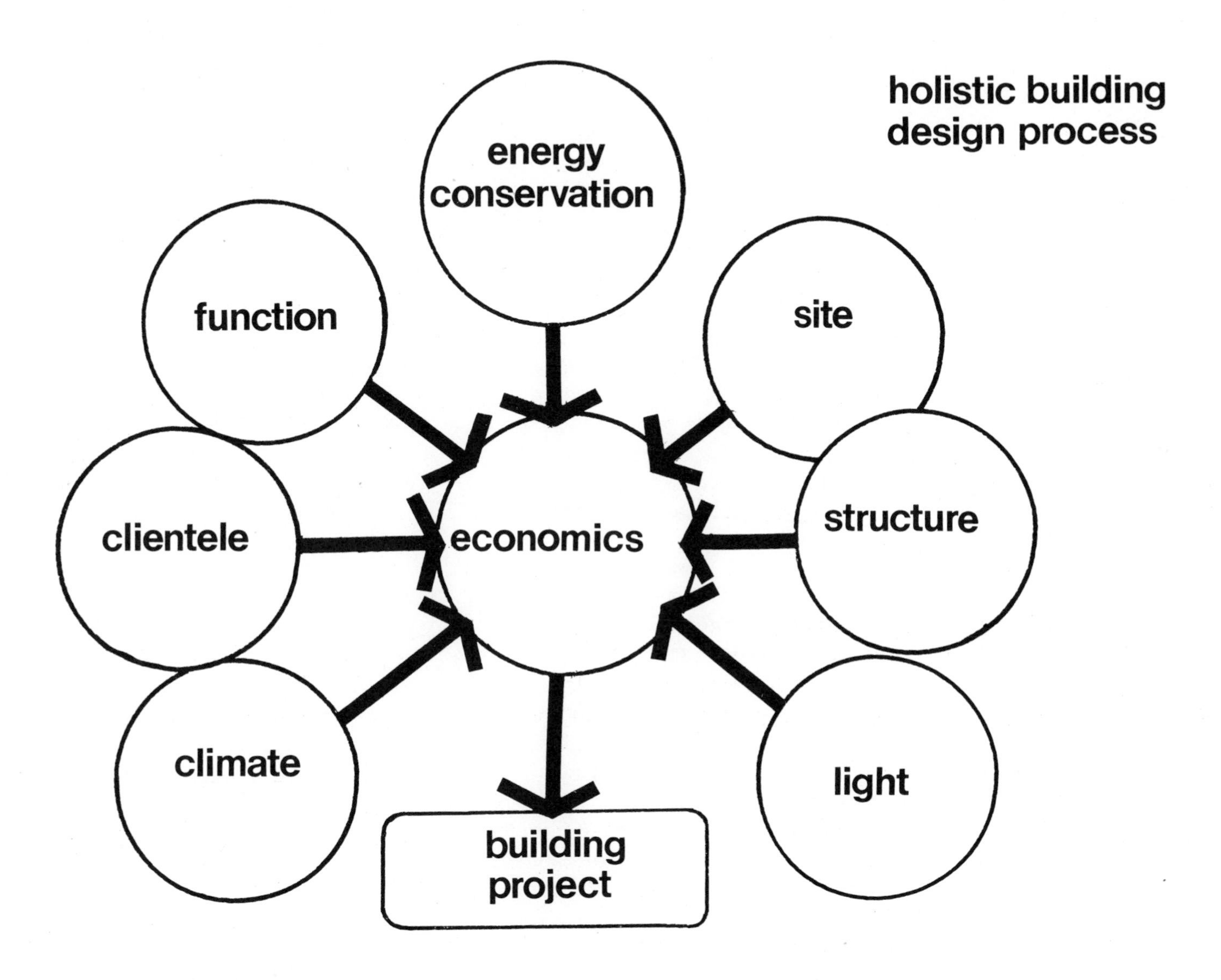

primary mechanical systems. (See the chart for placement of these terms.) All elements and openings of the exterior architectural envelope should optimally control the use of on-site sun, earth, air, and water energies.

Interior

Interior energy is largely neglected. All interior objects, materials, forms, and surfaces can control, store, or deflect energy flows in response to light and heat. Everything is energy, and every design either adds to or subtracts from comfort, space use, and the dimensions of human activity.

For the most part, we occupy indoor rather than outdoor space. Interior space and the happenings within it are generally personal, more secure, and intimate. Spatial divisions and spatial definition can greatly restrict or encourage energy flows, whereas electronic media (television, radio, telephone) can bring events of the world to interior space (see chart). Electronic control and service robots are a growing trend in homes and businesses.

The environmental ambience of the interior depends on light, color and texture, and form and decor (refer to the chart). Materials, objects, cabinetwork, and furniture possess a unique ability to change the energy transfers and sensory ambience of interior space through their rearrangement.

Indoor Air Quality

Air is our most basic life sustainer. Its quality, indoors and outdoors, affects our health and vitality. Air pollution is a major environmental problem for most cities. Outdoor air, smoking, industrial processes, food preparation, painting, cleaning, equipment, architectural and interior materials, and occupants produce indoor air pollution. Building codes prescribe ventilation requirements, which result principally in introducing polluted outdoor air into a building through inadequate filtration systems. It is illogical to ventilate buildings with polluted air or permit sources of unnecessary indoor pollution to exist. Smoking, in particular, should be eliminated or highly controlled.

Air can be purified by negative ionization, activated-charcoal, high-efficiency, chemical-absorbing (potassium permanganate), electronic, or bacteriological filtration—or even by growing plants. Oxygen can be provided by vegetation, oxygen-producing bacteria, or by electrolysis—using solar energy to separate the oxygen and hydrogen from the water. In the future, concentrated enrichment units could release oxygen, using preferential molecular sieves. Carbon dioxide can be absorbed or removed by green plants, ventilation, dissolution into water, or gas-absorption and filtration systems.

Retreat Spaces

Retreat spaces within a home or building extend protection from natural and other hazards. Air pollution, for example, can be a health hazard in both urban and rural areas, and retreat spaces may be designed to avoid or minimize its effects. While it might not be practical or economic to adequately control the air quality of an entire building, a retreat space with a highly controlled level of air quality can be biologically beneficial.

Intrusion and violence are growing concerns. Designing for defensive retreat with interception at the house perimeter can be more economic and protective when incorporated in the holistic design.

Fallout shelters provide a high level of security, and can serve as retreat spaces in emergency blackouts and as protection from natural hazards such as heavy storms and tornados. In addition to the thermal comfort required in retreat spaces, fallout-shelter design should allow for supportive energy, provisions, and an effective means of waste removal. The hazard's probable severity and duration should directly determine the level of protection.

Societal Responsibility

Every project has an impact upon community, national, and international resources. Every project also has an aesthetic impact upon its immediate environment. Architects, designers, planners, and developers must learn to understand the energy-form interactions between their projects and the projects' surroundings. The energy networks and rhythms of land planning and architectural design adhering to the laws of physics would create an urban holistic energy aesthetic.

Rhythm and Time

In the design process, it is good procedure to time ourselves to the rhythms of human life, the site, the architecture, and the interior. The rhythms of day and night, the seasonal solar path, the rhythms of wind and atmospheric change, and the rhythms of snow and rain are revealed as they intersect the forms, volume, and openings of the architecture. The prevalence of time becomes an energizing internal force. Earth topography has its rhythms. Nothing exists without time, yet we see it only by the manifestations of rhythm and change in our environment.

Variety

There is little danger of monotony in holistic energy design. The indigenous characteristics of each project, each person creatively involved, and each microclimate increase the probabilities of eliciting aesthetic interest and response. Both the dynamics and the limitations of the physical laws of natural order result in diverse meaning and form. Although contemporary architecture has produced a marvelous vocabulary of design, for the most part it blatantly disregards the structuring value of natural forces. Philosophically, we need to wed the life-giving flows of cosmic energy and the contemporary idioms of design.

Case Study Projects

The case studies of the author's architecture reflect only a direction of thought. A growing number of architects regard energy as a critical design element. The talent inherent in our design community represents a vast scope of creative possibility. Our near and distant futures desperately need the commitment to an architectural ethic that recognizes holistic energy flows and redundancies as predominant to design.

Holistic Research

Accepted scientific and technical research explores the manifestations of organic and inorganic matter, usually by isolating the phenomena being studied. The scientific method, however, excludes some contributing factors, thereby leaving the results open to further question. Much scientific and technical information extracted from studies of isolated phenomena is nonetheless useful as a guide and means of design-process analysis.

The holistic energy flows of the architecture, site, interior, and inhabitants are so complex, dynamic, and redundant that attempting to qualify and quantify the energy synergies and evaluate the holistic significance is very difficult. Generally, architecture is immutable. Most interior elements are movable; exterior vegetation grows. Such factors as the vagaries of climate and the uncertainties of human interaction are particularly troublesome. Thermal, gravitational, ionic, magnetic, radiative, and metabolic energies should all be part of the holistic energy equation.

Holistic design should be based on intuition, educated assumptions, and the observation and study of relevant past and present energy models and applied physics. In addition to these customary scientific and technical processes, which are of great value, holistic interactions of the total reality must be accounted for.

Traditionally, architecture is a "shelter"—from the climate, predators, and hostile persons; and for valuables and privacy. While these time-honored considerations remain of consummate value, holistic architecture would also optimally use the energies of the sun, earth, air, water, inhabitants, and architecture itself.

The First Law of Thermodynamics

Scientific research and architectural design are fundamentally concerned with the universal properties of matter and energy. The phenomena of heating and cooling are expressed within the laws of thermodynamics.

Within the universe, no energy is lost—it is only transferred or transformed as it exhibits individual form and indigenous properties.

The Second Law of Thermodynamics

Within the universe, all energy degrades, dissipates, and becomes less useful.

Passive and Active Solar Energy

The term *passive solar* is widely used to convey the concept of energy flowing naturally, without mechanical assistance. *Active solar*, in contrast, is defined as requiring mechanical assistance. From molecular, thermal, and illuminant standpoints, however, all energy is active. Passive energy, literally, does not exist. Only metaphorically can energy be termed passive or active. Technically, passive solar energy should be designated as *thermodynamic* solar energy, meaning that some of the sun's electromagnetic radiation converts to longer wavelength energy as it intercepts earthly objects and surfaces. Active solar energy might technically be called *mechanothermodynamic* solar energy (from the Greek *mechane*, meaning "machine"), the point being that active solar energy requires machines, devices, and controls.

Every square foot of site and architectural surface directly or indirectly receives solar electromagnetic radiation that provides daylight or thermal energy. Every square foot of site and external and internal architectural surface that does not optimally utilize this free beneficent energy is wasteful.

Air-Type Thermal Storage

Low-temperature direct and indirect solar thermal energy plus indoor heat can be stored in large masses such as concrete basements, horizontal or vertical gravel beds, or eutectic salts in insulated containers. Such systems function best between 21° C (70 F) and 38° C (100 F). Earth can be used as heat storage (wet earth absorbs heat better than dry earth). Higher-temperature direct solar-thermal energy 32° C to 93° C (90 F to 200 F) from thermosiphon or fan-powered solar collectors is most effectively stored in gravel storage bins (at 5 feet or more depth) or in eutectic-salts containers that are cubical, insulated, and controlled for time-interval heat loss. Heat transfer from air to contained fluids is less effective.

Important are the advisable lengths of time to store solar heat to sustain levels of indoor comfort.

Temperature, Wind, Humidity, Rain, Snow

Climate varies with the season, region, and microclimate of the site. The composite dynamics of these forces must be clearly understood to design effective holistic energy architecture.

In the design process, some energy-exchange fundamentals should be considered. Outdoor temperatures affect indoor comfort according to the energy exchange in the thin air film at the building's surface. Winds tend to remove this film. Cold winds produce a wind-chill factor upon the building and increase indoor heat loss. Roofs are particularly vulnerable to wind-chill losses because of their expanse and position as a cover of the warm ceiling portion of the building.

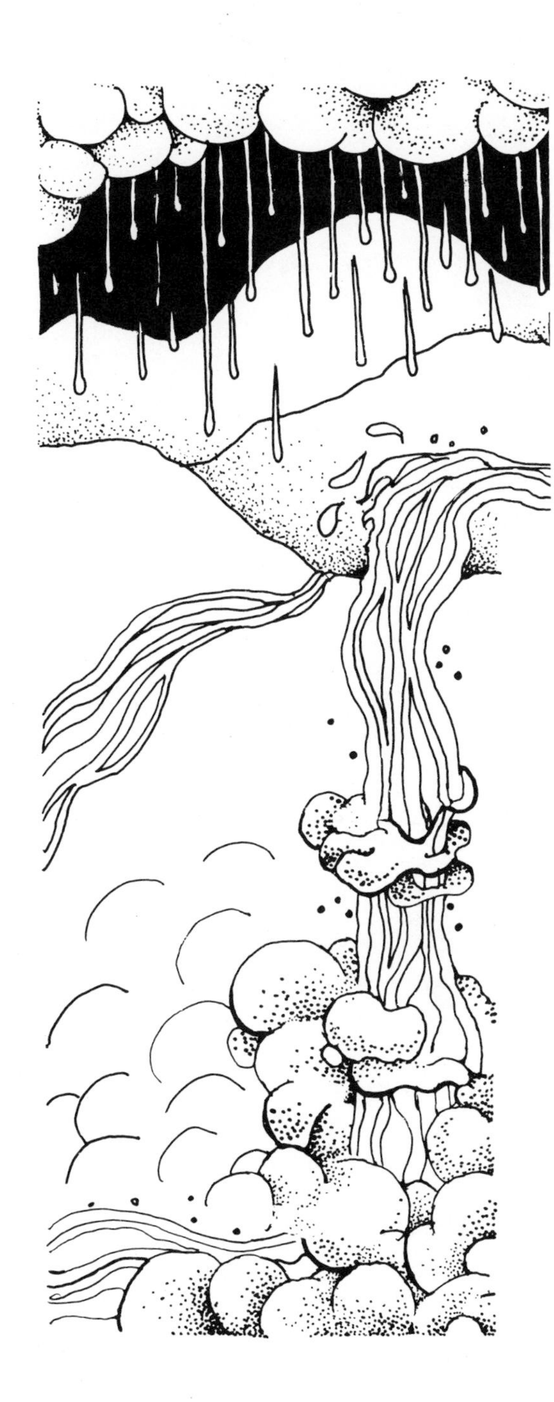

Temperature swings in warm and hot months from days to cooler nights provide an opportunity for cooling. When outdoor temperatures reside between 12.7° C and 23.8° C (55 F and 75 F)—during night or day—they should be directly used for indoor comfort. Higher indoor humidity makes winter air temperatures feel warmer as well as summer temperatures hotter.

Rain tends to ionize and wash the atmosphere of pollutants; it also cools the earth, buildings, and inhabitants by evaporation and sensible cooling. Snow provides a reflectance factor of 85 percent and acts as an insulative blanket over rooftops. The weight and avalanching of snow should be accounted for.

The New and the Old

This book has been written primarily about new architectural projects. Existing homes and buildings, however, need specific attention. Each existing building has its energy problems and personality. Many *Sun/Earth* concepts can be used to make existing buildings more energy effective and comfortable.

Needed Innovative Architectural Energy Development

- design of new and conversion of existing space for a wider range of use
- manufactured, transportable, energy-core modules for new residential construction
- adjustable architectural space to meet variable requirements over time
- down-sized and miniaturized habitat and occupational space
- manufactured, architecturally integrated elements for passive cooling, ventilating, air tempering, humidifying, dehumidifying, filtering, and heating
- manufactured, exterior doors that truly minimize energy loss and maximize security and summer ventilation
- glazing units that permit view and desired solar gains but minimize energy loss without movable insulation
- earth-technology construction techniques using on-site earth plus waste products (as stabilized, earth-insulative, or thermal mass interior or exterior walls)
- inexpensive, centralized computers to program the total heating, cooling, ventilating, and air-tempering aspects of "passive" systems
- tight, operable, insulated air intakes and inductive exhaust devices
- manufactured architectural components with prescribed energy attributes that permit flexibility in expansion or rearrangement to meet changing occupant needs
- integrated-energy kitchens that effectively use appliance, cooking, refrigeration, and dishwashing waste heat for food preparation and preservation
- low-cost, energy-efficient photovoltaic cells and batteries
- lower-cost, high-efficiency, solar collectors with high-density-Btu thermal storage systems for heating/cooling

Daylighting

Daylight reveals the external forms of architecture and penetrates the interior as useful light. Building design should balance daylight and illuminance of the interior for task functions, visual comfort, and ambience.

Sunspace

A sunspace is any isolated architectural space that is designed and located to receive an extensive amount of direct sunlight and daylight from skyvault irradiance. It can be an atrium, a greenhouse, a glazed portion of the building envelope, a sunroom, or any primarily passive-solar space that is part of the architecture. Sunspaces can serve other spaces with transient solar daylighting and heat.

The physical and thermal disadvantages of sloping glass for roofs and attached greenhouses should be considered with great caution.

Passive Cooling

Buildings can be passively cooled by the velocity of inductive interior air flow; outdoor breezes; evaporative techniques, including fountains and ponds of water; nocturnal cooling; thermal mass; earth coupling; earth-cooled air; natural ice formation; wind towers; thermal chimneys; porches; verandas; colonnades; arcades; interposed shading; vegetation; dehumidification; clear-sky temperature differential; and human migration to cooler places.

Calculations

Performance-analysis calculations are essential to a building's lighting and thermal design. Discussion of such methodologies are available from other published works (see Bibliography). Rules of thumb and generalized calculations can be applied for preliminary evaluations, but more definitive quantifications and qualifications are needed as the design process develops. Energy consultants are available for assistance in these areas, and computer read-outs can be effectively employed for energy projections and modeling. Use of small-scale models can be helpful for daylighting studies.

baseline criteria

HOLISTIC ENERGY DESIGN PROCESS

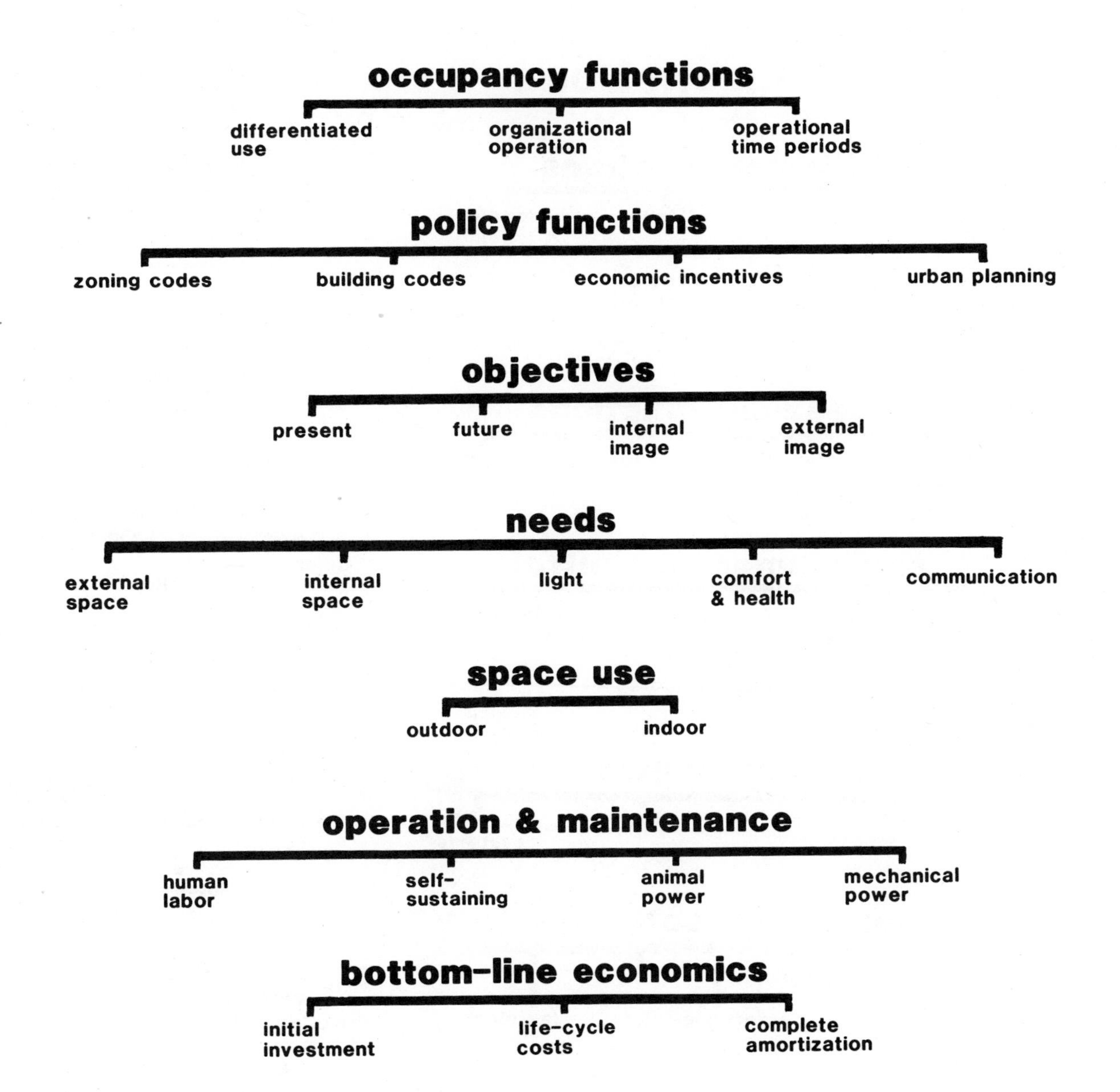

humanization

HOLISTIC ENERGY DESIGN PROCESS

outdoor environs

vehicular functions

types — on-site — off-site — unloading & loading

[cars/buses trucks bicycles] [drives & parking] [roads & stacking space]

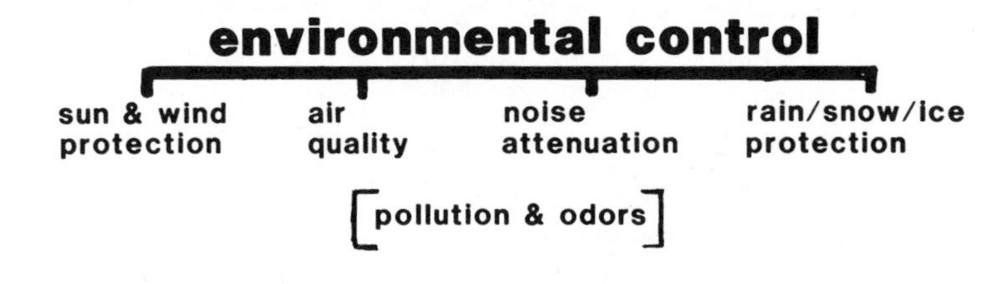

indoor environs

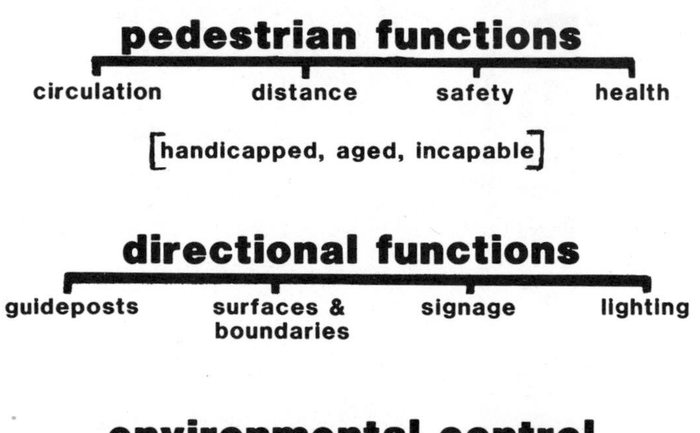

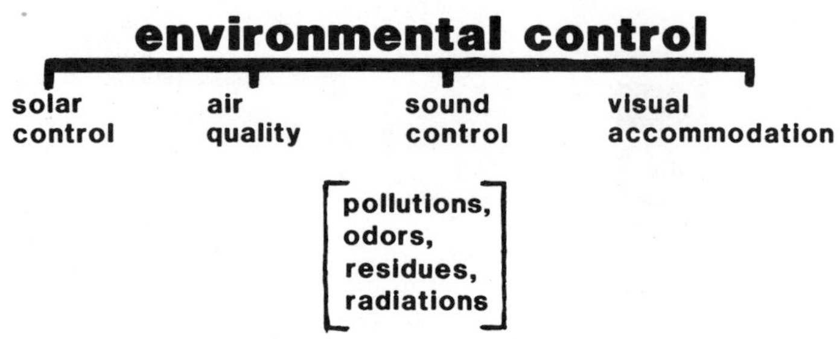

site

HOLISTIC ENERGY DESIGN PROCESS

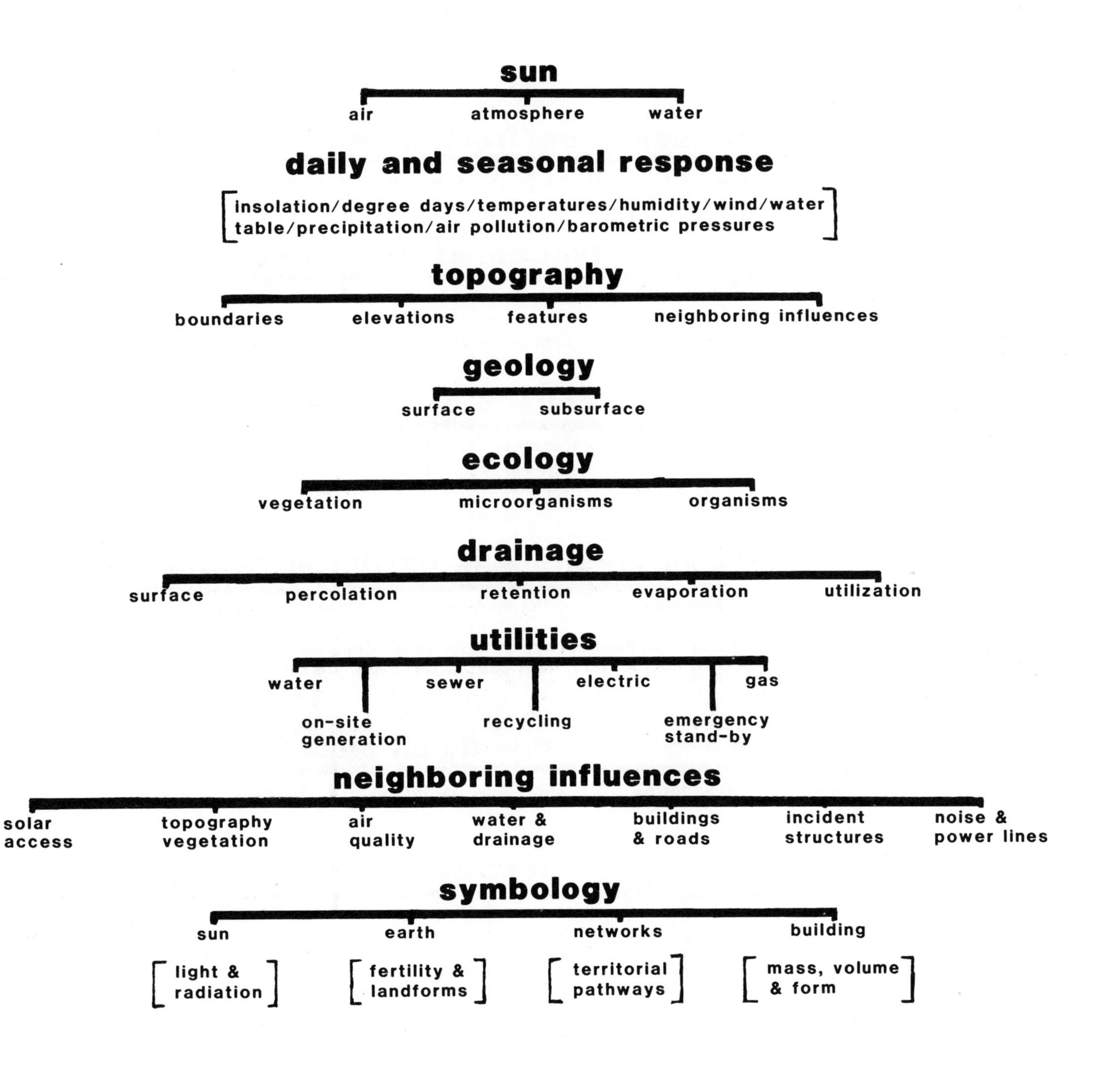

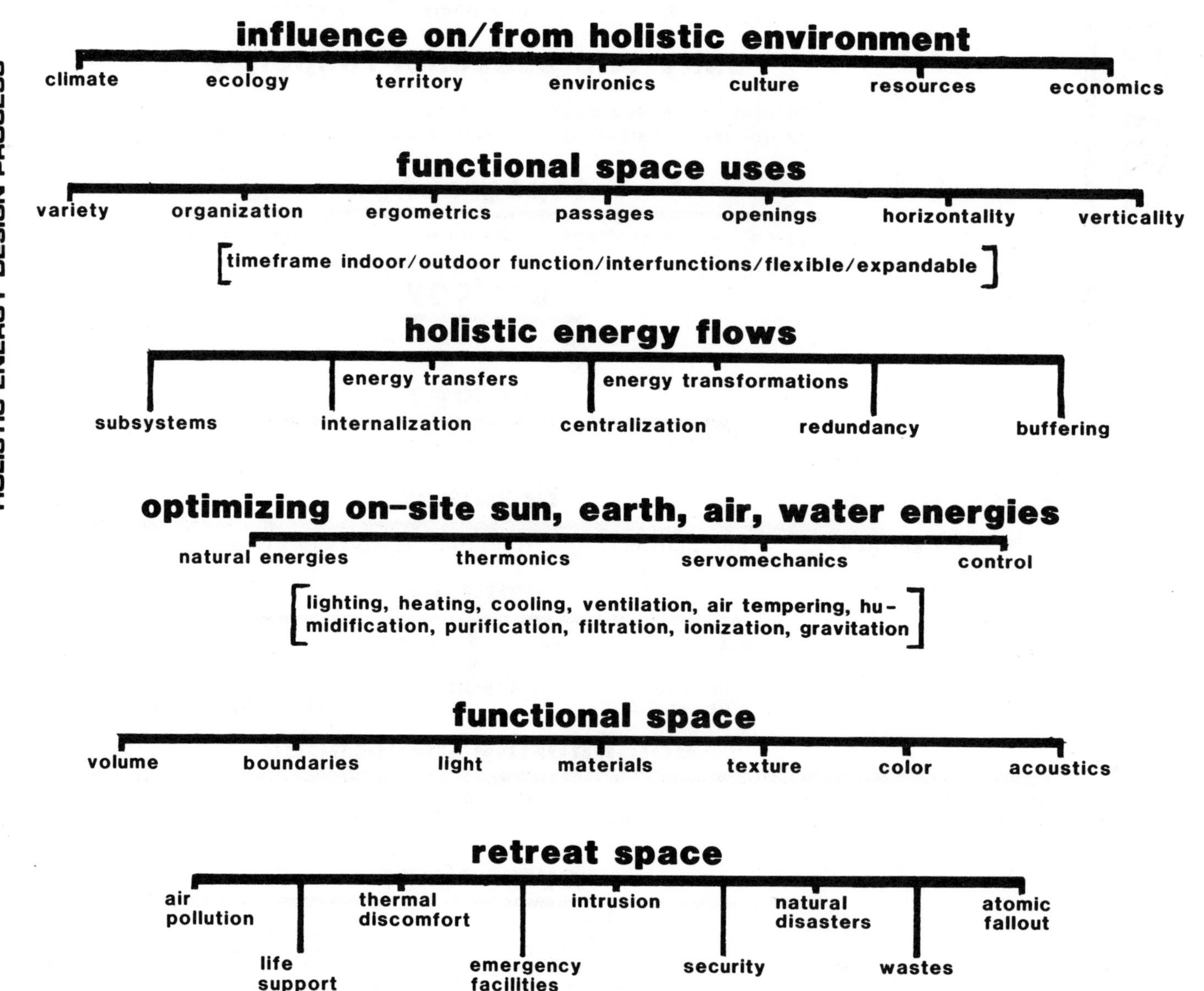
architecture
HOLISTIC ENERGY DESIGN PROCESS
influence on/from holistic environment
climate
ecology
territory
environics
culture
resources
economics
functional space uses
variety
organization
ergometrics
passages
openings
horizontality
verticality
[timeframe indoor/outdoor function/interfunctions/flexible/expandable]
holistic energy flows
energy transfers
energy transformations
subsystems
internalization
centralization
redundancy
buffering
optimizing on-site sun, earth, air, water energies
natural energies
thermonics
servomechanics
control
[lighting, heating, cooling, ventilation, air tempering, humidification, purification, filtration, ionization, gravitation]
functional space
volume
boundaries
light
materials
texture
color
acoustics
retreat space
air pollution
life support
thermal discomfort
emergency facilities
intrusion
security
natural disasters
wastes
atomic fallout

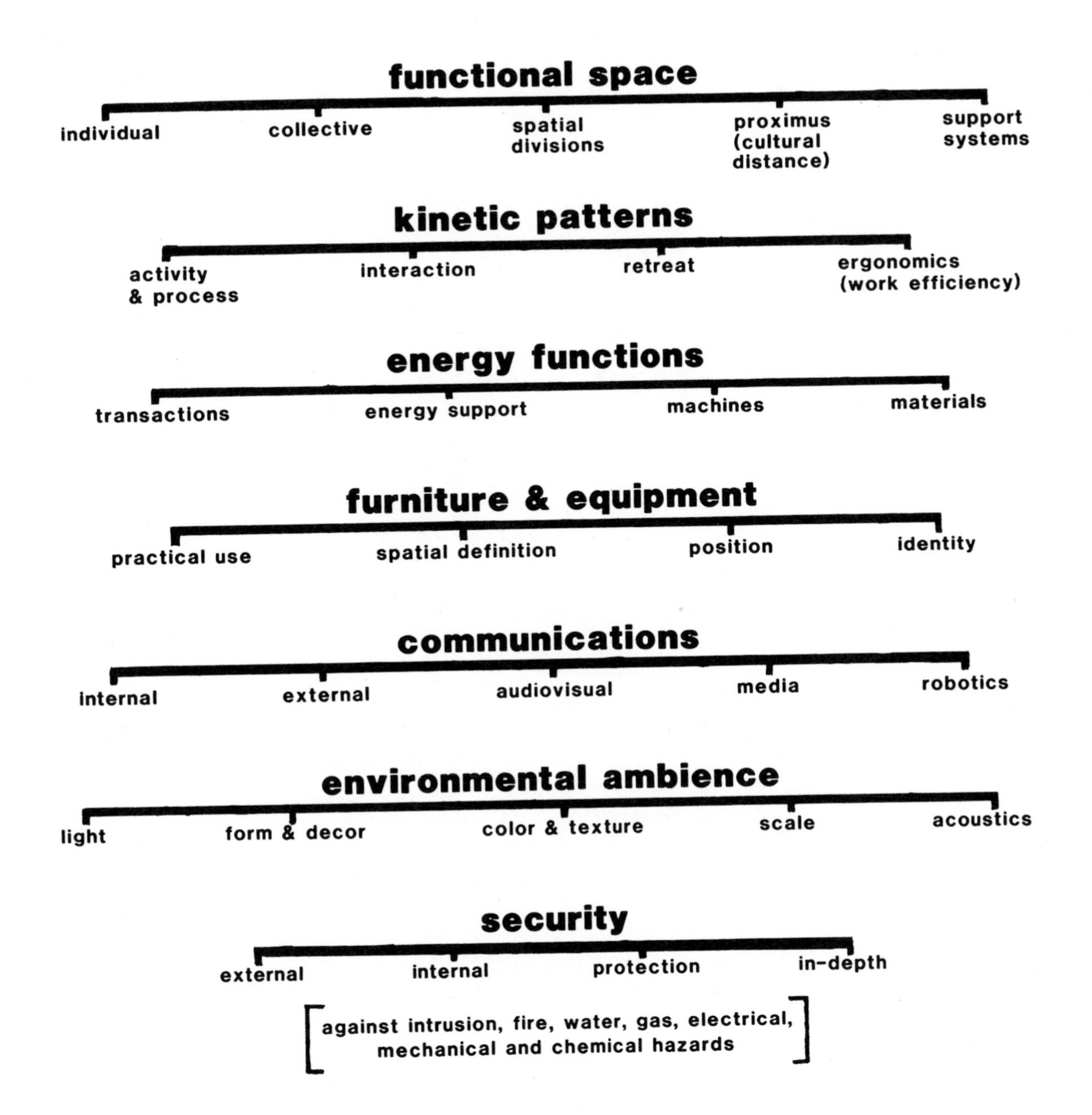
interior
HOLISTIC ENERGY DESIGN PROCESS
functional space
individual
collective
spatial divisions
proximus (cultural distance)
support systems
kinetic patterns
activity & process
interaction
retreat
ergonomics (work efficiency)
energy functions
transactions
energy support
machines
materials
furniture & equipment
practical use
spatial definition
position
identity
communications
internal
external
audiovisual
media
robotics
environmental ambience
light
form & decor
color & texture
scale
acoustics
security
external
internal
protection
in-depth
[against intrusion, fire, water, gas, electrical, mechanical and chemical hazards]

Appendix

It is interesting to see where energy is consumed in our society. The following lists some end uses of energy.

END USES OF ENERGY (consuming sector of the economy in parentheses)	Percent of Total
Fuel, excluding lubricants and greases (transportation)	24.9
Space heating (residential and commercial	17.9
Process steam (industrial)	16.7
Direct heat (industrial)	11.5
Electric drive (industrial)	7.9
Raw materials and feedstocks (commercial industrial and transportation)	5.5
Water heating (residential, commercial)	4.0
Air-conditioning (residential, commercial)	2.5
Refrigeration (residential, commercial)	2.2
Lighting (residential, commercial)	1.5
Cooking (residential, commercial)	1.3
Electrolytic processes (industrial)	1.2
Other	2.9
Total	100 percent

Energy Source	Energy Output	
.907 metric tons (1 ton) bituminous coal	7674 kilowatt-hours	27.7×10^9 joules (26.2×10^6 Btu)
159 liters (1 barrel) crude oil	1640 kilowatt-hours	5.9×10^9 joules (5.6×10^6 Btu)
159 liters (1 barrel) residual oil	1842 kilowatt-hours	6.64×10^9 joules (6.29×10^6 Btu)
3.7854 liters (1 gallon) gasoline	36 kilowatt-hours	132×10^6 joules (125×10^3 Btu)
3.7854 liters (1 gallon) No. 2 fuel oil	40 kilowatt-hours	147×10^6 joules (139×10^3 Btu)
.028 cubic meters (1 cubic foot) natural gas	.301 kilowatt-hours	1.09×10^6 joules (1031 Btu)
1 kw-hr electricity	1.0 kilowatt-hours	3.60×10^6 joules (3414 Btu)
.907 metric tons (1 ton) TNT	11657 kilowatt-hours	42.0×10^9 joules (39.8×10^6 Btu)
907,185 metric tons (1 megaton) nuclear bomb	11,657,000,000 kilowatt-hours	42.0×10^{15} joules (39.8×10^{12} Btu)
.035 ounces (1 gram) of matter completely converted to energy	24,985,354 kilowatt-hours	90.1×10^{12} joules (85.3×10^9 Btu)

The following conversions are useful for design purposes and will be especially helpful when the SI units of measurement are adopted in the United States.

Energy

1 British Thermal Unit (Btu) = 251.99 calories
= 1055.87 joules
= .00029287 kilowatt-hours
1 calorie = .003968 Btu
1 foot-pound = .324048 calories
1 joule = 1 watt sec
1 kilowatt-hour = 3414.43 Btu

Energy Density

1 calorie/square centimeter = 3.68669 Btu/square foot
1 Btu/square foot = .271246 calories/square centimeter
1 langley = 1 calorie/square centimeter

Power Density

1 cal./square centimeter/minute = 221.2 Btu/square foot/hour
1 watt/square centimeter = 3172 Btu/square foot/hour

Power

1 Btu/hour = 4.2 calories/minute
= .292875 watts
1 watt = 1 joule/sec

Flow Rate

1 cubic foot/minute = 471.947 cubic centimeter/second
1 liter/minute = .0353 cubic feet/minute
= .2642 gallons/minute

Metric Prefix	Common Usage	Scientific Notation
nano	1 billionth	10^{-9}
micro	1 millionth	10^{-6}
milli	1 thousandth	10^{-3}
centi	1 hundredth	10^{-2}
kilo	1 thousand	10^{3}
mega	1 million	10^{6}

Mass/Weight

1 pound = 16 ounces
= .45359 kilograms
1 ton = 907 kilograms
1 kilogram = 2.2046 pounds
1 metric ton = 1000 kilograms
= 2204.6 pounds

Velocity

1 foot/minute = .508 centimeter/second
1 mile/hour = 1.6093 kilometer/hour
1 kilometer/hour = .621 mph

Temperature

C = 5/9 (F − 32)
F = 9/5 (C) + 32

Length

1 mile = 5280 feet
= 1.6093 kilometers
= 1760 yards
1 kilometer = .621 miles
= 1000 meters
1 yard = .9144 meters
1 meter = 39.37 inches
= 3.28 feet
1 centimeter = .3937 inch
1 inch = 2.54 centimeters
1 foot = .3048 meter
1 angstrom = 1×10^{-8} centimeters

Area

1 square mile = 640 acres
= 2.59 square kilometers
1 square yard = .836 square meters
1 square foot = .0929 square meters
1 square inch = 6.4516 square centimeters
1 square centimeter = .155 square inches
1 square meter = 10.7639 square feet
= 1.196 square yards
1 square kilometer = .3861 square miles
1 acre = 43,560 square feet
= 4047 square meters

Volume-Liquid

1 gallon = 4 quarts
= 3.7854 liters
= 231 cubic inches
1 quart = .9463 liters
1 liter = 1000 cubic centimeters
= 1.0567 quarts
= .2642 gallons

Volume-Dry

1 cubic foot = 28.317 liters
1 cubic yard = .7645 cubic meters
1 cubic inch = 16.387 cubic centimeters
1 cubic centimeter = .06102 cubic inches
1 cubic meter = 35.3145 cubic feet
= 1000 liters
= 1.308 cubic yards

The following maps present the mean daily insolation on a horizontal surface for the months of June and December and year round. The unit of measurement is langleys, where 1 langley/day = 1 calorie/square centimeter/day (3.69 Btu/square foot/day).

The source of these maps is the United States Department of Commerce, Weather Bureau. Copies can be obtained from the Superintendent of Documents, United States Government Printing Office, Washington, D.C. 20402.

Mean Daily Solar Radiation in December (langleys)

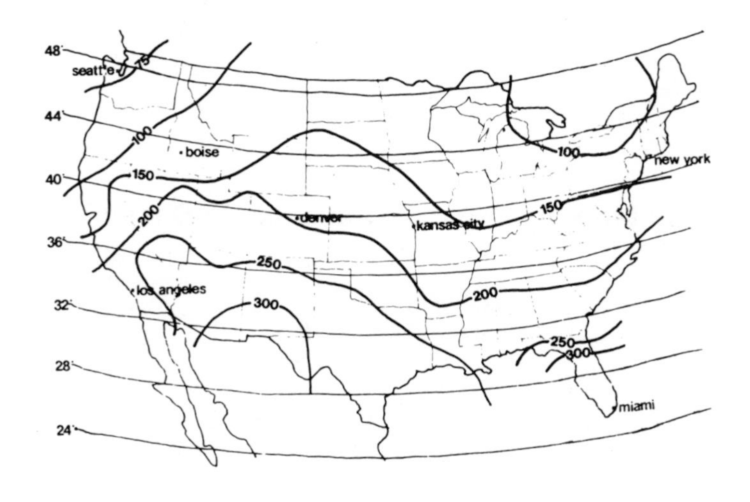

Mean Daily Solar Radiation in June (langleys)

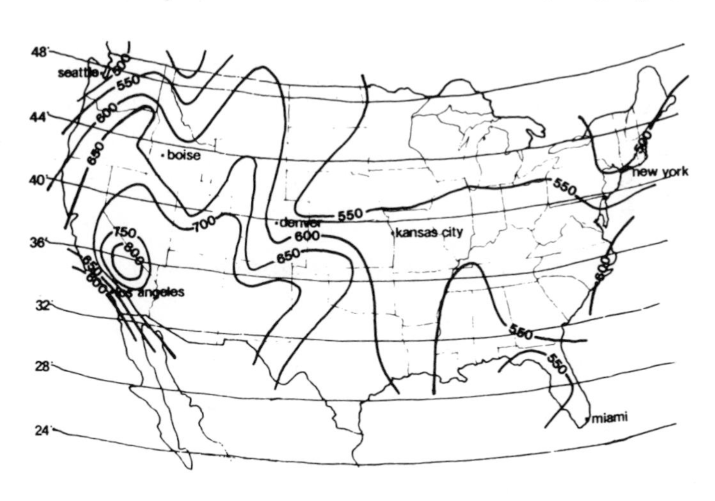

Annual Mean Daily Solar Radiation (langleys)

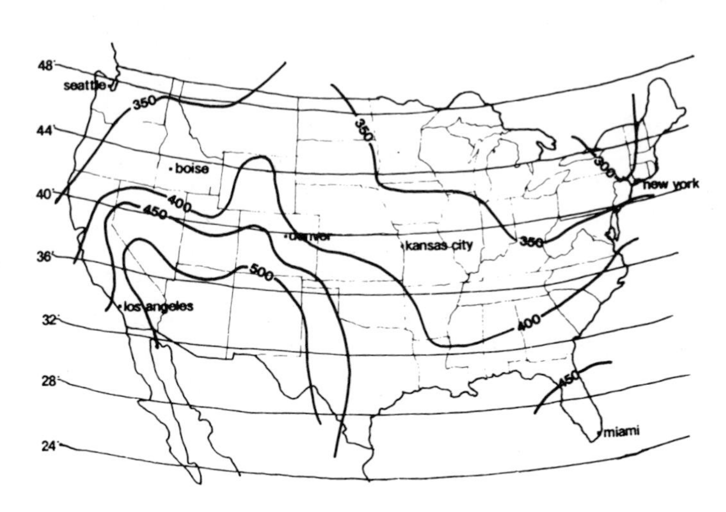

The following maps present the mean monthly hours of sunshine for the months of June and December and year round.

The source of these maps is the United States Department of Commerce, Weather Bureau. Copies can be obtained from the Superintendent of Documents, United States Government Printing Office, Washington, D.C. 20402.

Mean Monthly Hours of Sunshine in June

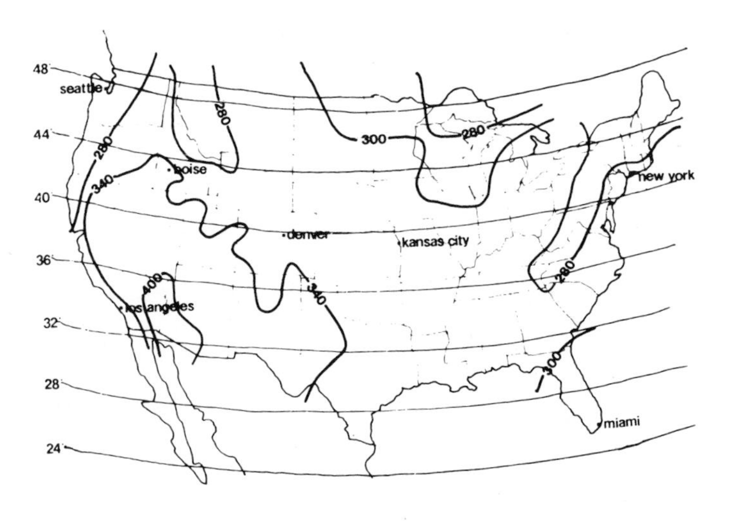

Mean Monthly Hours of Sunshine Annually

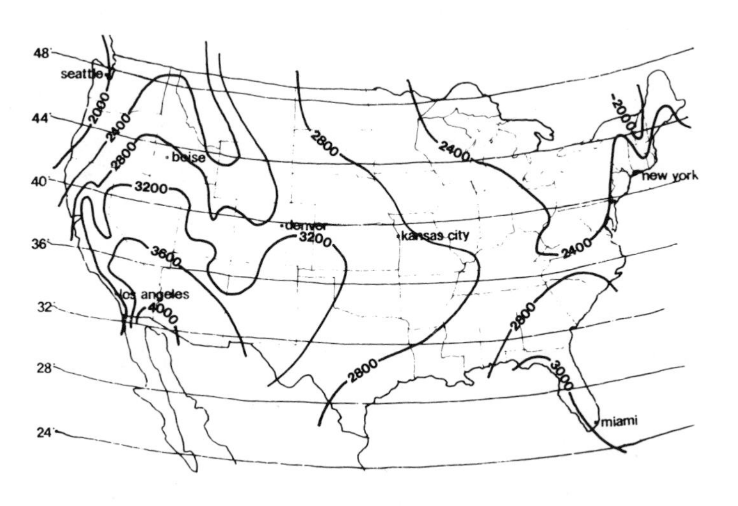

Mean Monthly Hours of Sunshine in December

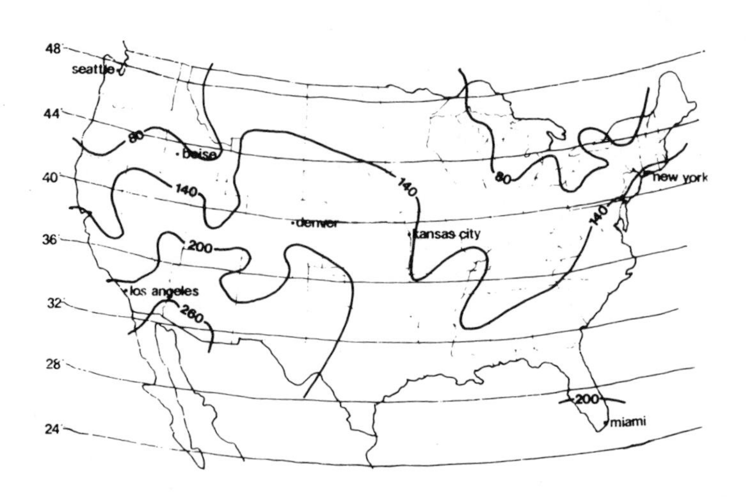

This map presents the average annual wind power (watts/square meter) that is available in the United States.

Average Annual Wind Power (watts/sq. meter)

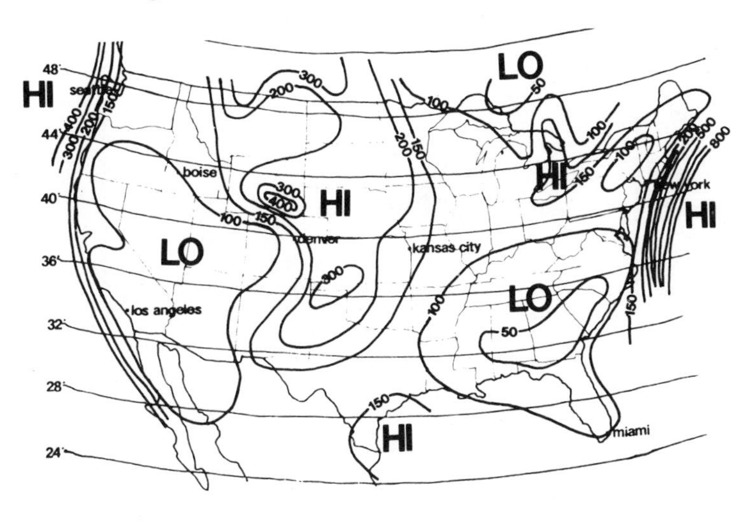

The following maps, based on mean daily insolation, present the relative suitability for winter and year-round solar-energy collection.

Relative Suitability for Solar Energy Collection (winter)

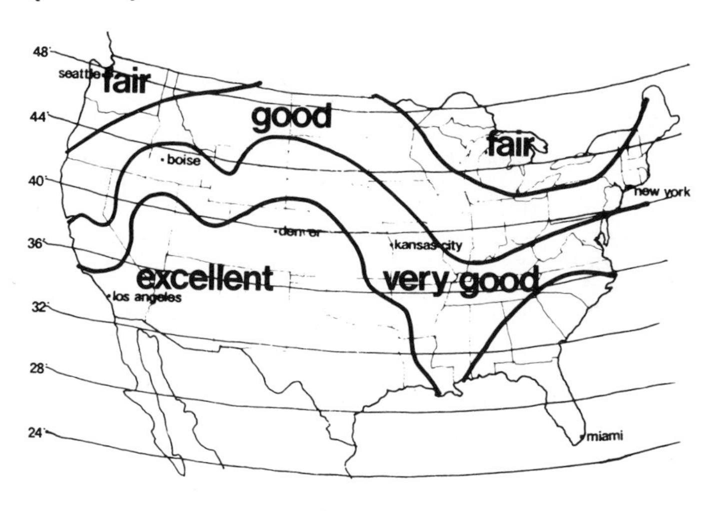

Relative Suitability for Solar Energy Collection (year-round)

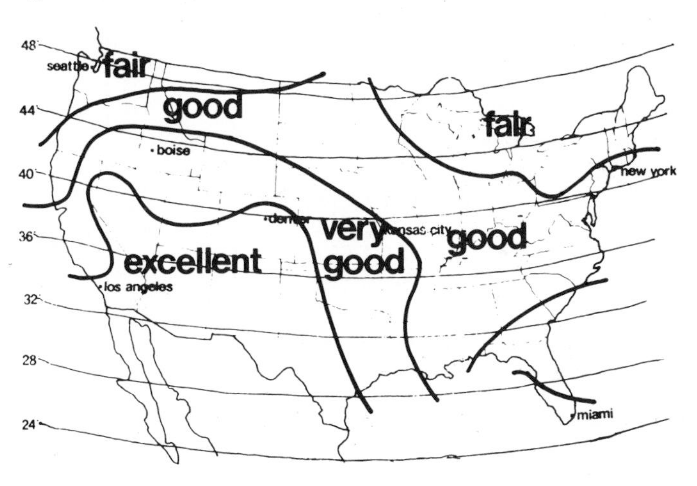

The following three charts present the average number of Btus per square foot per day striking surfaces tilted at various angles, in three geographic locations. The locations are Denver, Colorado; Nashville, Tennessee; and Ithaca, New York. Both climatic conditions and latitude of the locations were used to prepare the charts. These three cities were chosen for presentation because they represent divergent locations and also have unique quantities of available radiation. Data are available that can be used to prepare similar charts for most metropolitan areas in the United States.

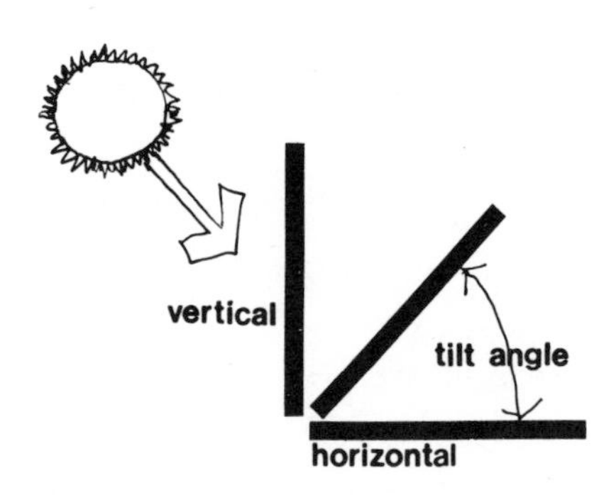

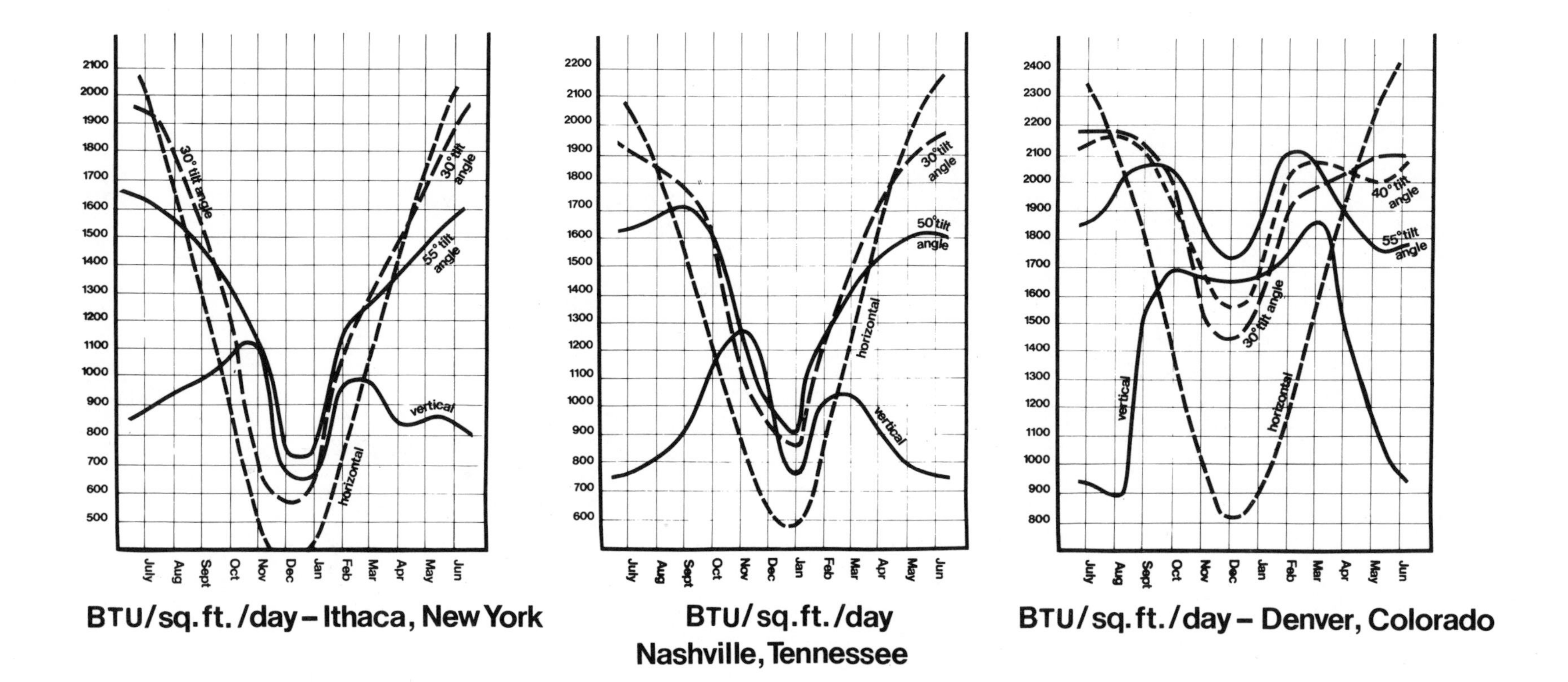

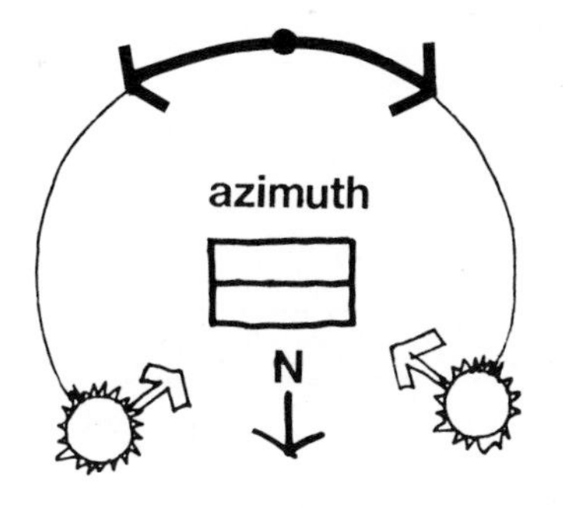

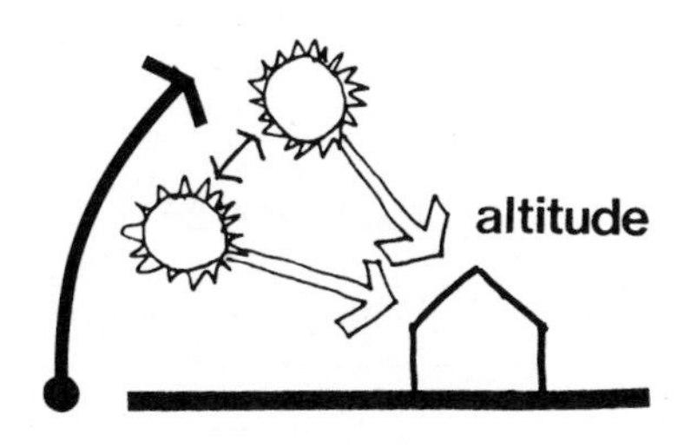

The following are charts of the sun's location in the sky relative to various northern latitudes. The altitude is measured in degrees above horizontal, and the azimuth is measured in degrees east or west of true south. The time is referenced to the sun, with noon occurring when the azimuth angle is zero degrees.

This information is valuable for calculating the recess depths of windows or for sizing wingwalls and overhangs where sun control to the interior is desired. It can be used for making most determinations relative to building components and the sun's position.

Latitude N 30°

SOLAR TIME AM	PM		JAN 21	FEB 21	MAR 21	APR 21	MAY 21	JUN 21	JUL 21	AUG 21	SEP 21	OCT 21	NOV 21	DEC 21
5	7	ALT												
		AZI												
6	6	ALT												
		AZI												
7	5	ALT	2	7	13	19	22	24	23	19	13	7		
		AZI	65	73	82	93	101	04	101	24	82	73		
8	4	ALT	14	19	26	32	35	37	36	32	26	20	14	11
		AZI	57	64	74	85	94	98	95	86	74	65	57	54
9	3	ALT	24	31	38	44	48	50	49	45	38	31	24	21
		AZI	47	54	63	76	87	92	88	77	63	54	47	44
10	2	ALT	32	40	49	57	61	63	61	57	49	40	32	29
		AZI	34	40	49	63	77	83	78	64	49	40	34	32
11	1	ALT	38	47	57	67	73	75	74	67	57	47	38	35
		AZI	18	22	28	40	57	67	59	41	28	22	18	17
12		ALT	40	49	60	72	80	83	81	72	60	50	40	37
		AZI	0	0	0	0	0	0	0	0	0	0	0	0

Latitude N 26°

SOLAR TIME AM	PM		JAN 21	FEB 21	MAR 21	APR 21	MAY 21	JUN 21	JUL 21	AUG 21	SEP 21	OCT 21	NOV 21	DEC 21
5	7	ALT												
		AZI												
6	6	ALT												
		AZI												
7	5	ALT	4	8	13	18	22	23	22	19	13	9	4	2
		AZI	65	74	83	94	102	106	103	95	83	74	66	62
8	4	ALT	16	21	27	32	35	36	35	32	27	21	16	14
		AZI	58	66	76	88	97	101	98	88	76	66	58	55
9	3	ALT	27	33	39	45	48	49	49	46	39	33	27	24
		AZI	48	56	66	80	91	96	92	81	66	56	48	45
10	2	ALT	36	43	51	58	62	63	62	59	51	43	36	33
		AZI	35	42	53	69	84	91	85	70	53	42	35	33
11	1	ALT	42	50	60	70	75	76	75	70	60	51	42	38
		AZI	19	24	31	47	70	83	72	49	31	24	19	18
12		ALT	44	53	64	76	84	87	85	76	64	54	44	41
		AZI	0	0	0	0	0	0	0	0	0	0	0	0

Latitude N 32°

SOLAR TIME AM	PM		JAN 21	FEB 21	MAR 21	APR 21	MAY 21	JUN 21	JUL 21	AUG 21	SEP 21	OCT 21	NOV 21	DEC 21
5	7													
6	6	ALT				6	10	12	11	7				
		AZI				100	107	110	108	101				
7	5	ALT	1	7	13	19	23	24	23	19	13	7	2	
		AZI	65	73	82	92	100	103	101	93	82	73	65	
8	4	ALT	13	19	25	31	35	37	36	32	25	19	13	10
		AZI	57	64	73	84	93	97	94	85	73	64	57	54
9	3	ALT	23	29	37	44	48	50	48	44	37	30	23	20
		AZI	46	53	62	74	85	89	86	75	62	53	46	44
10	2	ALT	31	39	47	56	61	62	61	56	47	39	31	28
		AZI	33	39	48	60	73	80	74	61	48	39	33	31
11	1	ALT	36	45	55	65	72	74	72	66	55	45	36	33
		AZI	18	21	27	38	52	61	53	38	27	21	18	16
12		ALT	38	47	58	70	78	82	79	70	58	48	38	35
		AZI	0	0	0	0	0	0	0	0	0	0	0	0

Latitude N 34°

SOLAR TIME AM	PM		JAN 21	FEB 21	MAR 21	APR 21	MAY 21	JUN 21	JUL 21	AUG 21	SEP 21	OCT 21	NOV 21	DEC 21
5	7	ALT												
		AZI												
6	6	ALT												
		AZI												
7	5	ALT		6	12	19	23	25	23	19	12	6		
		AZI		73	81	92	99	103	100	92	81	73		
8	4	ALT	11	18	24	31	36	37	36	32	24	18	12	9
		AZI	56	63	72	83	91	95	92	83	72	63	56	54
9	3	ALT	21	28	36	43	48	49	48	44	36	28	21	18
		AZI	45	52	61	72	82	87	83	73	61	52	46	43
10	2	ALT	29	37	46	55	60	62	60	55	46	37	29	26
		AZI	32	38	46	58	70	76	71	50	46	38	33	31
11	1	ALT	34	43	53	64	71	73	71	64	53	43	34	31
		AZI	17	20	26	35	47	55	49	36	26	20	17	16
12		ALT	36	45	56	68	76	79	77	68	56	46	36	33
		AZI	0	0	0	0	0	0	0	0	0	0	0	0

Latitude N 36°

SOLAR TIME AM	PM		JAN 21	FEB 21	MAR 21	APR 21	MAY 21	JUN 21	JUL 21	AUG 21	SEP 21	OCT 21	NOV 21	DEC 21
5	7	ALT						2						
		AZI						117						
6	6	ALT				7	12	14	12	7				
		AZI				99	106	109	106	99				
7	5	ALT		6	12	19	24	25	24	19	12	6		
		AZI		72	81	91	98	101	98	91	81	72		
8	4	ALT	10	17	24	31	36	38	36	31	24	17	10	8
		AZI	56	62	71	81	90	93	90	81	71	62	56	53
9	3	ALT	20	27	35	42	48	49	48	42	35	27	20	17
		AZI	44	51	59	70	80	84	80	70	59	51	44	42
10	2	ALT	27	35	45	54	60	62	60	54	45	35	27	24
		AZI	32	36	44	55	67	72	67	55	44	36	32	30
11	1	ALT	32	41	52	62	69	72	69	62	52	41	32	28
		AZI	17	20	24	33	43	49	43	33	24	20	17	15
12		ALT	34	43	54	66	74	77	74	66	54	43	34	30
		AZI	0	0	0	0	0	0	0	0	0	0	0	0

Latitude N 38°

SOLAR TIME AM	PM		JAN 21	FEB 21	MAR 21	APR 21	MAY 21	JUN 21	JUL 21	AUG 21	SEP 21	OCT 21	NOV 21	DEC 21
5	7	ALT												
		AZI												
6	6	ALT				7	12	14	13	6				
		AZI				99	106	109	106	100				
7	5	ALT		5	12	19	24	26	24	19	12	5		
		AZI		72	81	90	98	101	98	91	81	72		
8	4	ALT	9	16	23	31	36	37	36	31	23	16	9	7
		AZI	56	62	70	80	89	92	89	81	70	62	56	53
9	3	ALT	18	26	34	42	47	49	48	42	34	26	18	15
		AZI	44	50	58	69	78	82	79	70	58	51	45	42
10	2	ALT	26	34	43	52	58	61	59	53	43	34	26	22
		AZI	31	36	43	53	64	69	65	54	43	36	31	30
11	1	ALT	30	39	50	60	68	71	68	61	50	40	31	27
		AZI	16	19	24	31	40	46	41	31	24	19	16	15
12		ALT	32	41	52	64	72	75	73	64	52	42	32	29
		AZI	0	0	0	0	0	0	0	0	0	0	0	0

Latitude N 40°

SOLAR TIME AM	PM		JAN 21	FEB 21	MAR 21	APR 21	MAY 21	JUN 21	JUL 21	AUG 21	SEP 21	OCT 21	NOV 21	DEC 21
5	7	ALT					2	4	2					
		AZI					115	117	115					
6	6	ALT				7	13	15	13	8				
		AZI				99	106	108	106	100				
7	5	ALT		4	11	19	24	26	24	19	11	5		
		AZI		72	80	90	97	100	97	90	80	72		
8	4	ALT	8	15	23	30	35	37	36	31	23	15	8	6
		AZI	55	62	70	79	87	91	88	80	70	62	55	53
9	3	ALT	17	24	33	41	47	49	47	42	33	25	17	14
		AZI	44	50	57	67	76	80	77	68	57	50	44	42
10	2	ALT	24	32	42	51	58	60	58	52	42	32	24	21
		AZI	31	35	42	51	61	66	62	52	42	36	31	29
11	1	ALT	28	37	48	59	66	69	67	59	48	38	29	25
		AZI	16	19	23	29	37	42	78	30	23	19	16	15
12		ALT	30	39	50	62	70	74	71	62	50	40	30	27
		AZI	0	0	0	0	0	0	0	0	0	0	0	0

Latitude N 42°

SOLAR TIME AM	PM		JAN 21	FEB 21	MAR 21	APR 21	MAY 21	JUN 21	JUL 21	AUG 21	SEP 21	OCT 21	NOV 21	DEC 21
5	7	ALT												
		AZI												
6	6	ALT				8	13	15	14	8				
		AZI				99	105	108	106	99	90			
7	5	ALT		4	11	19	24	26	25	19	11	4		
		AZI		72	80	89	96	99	96	89	80	72		
8	4	ALT	7	14	22	30	35	37	36	30	22	14	7	4
		AZI	55	61	69	78	86	89	86	79	69	61	55	53
9	3	ALT	15	23	32	40	46	48	47	41	32	23	16	12
		AZI	44	49	56	66	74	78	75	66	56	49	44	42
10	2	ALT	22	30	40	50	56	59	57	50	40	31	22	19
		AZI	30	35	41	50	58	63	59	50	41	35	31	29
11	1	ALT	26	35	46	57	65	68	65	58	46	36	27	23
		AZI	16	18	22	28	35	39	35	28	22	18	16	15
12		ALT	28	37	48	60	68	71	69	60	48	38	28	25
		AZI	0	0	0	0	0	0	0	0	0	0	0	0

Latitude N 44°

SOLAR TIME AM	PM		JAN 21	FEB 21	MAR 21	APR 21	MAY 21	JUN 21	JUL 21	AUG 21	SEP 21	OCT 21	NOV 21	DEC 21
5	7	ALT					4	6	4					
		AZI					115	117	115					
6	6	ALT				8	14	16	14	8				
		AZI				98	105	107	105	98				
7	5	ALT		3	11	19	24	27	24	19	11	3		
		AZI		72	80	88	95	98	95	88	80	72		
8	4	ALT	6	13	22	30	35	37	35	30	22	13	6	3
		AZI	55	61	68	77	85	88	85	77	68	61	55	53
9	3	ALT	13	22	31	40	46	48	46	40	31	22	13	11
		AZI	43	48	55	64	72	76	72	64	55	48	43	41
10	2	ALT	20	28	38	49	51	58	51	49	38	28	20	17
		AZI	30	34	40	48	56	60	56	48	40	34	30	29
11	1	ALT	24	33	44	55	63	66	63	55	44	33	24	21
		AZI	15	18	21	27	33	36	33	27	21	18	15	15
12		ALT	26	35	46	58	66	69	66	58	46	35	26	22
		AZI	0	0	0	0	0	0	0	0	0	0	0	0

Latitude N 46°

SOLAR TIME AM	PM		JAN 21	FEB 21	MAR 21	APR 21	MAY 21	JUN 21	JUL 21	AUG 21	SEP 21	OCT 21	NOV 21	DEC 21
5	7	ALT												
		AZI												
6	6	ALT				8	14	17	15	9				
		AZI				98	104	107	105	99				
7	5	ALT		2	10	19	25	27	25	19	10	3		
		AZI		72	79	87	94	97	94	88	79	72		
8	4	ALT	5	12	20	29	35	37	35	30	20	12	5	2
		AZI	55	60	67	76	83	86	84	76	67	61	55	53
9	3	ALT	12	20	29	39	45	47	45	39	29	21	13	9
		AZI	43	48	54	63	70	74	71	63	54	48	43	41
10	2	ALT	19	27	37	47	54	57	55	48	37	27	19	15
		AZI	30	33	39	46	53	57	54	47	39	34	30	28
11	1	ALT	23	32	42	53	61	64	62	54	42	32	23	19
		AZI	15	17	20	25	30	33	31	25	20	17	15	15
12		ALT	24	33	44	56	64	67	65	56	44	34	24	21
		AZI	0	0	0	0	0	0	0	0	0	0	0	0

Latitude N 48°

SOLAR TIME AM	PM		JAN 21	FEB 21	MAR 21	APR 21	MAY 21	JUN 21	JUL 21	AUG 21	SEP 21	OCT 21	NOV 21	DEC 21
5	7	ALT					5	8	6					
		AZI					114	117	115					
6	6	ALT				9	15	17	15	9				
		AZI				98	104	106	104	98				
7	5	ALT		2	10	19	25	27	25	19	10	2		
		AZI		72	79	87	93	96	94	87	79	72		
8	4	ALT	4	11	20	29	35	37	35	29	20	11	4	
		AZI	53	60	67	75	82	85	82	75	67	60	55	
9	3	ALT	11	19	28	38	44	47	45	38	28	19	11	8
		AZI	43	47	53	61	68	72	69	62	53	47	43	41
10	2	ALT	17	26	35	46	53	56	54	46	35	26	17	14
		AZI	29	33	38	45	51	55	52	45	38	33	30	28
11	1	ALT	21	30	40	52	60	63	60	52	40	30	21	17
		AZI	15	17	20	24	29	31	29	24	20	17	15	14
12		ALT	22	31	42	54	62	65	63	54	42	32	22	19
		AZI	0	0	0	0	0	0	0	0	0	0	0	0

Following is a list of some of the most common household appliances and the average wattage of each when in use. This list can be used to estimate the amount of energy and ultimately the amount of money that is spent each year for their operation. The procedure is to first estimate the number of hours per year each appliance is operated, use the average wattage from the list to determine the annual number of kilowatt-hours (kwh) of electricity consumed by the appliance, and then with the cost per kwh of electricity figure the yearly cost of operation.

For example, estimate that a broiler is used for three hours per week, then the average yearly usage is $3 \times 52 = 156$ hours. During operation, the broiler requires 1430 watts. For the year the total consumption is $1430 \times 156 =$ 223,080 watt-hours. Dividing by 1000, the consumption is 223,080/1000 = 223 kwh.

If the price for electricity is 8¢/kwh, then the annual cost to operate the broiler is $17.84. Adding the costs to run individual appliances will yield the total annual cost for operating all household appliances.

Appliance	Average Wattage
Air cleaner	50
Air conditioner	1566
Blanket, electric	180
Blender	350
Bottle sterilizer	500
Bottle warmer	500
Broiler	1430
Carving knife	90
Clock	2
Clothes dryer, electric heat	4856
Clothes dryer, gas heat	325
Coffee pot	890
Corn popper	460-550
Curling iron	10-20
Deep fryer	1440
Dehumidifier	300-500
Dishwasher	1200
Disposal	375
Drill, electric	250
Egg cooker	510
Fan, attic	370
Fan, circulating	88
Fan, roll away	171
Fan, window	200
Food warming tray	350
Foot warmer	75
Floor polisher	300
Freezer, .42 cubic meter (15 cubic feet)	340
Freezer, .42 cubic meter (15 cubic feet), frost-free	440
Frying pan	1190
Germicidal lamp	20
Griddle	700
Grill	1000
Hair dryer	400
Heat lamp, infrared	250
Heater, portable	1300
Heating pad	70
Hot plate	1250
Humidifier	177
Iron, hand	1100
Knife sharpener	125
Mixer, food	120
Movie projector	500
Oven, microwave	1450
Radio, table	60
Radio, record player	100
Range, with oven	12200
Refrigerator/freezer .40 cubic meter (14 cubic feet)	320
Refrigerator/freezer frost-free, .40 cubic meter (14 cubic feet)	610
Roaster	1300
Rotisserie	1400
Sandwich grill	1160
Sewing machine	75
Shaver	12
Stereo system	200
Sun lamp	300
Toaster	1150
Toothbrush, electric	7
Trash compactor	400
TV—black/white, tube type	160
TV—black/white, solid state	55
TV—color, tube type	300
TV—color, solid state	200
Typewriter, electric	40
Vacuum cleaner	600
Vaporizer	300
Vibrator	40
Waffle iron	1100
Washing machine, automatic	500
Washing machine, nonautomatic	280
Waste disposer	440
Water heater	2470
Water heater, quick recovery	4474
Water pump, electric	450

Glossary

absorptance: a ratio of the radiation absorbed by a surface to the radiation incident on that surface.

absorption refrigeration: a cooling system that uses heat as its primary source of energy and evaporation as the cooling means.

air change: the replacement of the air contained within an enclosed space within a given period of time.

air-conditioning: the process of treating air to control its temperature, humidity, flow, cleanliness, and odor.

alternating current (AC): electrical current in which the direction of electron flow is reversed at regular intervals. In the United States, 60 hertz (cycles per second) is the standard frequency. In Europe, 50 hertz is standard.

altitude: angle of the sun above the horizon calculated in a vertical plane.

ambient temperature: the temperature of the outside air or air surrounding a space or building.

ampere (amp): a measure of the quantity of electric current flowing in a circuit. One volt applied across a resistance of one ohm will cause one amp to flow.

auxiliary furnace: a supplementary heating unit used to provide heat to a space when its primary heat source cannot do so adequately.

azimuth: measurement of the sun's position in plan view (from above), expressed as angular distance between true south and the point at the horizon directly below the sun.

barrel: a measure of a quantity of fluid, usually equal to 42 U.S. gallons, 5.6 cubic feet, or 159 liters.

battery: a device to store electrical energy as chemical potential energy. Within limits, most can be repeatedly charged and discharged.

berm: a man-made mound or small hill of earth.

bioconversion: conversion of solar energy to fuel by the natural process of photosynthesis.

biofuels: renewable fuels and energy sources derived from organic material (wood, methane, etc.).

biomass: a renewable resource, generally plant material and/or organic waste, that can be converted to provide a variety of fuels and energy products.

biosphere: the zone of air, land, and water, above and below the earth's surface, which is occupied by plants and animals.

blackout: a complete shutoff of electrical energy from a power-generating source caused by overloads, power shortages, or outages resulting from equipment or transmission breakdown.

boiler: a device used to heat water or to produce steam for space heating or other uses, including power generation.

British thermal unit (Btu): the amount of heat necessary to raise the temperature of one pound of water one degree Fahrenheit.

brownout: a reduction in line voltage planned to alleviate overloads on power generating equipment. Reduced voltage diminishes the brightness of incandescent lamps.

building envelope: exterior components of construction that enclose an interior space.

calorie: the amount of heat necessary to raise the temperature of one gram of water one degree Celsius.
change of state (phase change): the change from the solid, liquid, or gaseous state to either of the other two.
coefficient of performance (COP): the ratio of the energy output of a device, such as a heat pump, to the energy input.
collector tilt: the angle at which a solar collector is inclined with respect to a horizontal plane.
color rendition: the effect of a light source on an object's perceived color.
color temperature: the measurement in degrees Kelvin of the color output of a light source.
comfort zone: the range of temperatures and humidities over which most adults feel comfortable under normal living and working conditions.
concentrator: a device used to intensify the solar radiation striking a surface.
condensation: the process of changing a vapor into a liquid by the extraction of heat.
condenser: a component of a system in which a working fluid undergoes a change of state from gas to liquid by the rejection of latent heat to a cooling medium, producing a heating effect.
cooling load: the amount of heat that must be removed from a building to maintain a comfortable temperature, measured in Btu/hr or tons of air-conditioning. (1 ton = 12,000 Btu/hour or 3515 watts).
cooling pond: a body of water that dissipates heat by evaporation, convection, and radiation.
Crooke's radiometer: a partially evacuated hollow glass sphere containing vanes that are black on one side and silvered on the other side. These spin in the presence of thermal radiation caused by differing rates of radiation absorption between the silvered and the blackened surfaces.
crude oil: the natural mixture of liquid hydrocarbons extracted as petroleum from under the earth's surface. Also may include oil extracted from tar sands and oil shale.
daylighting: the use of natural light for illumination
degree day, heating: a unit used to estimate the heating-fuel consumption and the nominal heating load of a building. For any one day, when the mean temperature is less than 18.3° C (65 F), there exist as many degree days as degrees Celsius × 1.8 (degrees F) difference in temperature between the mean temperature for the day and 18.3° C (65 F). The sum of the degree days constitutes the annual degree day heating requirements.
design conditions: selected indoor and outdoor wet-bulb and dry-bulb temperatures for a specific location that determine the maximum heating and cooling loads of a building.
direct current (DC): electrical current in which the direction of electron flow never changes.
direct gain: solar radiation directly intercepting or entering a building.
disability glare: reflected light that impairs a vieer's ability to accurately discern an object, perceive color, or read printed matter.
dry-bulb temperature: the local air temperature as indicated by a dry-temperature measuring sensor (the one with which we are most familiar).
duct: a conduit or tube through which air or other gases flow.
economizer cycle: a cooling mode that uses cool outdoor air to offset heat gains in building rather than using an energy-consuming cooling device.
efficiency, thermal: the ratio of the useful heat at the point of use to the thermal energy input for a designated time period; expressed as percent.
effluent: discharged waste suspended in liquid or gas.
electric-demand limiter: a device that selectively switches off electrical equipment whenever total electrical demand rises beyond a predetermined level.
electric-resistance heat: the conversion of electrical energy to heat energy, using an electrical resistor. May be applied to space heating or water heating. 1Kwh = 3,413 Btu.
electricity: the interaction between particles of positive and negative charge. Utilized as a flow of electrons, producing electric current.
electromagnetic spectrum: the entire range of wavelengths of electromagnetic radiation, extending from gamma rays to the longest radio waves.
emittance: a ratio of the amount of energy radiated by a surface to that of a black body (a perfect absorber of heat that emits none) at the same temperature.
energy: the capacity for doing work; taking a number of forms that may be transformed from one into another, such as thermal (heat), mechanical (work), electrical, and chemical; in customary units, measured in kilowatt hours (kwh) or British thermal units (Btu); in SI units, measured in joules (J), where 1 joule = 1 watt-second.
direct energy: energy used in its most immediate form, that is, natural gas, electricity, oil.
indirect energy: energy that is converted into goods that are then consumed. Examples: food through photosynthesis, fibers, plastics, chemicals.
net energy: the energy remainder or deficit after the energy costs of extracting, concentrating, and distribution are subtracted.
net reserves: an estimate of the net energy that can be delivered from a given energy resource.
energy storage: the ability to retain energy by converting it to a form (gravitational potential, chemical, etc.) from which it can be retrieved for useful purposes.
ethyl alcohol: a colorless, volatile, flammable liquid (C_2H_5OH) that is produced by the fermentation of grains and fruits.
eutectic salt: a combination of two (or more) mutually soluble materials, requiring a large amount of heat to melt or an equally large amount of heat to solidify, that melts or freezes at constant temperature and with constant composition. The phase-change temperature is the lowest achievable with these materials.
evaporator: a component of a system in which a working fluid undergoes a change of state from liquid to gas, taking on latent heat and thereby providing a cooling effect.
fan coil: a heating or cooling device that forces air through heating or cooling coils.
fission: the splitting of an atomic nucleus, releasing large amounts of energy. Usually the uranium-235 atom is split, producing heat for the generation of electricity with steam.

flow rate: the volume or weight per unit time of a fluid flowing through an opening or duct.
flue: the exhaust channel through which gas and fumes produced by combustion exit from a building.
footcandle: unit of measurement of the intensity of light on a surface that is everywhere one foot from a uniform point source of light from a standard (sperm whale oil) candle and equal to one lumen per square foot.
fossil fuel: a class of organic compounds formed by the decay of matter under the influence of heat and pressure over millions of years, that when burned liberate large quantities of energy; exemplified by coal, oil, and natural gas.
frostline: the depth of frost penetration in the earth. This depth varies from one geographic location to another.
fuel: any substance that can be burned to produce heat.
fuel cell: a device in which hydrogen and oxygen are combined in an electrochemical reaction to generate electricity and produce water as a by-product.
fusion (nuclear): the release of energy by the formation of a heavier nucleus from two lighter ones.
geothermal energy: heat energy contained in large underground reservoirs of steam and hot water, produced by molten material from the earth's interior.
heat exchanger: a device used to transfer heat from one temperature level to another.
heat gain: an increase in the amount of heat contained in a space, resulting from solar radiation and the heat given off by people, lights, equipment, machinery, and other sources.
heat loss: a decrease in the amount of heat contained in a space, resulting from heat flow through walls, windows, roof, and other components of the building envelope and from the infiltration of cold outdoor air.
heat pipe: a closed pipe containing a liquid and a wick that will transfer heat from one end to the other end without any input of work.
heat pump: a reversible refrigeration system that delivers more heat energy to the end use than is input to the compressor. The additional energy input results from the absorption of heat from a low temperature source.
heat sink: a body or substance that is capable of accepting or rejecting heat.
heat wheel: a device used in ventilating systems that tends to bring incoming air into thermal equilibrium with exiting air. As a result, hot summer air is cooled and cold winter air is warmed.
heliodon: a device used to simulate the effect the sun's position has on models of buildings and other objects; used primarily to conduct shadowing studies.
heliostat: an instrument consisting of a mirror mounted on an axis, mechanically rotated to steadily reflect the sun in one direction.
heliotropism: property of being able to follow the sun's apparent motion across the sky.
horsepower: a standard unit of power equal to 746 watts, 2545 Btu per hour, or 550 foot-pounds per second.
humidistat: a device to continuously monitor the humidity in a space and activate the humidity-control equipment to ensure that the humidity is maintained at a preset level.
HVAC: abbreviation for heating, ventilating, and air conditioning.
hydroelectric plant: an electrical power generating plant in which the kinetic energy of water is converted to electrical energy using a turbine generator.
hydronic heating: a heating system in which liquid is used for heat transport.
illumination: light provided to interior and exterior spaces.
indirect gain: solar thermal gains from diffuse skyvault irradiance as well as other surfaces external to the building.
infiltration: the uncontrolled flow of air into a building through cracks, openings, doors, or other areas that allow air to penetrate.
insolation (*In*cident *Sol*ar Radi*ation*): the amount of solar radiation striking a surface during a specified period of time.
kilowatt (Kw): a unit of power equal to 1000 watts.
kilowatt hours (Kwh): a unit of work equal to the consumption of 1000 watts in one hour.
langley: a measure of irradiation in terms of langleys per minute, where one langley equals 1 calorie per square centimeter. Named in honor of American astronomer Samuel P. Langley.
latent heat: a change in heat content caused by a change in state that occurs without a corresponding change in temperature.
load: the demand on the operating resources of a system.
lumen: a unit measure of the light output of a lamp, where one lumen provides an intensity of one footcandle at a distance of one foot from the light source.
lux: a measure of light intensity on a plane, denoting lumens per square meter.
megawatt: a unit of power equal to one million watts or one thousand kilowatts.
methane: a colorless, odorless, flammable gaseous hydrocarbon (CH_4), which is the product of the decomposition of organic matter. It is the major component of natural gas.
peak load: the maximum energy demand placed on a system.
photolysis: chemical decomposition caused by the action of solar radiant energy.
photosynthesis: the production of chemical compounds in plants using solar radiant energy, specifically the production of carbohydrates by plants containing chlorophyll using sunlight, carbon dioxide, and water.
phototropism: growth or movement in response to a light source.
plenum: a compartment for the passage and distribution of air.
power: the rate at which work is performed or energy expended.
quad: Equals 10^{15} Btu (one quadrillion Btu), or 10.558×10^{18} joules.
raw energy source: an original unrefined source of energy.
recovered energy: energy utilized that would otherwise be wasted.
recycle: the process of recovering resources from waste material for reprocessing and reuse.

refinery: a chemical processing plant in which crude oil is separated into more useful hydrocarbon compounds.
reflectance: the ratio of the amount of radiation reflected by a surface to the amount of radiation incident on the surface.
R-value: the measure of a material's resistance to heat flow. The higher the R-value, the greater the resistance.
selective surface: pertaining to solar collectors, the surface produced by the application of a coating to an absorber plate, with spectral selective properties, which maximizes the absorption of incoming solar radiation (0.3 to 3.0 micron range) and emits much less radiation (3.0 to 30.0 micron range) than an ordinary black surface at this same temperature would emit.
sensible heat: heat that produces a temperature change.
Skyshaft: a multichambered Plexiglas device that penetrates a roof used to provide natural interior illumination with a minimum of heat loss or gain.
solar cell: also, photovoltaic cell. A device employing crystals (silicon, for example) that, when exposed to solar radiation, generates an electric current.
solar cooker: a device for cooking that uses the sun as an energy source.
solar energy: energy in the form of electromagnetic radiation received from the sun.
solar power farm: an installation for generating electricity on a large scale using solar energy, consisting of an array of solar collectors, steam or gas turbines, and electrical generators.
space heating: interior heating of a building or room.
stratification: the tendency of heated air or fluid to rise and arrange itself in layers. The top layer is warmer than the bottom.
sun time: time of day at a specific location as determined by the position of the sun.
sun tracking: the ability to follow the apparent motion of the sun across the sky.
therm: a quantity of heat equal to 100,000 Btu, or 1055.87×10^5 joules.
thermal transmission: the passage of heat through a material.
thermometer: an instrument for measuring temperature.
thermostat: a device to continuously monitor the temperature in a space and activate temperature-control equipment to ensure that the temperature in the space is maintained at a preset level.
ton (of air conditioning): the thermal refrigeration energy required to create one ton of ice (907 kg) in one day. Equals 12,000 Btu/hour (3515 watts).
torque: a turning or twisting force.
U-value: a coefficient that indicates the rate at which energy (Btu/hour) passes through a component for every degree (Fahrenheit) of temperature difference between one side and the other under steady state conditions. The reciprocal of R-value.
vapor barrier: a component of construction that is impervious to the flow of moisture, used to prevent moisture travel to a point where it may condense.
vent: any penetration through the building envelope that is specifically designed for the flow of air into or from the building.
ventilation air: outside air that is intentionally caused to enter an interior space.
weatherstripping: foam, metal, or rubber strips used to form an airtight seal around windows, doors, or openings to reduce infiltration.
wet-bulb temperature: the local air temperature as indicated by a wet-temperature measuring sensor.
wind energy: the kinetic energy of air motion over the earth's surface caused by the sun's heating of the atmosphere.
wind machine: any one of a number of devices used to convert the kinetic energy of wind to another form of energy for useful purposes.
work: the transfer of energy from one physical system to another. In mechanics, the transfer of energy to a body by the application of force.

Bibliography

The annotated bibliography that follows contains the references used in the preparation of this book, plus other publications that can be of value for exploring concepts beyond the scope of those presented here. The bibliography is divided into major categories to facilitate access to publications on specific topics, with comments included on each entry to expedite the selection process.

General Energy and Alternative Energy Sources

Abelson, Philip H. *Energy: Use, Conservation, and Supply*. Washington, DC: American Association for the Advancement of Science, 1974.
Twenty-four articles on new and developing energy approaches.

Alves, Ronald. *Living with Energy*. New York: Viking Press, Penguin Books, 1978.
Descriptions of many energy and life-style alternatives (appropriate technology). Includes bibliography and resource listing.

The Biosphere. A *Scientific American* Book. San Francisco: W.H. Freeman and Company, 1970.
A collection of eleven articles on the earth's biosphere.

Clark, Peter. *Natural Energy Workbook*. Berkeley: Visual Purple, 1974.
Presents basic concepts related to solar, wind, and other alternative-energy source systems.

Clark, Wilson. *Energy for Survival: The Alternative to Extinction*. Garden City, NY: Doubleday and Company, Anchor Press, 1975.
This book not only examines the advantages and disadvantages of fuels currently in use, but also details numerous natural energy principles. The energy policy and habits of American society are examined.

Davis, Albert J. and Schubert, Robert P. *Alternative Natural Energy Sources in Building Design*. Second edition, revised. New York: Van Nostrand Reinhold Company, 1981.
Discusses alternative energy solutions and applications, and energy-conservation techniques.

Energy: AIA Energy Notebook. Washington, DC: American Institute of Architects, 1975—.
Ongoing notebook with references to books, articles, and regulation involving energy and its use.

Energy and Power. A *Scientific American* Book. San Francisco: W.H. Freeman and Company, 1971.
A collection of eleven *Scientific American* articles on the general use of energy and its effect on various cultures.

Gabel, Medard. *Energy, Earth, and Everyone: A Global Energy Strategy for Spaceship Earth*. San Francisco: Straight Arrow Books, 1975.
Developed from the World Game Workshop sponsored by Earth Metabolic Design of Connecticut. Presents an analysis of existing and alternative energy sources worldwide. Includes a foreward by R. Buckminster Fuller.

Gay, Larry. *The Complete Book of Heating with Wood.* Charlotte, VT: Garden Way Publishing Company, 1974.
Information on finding and preparing wood for burning and explanation of numerous types of stoves.

Hammond, Allen L.; Metz, William D.; and Maugh, Thomas H., II. *Energy and the Future*. Washington, DC: American Association for the Advancement of Science, 1973.
Analysis of various energy sources and the implications of their use.

Hayes, Denis. *Rays of Hope: The Transition to a Post-Petroleum World*. A Worldwatch Institute Book. New York: W.W. Norton and Company, 1977.
The author, former director of the Solar Energy Research Institute, describes the energy problem with a global perspective and outlines the transition to increased conservation and development of alternative and renewable energies.

Hunt, V. Daniel. *Energy Dictionary*. New York: Van Nostrand Reinhold Company, 1979.
A compilation of more than four thousand terms associated with the energy field.

Kreith, Frank. *Principles of Heat Transfer*. Third edition. New York: Intext Educational Pubs., 1973.
A textbook presenting engineering and technical aspects of thermal conduction, radiation, and convection.

Loftness, Robert L. *Energy Handbook*. New York: Van Nostrand Reinhold Company, 1978.
A large collection of data and information on energy and energy sources. Useful for in-depth research.

Lovins, Amory. *Soft Energy Paths: Toward a Durable Peace*. Cambridge, MA: Ballinger Publishing Company, 1972.
A detailed rationale for an immediate start towards conversion to solar and "soft" energy technologies.

Man and the Ecosphere. A *Scientific American* Book. San Francisco: W.H. Freeman and Company, 1971.
A collection of twenty-seven articles on man's ecosphere and the environmental effect of human actions.

Mother Earth News. Handbook of Homemade Power. New York: Bantam Books, 1974.
Many short articles and interviews containing ideas, facts, and tips on the use of renewable resources and alternative energies.

Portola Institute. *Energy Primer: Solar, Water, Wind, and Bio-fuels*. Menlo Park, CA: Portola Institute, 1974.
A collection of articles on solar, wind, water, architecture, and integrated energy systems, with a *Whole Earth Catalog*-style listing of energy-related hardware and information.

Prenis, John, ed. *Energy Book #1: Natural Sources and Backyard Applications*. Philadelphia: Running Press, 1975.
Collection of articles and information on wind, solar, and biosystems energy generation.

____. *Energy Book #2: More Natural Sources and Backyard Applications*. Philadelphia: Running Press, 1977.
Collection of articles by a variety of authors on the alternative energy field, including solar, wind, and methane technologies.

Schumacher, E.F. *Small is Beautiful: Economics as if People Mattered*. New York: Harper and Row, Publishers, 1973.
A pioneering book introducing the concept of appropriate technology, "technology with a human face."

Shelton, Jay W. and Shapiro, Andrew B. *The Woodburner's Encyclopedia*. Waitsfield, VT: Vermont Crossroads Press, 1976.
A comprehensive book dealing with wood as an alternative energy resource; includes a manufacturers listing, specifications chart, and chapters on economics and safety.

Skurka, Norma and Naar, John. *Design for a Limited Planet: Living with Natural Energy*. New York: Ballantine Books, 1976.
Brief descriptions of a great many alternative energy residences.

Sofer, Samir S. and Zaborsky, Oskar R., eds. *Biomass Conversion Processes for Energy and Fuels*. New York: Plenum Press, 1981.
Introductory textbook describing biomass sources, conversion processes, and technical and economic considerations needed for effective conversion.

Solar Energy Research Institute. *A New Prosperity: Building a Sustainable Energy Future*. Andover, MA: Brick House Publishing Company, 1981.
The report on solar energy and conservation produced by the Solar Energy Research Institute and suppressed by the Department of Energy concludes it is possible to engineer a series of events that would result in a full-employment economy, increased worker productivity and national income, and reduced national energy consumption by nearly twenty-five percent, with twenty to thirty percent of this reduced demand being supplied by renewable resources.

Stobaugh, Robert and Yergin, Daniel, eds. *Energy Future: Report of the Energy Project of the Harvard Business School*. New York: Random House, 1979.
Analysis of projected future trends and solutions to the energy crisis.

Stoner, Carol H., ed. *Producing Your Own Power*. Emmaus, PA: Rodale Press, 1974.
Describes small-scale power production from alternative energy sources; e.g., wind, water, wood, methane, and sun.

United States Council on Environmental Quality. *The Costs of Sprawl: Environmental and Economic Costs of Alternative Residential Development Patterns at the Urban Fringe*. Two volumes. Washington, DC: United States Government Printing Office, 1974.
A report on the effects of low-density sprawl on energy, the environment, and human life.

Vale, Brenda and Vale, Robert. *The Autonomous House: Designing and Planning for Self-Sufficiency*. New York: Universe Books, 1975.
Use of solar, wind, methane generation, waste and water reclamation, and other alternative energy technologies to achieve self-sufficient living.

Voegli, Henry E. and Tarrant, John J. *Survival 2001: Scenario from the Future*. New York: Van Nostrand Reinhold Company, 1975.
A book that presents alternative energy sources: wind, solar, water, wave, and ocean power, from a futuristic standpoint.

Wilson, Mitchell. *Energy*. New York: Time, Inc., 1963.
General energy book covering principles and application of various energy sources.

Energy Conservation

Browne, Dan. *Alternative Home Heating*. New York: Holt, Rinehart, and Winston, 1980.

A guide to residential alternative heating systems. Describes solar systems, wood-burning stoves and fireplaces, and heat pumps.

Clegg, Peter. *New Low-Cost Sources of Energy for the Home*. Charlotte, VT: Garden Way Publishing Company, 1975.
Describes use of solar, wind, water, wood heating, and water and waste systems as sources of residential energy.

Clovis Heimsath Associates. *Energy Conservation in Tract Housing*. Washington, DC: American Institute of Architects, 1974.
Examines the effect of incorporating energy-conserving modifications into existing tract housing.

Dubin, Fred S. and Long, Chalmers G., Jr. *Energy Conservation Standards for Building Design, Construction, and Operation*. New York: McGraw-Hill Book Company, 1979.
Basic outline of concepts and directions needed for energy-efficient building design.

Farallones Institute. *The Integral Urban House: Self-Reliant Living in the City*. San Francisco: Sierra Club Books, 1979.
Applications of appropriate technology and energy-conservation systems to the urban environment.

Griffin, C.W., Jr. *Energy Conservation in Buildings: Techniques for Economical Design*. Washington, DC: Construction Specifications Institute, 1974.
Presents principles of mechanical systems and building design necessary for energy-conserving structures.

Kern, Ken. *The Owner-Built Home*. New York: Charles Scribner's Sons, 1975.
Presents alternatives to the conventional contractor-built residence, with ideas on how to use the surroundings of a house to the best advantage.

Kern, Barbara and Kern, Ken. *The Owner-Built Homestead*. New York: Charles Scribner's Sons, 1977.
Gives information and advice on self-reliant living through small-scale intensive farming.

Knowles, Ralph. *Energy and Form: An Ecological Approach to Urban Growth*. Cambridge, MA: MIT Press, 1974.
A worthwhile book on the historical analysis of primitive architecture and design concepts. Study includes the development of forms that respond to natural and climatic stimuli.

Langdon, William K. *Movable Insulation*. Emmaus, PA: Rodale Press, 1980.
A guide for homeowners, illustrating many different movable-insulation systems for reducing heating and cooling losses through windows.

Leckie, Jim; Masters, Gil; Whitehouse, Harry; and Young, Lily. *More Other Homes and Garbage: Designs for Self-Sufficient Living*. Revised edition. San Francisco: Sierra Club Books, 1981.
Expanded, revised, and updated version of the original, 1975 book explaining appropriate technology and alternative energy sources for decentralized application to residences. A bibliography appears at the end of each chapter.

Morrison, James W. *The Complete Energy Savings Handbook for Homeowners*. New York: Harper and Row, Publishers, 1977.
Describes residential energy-conservation measures, including a section on tax credits.

Parks, Alexis. *People Heaters: A People's Guide to Keeping Warm in Winter*. Andover, MA: Brick House Publishing Company, 1981.
Discusses how to live without central heating systems. Ideas about clothing, beds, saunas, greenhouses, and other devices to heat people instead of spaces.

Schoen, Richard; Hirshberg, Alan; and Weingart, Jerome. *New Energy Technologies for Buildings*. Cambridge, MA: Ballinger Publishing Company, 1975.
A report on new energy-conserving technologies for buildings, barriers. Incentives for their use, recommendations.

Steadman, Philip. *Energy, Environment, and Building*. New York: Cambridge University Press, 1975.
Includes articles on solar and wind energy, plus methods for energy conservation in buildings. Numerous examples and excellent descriptions of energy-conserving buildings in the United States.

Stein, Richard G. *Architecture and Energy*. Garden City, NY: Doubleday and Company, Anchor Press, 1977.
Describes current energy-use patterns in architecture and gives examples of energy-conserving design practices.

Stoner, Carol H., ed. *Goodbye to the Flush Toilet*. Emmaus, PA: Rodale Press, 1977.
Presents alternatives to the conventional flush toilet. Describes composting toilets, and grey-water treatment and other techniques for home water conservation.

United States Department of Commerce, National Bureau of Standards. *Energy Conservation Program Guide for Industry and Commerce*. NBS Handbook 115. Prepared by R.R. Gatts, R.G. Massey, and John C. Robertson. Washington, DC: United States Government Printing Office, 1974.
Proposes new standards to American industry and commerce for decreasing their total energy consumption.

____. *Technical Options for Energy Conservation in Buildings*. NBS Technical Note 789. Washington, DC: United States Government Printing Office, 1973.
Investigates energy-conserving options for new and existing buildings. Several energy-conservation features are discussed.

____. *Window Design Strategies to Conserve Energy*. NBS Building Science Series 104. Prepared by Robert S. Hastings and Richard W. Crenshaw. Washington, DC: United States Government Printing Office, 1977.
Discusses thirty-three window design strategies, covering six energy functions of windows: solar heating, daylighting, shading, insulation, airtightness, and ventilation. Lists advantages and disadvantages of each strategy and gives references.

United States Department of Energy. *Minimum Energy Dwelling (MED) Workbook: An Investigation of Techniques and Materials for Energy Conscious Design*. Prepared by Burt Hill Kosar Rittlemann Associates. Springfield, VA: National Technical Information Service, United States Department of Commerce, 1977.
Presents issues dealing with energy conservation in dwellings. Includes an extensive glossary.

United States General Services Administration. *Energy Conservation Design Guidelines for New Office Buildings*. Second edition. Washington, DC: United States Government Printing Office, 1975.
Energy-conservation guidelines for new federal office buildings are presented. Most of the energy-consuming aspects of office buildings are analyzed.

Villeco, Marguerite, ed. *Energy Conservation in Building Design*. Washington, DC: American Institute of Architects, 1974.
Includes design methods for maximizing energy efficiency in building design. Also included are analyses of energy uses in existing buildings.

Walker, Howard V. *Energy Conservation Design Resource Handbook*. Ottawa, Ontario, Canada: Royal Architectural Institute of Canada, 1979.
A comprehensive handbook supplying answers to design questions about energy use; concise style. All data are presented in metric form.

Architecture

American Society of Heating, Refrigerating, and Air-Conditioning Engineers. *ASHRAE Handbook and Product Directory: 1974 Applications*. New York: ASHRAE, 1974.
Presentation of technical data and information on heating, ventilating, and air-conditioning systems, units, and components.

____. *Handbook of Fundamentals*. New York: ASHRAE, 1972.
Authoritative reference on theory and basic information needed for development, system design, and application of heating, ventilating, and air-conditioning equipment.

American Society of Landscape Architects Foundation. *Landscape Planning for Energy Conservation*. Reston, VA: Environmental Design Press, 1977.
Description of climatic analysis and its impact; discussion of energy-conserving siting and design.

Arizona State University, College of Architecture. *Earth Integrated Architecture*. Tempe, AZ: Arizona State University, College of Architecture, 1974.
A notebook covering the applications and implementation of below-grade architectural principles.

Aronin, Jeffrey Ellis. *Climate and Architecture*. New York: Reinhold Publishing Corp., 1953.
An introduction to the effects of macroclimate and microclimate and their application to the design of buildings and towns.

Butler, Lee Porter. *Ekose'a Homes: Natural Energy-Conserving Designs*. San Francisco: Ekose'a, 1978.
Describes the concepts of the double-skin, or "double envelope," home that uses gravity convection to maintain thermal equilibrium. Ten designs are illustrated.

Campbell, Stu. *The Underground House Book*. Charlotte, VT: Garden Way Publishing Company, 1980.
A fairly detailed guide to underground housing, dealing with design considerations, economics, water, and site conditions.

Ching, Francis D.K. *Architecture: Form, Space, and Order*. New York: Van Nostrand Reinhold Company, 1979.
Examines the principles of form, space, and order in architectural design. Discusses point, line, plane, volume, proportion, circulation, and the interdependence of form and space.

Dresser, Peter van. *A Landscape for Humans*. Albuquerque, NM: Biotechnic Press, 1972.
Examines the rationale behind regional and local planning concepts and the implications of more human-oriented planning processes.

Evans, Benjamin H. *Daylight in Architecture*. New York: McGraw-Hill Book Co., 1981.
Explains the architectural design potential of daylight, describing principles and techniques for implementation and giving numerous case studies. Use of scale models for design evaluation is discussed, as is integration of daylighting with other design concerns.

Givoni, Baruch. *Man, Climate, and Architecture*. Second edition. New York: Van Nostrand Reinhold Company, 1981.
Presents an in-depth discussion of the relationship of the psychological, physical, and architectural aspects of man, climate, and architecture to interior and exterior building design.

Grillo, Paul Jacques. *Form, Function, and Design*. New York: Dover Publications, 1960.
Presents an analysis of design through a series of variables, including climate, orientation, natural archetypes, proportion, energy, mass, and motion.

Halprin, Lawrence. *The RSVP Cycles: Creative Processes in the Human Environment*. New York: George Braziller, 1969.
Presents design analysis in response to human needs and the environment.

Hancocks, David. *Master Builders of the Animal World*. New York: Harper and Row, Publishers, 1973.
Examines the design habits and dwellings developed by animals in response to their individual climates and needs.

Kahn, Lloyd, ed. *Shelter*. Bolinas, CA: Shelter Publications, 1973.

____. *Shelter II*. Bolinas, CA: Shelter Publications, 1978.
This book and the previous entry present many examples of indigenous buildings and give information about hand-built housing and construction techniques.

Laseau, Paul. *Graphic Thinking for Architects and Designers*. New York: Van Nostrand Reinhold Company, 1980.
Discusses graphic thinking in relation to community and public design processes and describes sketching techniques and graphic language.

Lynch, Kevin. *Site Planning*. Second edition. Cambridge, MA: MIT Press, 1971.
Illustrates principles of climate-responsive site planning and site-oriented building organization.

McGuiness, William J. and Stein, Benjamin. *Mechanical and Electrical Equipment for Buildings*. New York: John Wiley and Sons, 1974.
A textbook covering the design, techniques, and construction of mechanical systems for buildings.

McHarg, Ian L. *Design with Nature*. Garden City, NY: Doubleday and Company, Natural History Press, 1969.

Presents an ecological approach to planning and development. Used as a textbook.
Markus, T.A. and Morris, E.N. *Buildings, Climate, and Energy.* New York: Pitman Publishing, 1980.
An overview of the energy implications of building design. Discusses human response and thermal comfort, climatic factors, heat gain and loss, economics, and building configuration.
Olgyay, Victor. *Design with Climate: Bioclimatic Approach to Architectural Regionalism*. Princeton, NJ: Princeton University Press, 1963.
Describes the design of buildings to respond to the natural forces of climate and the sun. Examples of many different project sizes and applications.
Rubenstein, Henry M. *A Guide to Site and Environmental Planning.* New York: John Wiley and Sons, 1969.
An analysis of landscaping, site planning, and environmental concerns.
Shurcliff, William A. *Super Insulated Houses and Double Envelope Houses.* Andover, MA: Brick House Publishing Company, 1981.
Describes and gives specific examples of each type of house, with detailed analysis of construction, costs, and ventilation.
____. *Thermal Shutters and Shades*. Andover, MA: Brick House Publishing Company, 1980.
Presentation and critique of a great variety of movable insulation designs.
Sizemore, Michael M.; Ogden, Harry; and Ostrander, William S. *Energy Planning for Buildings.* Washington, DC: American Institute of Architects, 1979.
Discusses building energy performance and evaluation of energy-conscious design and redesign.
United States Department of the Interior, National Park Service. *Plants/People/And Environmental Quality: A Study of Plants and Their Environmental Functions.* Prepared by Gary O. Robinette. Washington, DC: United States Government Printing Office, 1972.
Describes environmentally responsive landscape design and planting, as well as the environmental aspects of site planning and organization.
University of Minnesota, Underground Space Center. *Earth Sheltered Community Design*. New York: Van Nostrand Reinhold Company, 1981.
Explores the prospects of designing entire communities of earth-sheltered housing and discusses issues of development, layout, site, and design. Ten case studies are presented.
____. *Earth Sheltered Homes: Plans and Designs.* New York: Van Nostrand Reinhold Company, 1981.
Detailed construction information, plans, and energy data are presented for twenty-three earth-sheltered homes.
____. *Earth Sheltered Housing Design: Guidelines, Examples, and References.* New York: Van Nostrand Reinhold Company, 1981.
A comprehensive manual on underground housing design; plans and sections of twenty earth-sheltered homes.
Wagner, Walter F., Jr., ed. *Energy-Efficient Buildings.* New York: McGraw-Hill Book Company, 1980.
Case studies of energy-efficient building designs and conservation techniques, including solar and underground buildings.

Geothermal

Armstead, H. Christopher, ed. *Geothermal Energy: Review of Research and Development*. Paris: Unesco Press, 1974.
Covers the basic theories and exploration techniques of geothermal resources.
Burrows, William. "Utilization of Geothermal Energy in Rotorua, New Zealand." *Multipurpose Use of Geothermal Energy*, P.J. Lienau and J.W. Lund, eds. Klamath Falls, OR: Oregon Institute of Technology (1974): 43–59.
Describes geothermal heating systems in a New Zealand city.
Kruger, Paul and Otte, Carol, eds. *Geothermal Energy: Resources, Production, Stimulation*. Stanford, CA: Stanford University Press, 1979.
A collection of eighteen articles on the use and exploration of geothermal resources.
McFarland, Robert D. *Geothermal Reservoir Models: Crack Plane Model*. Los Alamos, NM: Los Alamos Scientific Laboratory, 1975.
Describes hot-rock geothermal reservoir models and experiments by the Los Alamos Scientific Laboratory.
Purvine, W.D. "Utilization of Thermal Energy at Oregon Institute of Technology, Klamath Falls, Oregon." *Multipurpose Use of Geothermal Energy*, edited by P.J. Lienau and J.W. Lund. Klamath Falls, OR: Oregon Institute of Technology (1974) 179–191.
Discusses how the Oregon Institute of Technology campus was relocated to make use of geothermal hot water for space heating.
Smith, Morton C. "Los Alamos Dry Geothermal Source Demonstration Project." *Proceedings of the Geothermal Power Development Conference*. Los Alamos, NM: Los Alamos Scientific Laboratory (1974): 29–34.
Geothermal project description at Los Alamos Scientific Laboratory.
Smith, Morton C., et al. *Man-Made Geothermal Reservoirs.* Los Alamos, NM: Los Alamos Scientific Laboratory, 1975.
Geothermal exploration experiments by Los Alamos Scientific Laboratory.
Storey, David M. "Geothermal Drilling in Klamath Falls, Oregon." In *Multipurpose Use of Geothermal Energy*, P.J. Lienau and J.W. Lund, eds. Klamath Falls, OR: Oregon Institute of Technology (1974): 192–200.
Discusses geothermal well-drilling methods.

Solar

Anderson, Bruce. *Solar Energy Fundamentals in Building Design*. New York: McGraw-Hill Book Company, 1977.
Outgrowth of Anderson's MIT thesis. A good reference book on energy-conserving design and solar applications.
Anderson, Bruce with Michael Riordan. *The Solar Home Book.* Palo Alto, CA: Cheshire Books, 1976.
Describes the fundamentals of solar-energy use for residential applications.
Anderson, Bruce and Wells, Malcolm. *Passive Solar Energy: The Homeowner's Guide to Natural Heating and Cooling*. Andover, MA: Brick House Publishing Company, 1981.

A nontechnical introduction to the basics of passive-solar energy. Discusses heating, cooling, construction details, and tax credits.

Arizona State University, College of Architecture. *Solar-Oriented Architecture*. Washington, DC: American Institute of Architects, 1975.
A survey of solar-heated and -cooled buildings in the United States. Includes descriptions of basic solar principles.

Association for Applied Solar Energy. *Proceedings of the World Symposium on Applied Solar Energy*. Menlo Park, CA: Stanford Research Institute, 1956.
Interesting book on applications of solar energy in the 1950s and earlier.

Baer, Steve. *Sunspots: Collected Facts and Solar Fiction*. Second edition. Albuquerque, NM: Zomeworks Corporation, 1977.
A look at nonconventional solar energy applications, with social commentary and "solar fiction."

Bainbridge, David; Corbett, Judy; and Hofacre, John. *Village Homes' Solar House Design*. Emmaus, PA: Rodale Press, 1979.
Collection of forty-three energy-conscious solar designs of homes in the Village Homes subdivision of Davis, California.

Beckman, William; Klein, S.A.; and Duffie, J.A. *Solar Heating Design by the F-Chart Method*. New York: John Wiley and Sons, 1977.
Presentation of a method for combining a solar collector and conventional furnace or heater into one system to supply to entire heating load.

Brinkworth, Brian J. *Solar Energy for Man*. New York: John Wiley and Sons, Halsted Press, 1972.
Review of basic energy principles, applications, and methods of solar energy collection.

Butti, Ken and Perlin, John. *A Golden Thread: 2,500 Years of Solar Architecture and Technology*. Palo Alto, CA: Cheshire Books, 1980.
Carefully researched history of solar energy, with illustrations of past applications and inventions.

Carriere, Dean. *Solar Houses for a Cold Climate*. New York: Charles Scribner's Sons, 1980.
Detailed introduction to solar energy for residential applications.

Cheremisinoff, Paul N. and Regino, Thomas C. *Principles and Applications of Solar Energy*. Ann Arbor, MI: Ann Arbor Science Publishers, 1978.
Overview description of solar-energy use and conversion.

Clegg, Peter and Watkins, Derry. *The Complete Greenhouse Book: Building and Using Greenhouses from Cold Frames to Solar Structures*. Charlotte, VT: Garden Way Publishing Company, 1978.
A very complete book emphasizing solar greenhouses and organic indoor gardening.

Davis, Norah Deakin and Lindsey, Linda. *At Home in the Sun: An Open-House Tour of Solar Homes in the United States*. Charlotte, VT: Garden Way Publishing Company, 1979.
Brief descriptions of thirty-one solar homes.

Dean, Thomas Scott. *How to Solarize Your House: A Practical Guide to Design and Construction for Solar Heating*. New York: Charles Scribner's Sons, 1980.
Fairly technical guide to home solar heating, with emphasis on active systems.

DeKorne, James B. *The Survival Greenhouse: An Eco-System Approach to Home Food Production*. Second edition. Culver City, CA: Peace Press, 1978.
Describes one family's experience with building and maintaining a food-producing greenhouse.

Edmund Scientific Co.; Homan, Bob; Homan, Nancy; Thomason, Dr. Harry; and Wells, Malcolm. *Solaria: On the Threshold of Environmental Renaissance*. Barrington, NJ: Edmund Scientific Company, 1975.
Covers the background, design, and construction of "Solaria," an earth-sheltered solar house.

Ericson, Katherine. *The Solar Jobs Book*. Andover, MA: Brick House Publishing Company, 1980.
Explains how to find a job in solar energy and where to obtain a solar education. Extensive appendix and bibliography.

Gropp, Louis. *Solar Houses: 48 Energy-Saving Designs*. New York: Random House, Pantheon Books, 1978.
Provides a look at selected active, passive, and underground houses nationwide.

Herdeg, Walter. *The Sun in Art*. Zurich: Graphis Press, 1968.
Examples of man's use of the sun in art throughout the ages.

International Solar Energy Society, American Section. *AS/ISES 1980: Proceedings of the 1980 Annual Meeting, American Section of the International Solar Society*. Gregory E. Franta and Barbara H. Glenn, eds. Two vols. Newark, DE: American Section of the International Solar Energy Society, 1980.
Contains three hundred twenty-five papers covering most aspects of solar-energy conversion and use.

———. *Passive Cooling: Proceedings of the International Passive and Hybrid Cooling Conference*. Arthur Bowen, Eugene Clark, and Kenneth Labs, eds. Newark, DE: American Section of the International Solar Energy Society, 1981.
Contains one hundred forty-four papers covering most aspects of passive cooling of buildings in different climate regions.

———. *Passive Solar State of the Art: Proceedings of the 2nd National Passive Solar Conference*. Don Prowler, ed. Three vols. Newark, DE: American Section of the International Solar Energy Society, 1978.
Contains one hundred fifty-six papers covering most aspects of passive solar heating and cooling of buildings.

———. *Passive Solar Takes Off: Proceedings of the 3rd National Passive Solar Conference*. Harry Miller, Michael Riordon, and David Richards, eds. Newark, DE: American Section of the International Solar Energy Society, 1979.
Contains one hundred sixty-two papers on passive-solar heating and cooling.

———. *Proceedings of the 5th National Passive Solar Conference*. John Hayes and Rachel Snyder, eds. Two vols. Newark, DE: American Section of the International Solar Energy Society, 1980.
Contains two hundred ninety papers covering passive solar heating and cooling of buildings.

____. *Proceedings of the 4th National Passive Solar Conference*. Gregory E. Franta, ed. Newark, DE: American Section of the International Solar Energy Society, 1979.
Contains two hundred papers covering passive solar heating and cooling of buildings.

____. *Proceedings of the 1978 Annual Meeting of the American Section of the International Solar Energy Society, Inc.* Karl W. Böer and Gregory E. Franta, eds. Two vols. Newark, DE: American Section of the International Solar Energy Society, 1977.
Contains three hundred twenty-five papers covering most aspects of solar energy conversion and use.

____. *Proceedings of the 1977 Annual Meeting, American Section of the International Solar Energy Society*. Charles Beach and Edward Fordyce, eds. Three vols. Cape Canaveral, FL: American Section of the International Solar Energy Society, 1977.
Contains two hundred ten papers covering solar-energy conversion and use.

____. *Proceedings of the 6th National Passive Solar Conference*. John Hayes and William A. Kolar, eds. Newark, DE: American Section of the International Solar Energy Society, 1981.
Contains one hundred eighty-five papers concerning passive-solar heating and cooling of buildings.

____. *Sharing the Sun: Solar Technology in the Seventies.* Karl W. Böer, ed. Ten vols. Cape Canaveral, FL: American Section of the International Solar Energy Society, 1976.
Contains three hundred fifty papers covering most aspects of solar energy conversion and use.

____. *Sun II: Proceedings of the International Solar Energy Society Silver Jubilee Congress*. Karl W. Böer and Barbara H. Glenn, eds. Three vols. New York: Pergamon Press, 1979.
Contains five hundred sixteen papers from the 1979 annual meeting.

Joint Venture and Friends. *Here Comes the Sun 1981*. Washington, DC: American Institute of Architects, 1975.
Describes principles of solar collection and its design implications in multifamily housing.

Kreider, Jan F. *The Solar Heating Design Process: Active and Passive Systems*. New York: McGraw-Hill Book Company, 1982.
An authoritative design reference for solar space and water heating systems.

Kreider, Jan F. and Kreith, Frank. *Solar Heating and Cooling: Engineering, Practical Design, and Economics*. Washington, DC: Hemisphere Publishing Corporation, 1975.
A definitive text on solar energy, heat-transfer principles, and solar heating and cooling of buildings.

Kreider, Jan F. and Kreith, Frank, eds. *Solar Energy Handbook*. New York: McGraw-Hill Book Company, 1981.
A comprehensive collective of data and procedures concerning the forms of solar conversion.

Kreith, Frank and Kreider, Jan F. *Principles of Solar Engineering*. New York: McGraw-Hill Book Company, 1978.
Describes basic engineering principles used in solar-energy-collection systems.

Lebens, Ralph M. *Passive Solar Heating Design*. London: Applied Science Publishers, 1980.
A workbook of passive-solar-design methods for the building designer. Includes a simulation program designed for a hand-held programmable calculator.

Libbey-Owens-Ford Company. *Sun Angle Calculator*. Toledo, OH: Libbey-Owens-Ford Co., 1974.
Discussion of a device designed to determine the position of the sun during the year or day at many different latitudes.

Löf, George O.G. and Ward, D.S. *Performance of a Residential Solar Heating and Cooling System*. Fort Collins, CO: Solar Energy Applications Laboratory, Colorado State University, 1975.
An analysis of the National Science Foundation/CSU solar home.

Lucas, Ted. *How to Build a Solar Heater*. Revised edition. New York: Crown Publishers, 1980.
Discusses solar applications for space heating and cooling as well as water heating. Includes list of manufacturers and a bibliography.

McCallagh, James C., ed. *The Solar Greenhouse Book*. Emmaus, PA: Rodale Press, 1978.
An extensive treatment of all aspects of solar greenhouses: design, construction, management.

McDaniels, David K. *The Sun: Our Future Energy Source*. New York: John Wiley and Sons, 1979.
Introductory text to solar energy and solar applications.

Massdesign. *Solar Heated Houses for New England and Other North Temperature Climates*. Third edition. Washington, DC: American Institute of Architects, 1975.
Includes examples of solar houses suited for the New England area. Also included is a sample computer program to determine the availability of and the requirement for solar energy in houses.

Mazria, Edward. *Passive Solar Energy Book*. Emmaus, PA: Rodale Press, 1979.
Guide to passive-solar applications for homeowners, architects, and builders. Gives twenty-seven "rules of thumb" for sizing and designing passive systems for homes, greenhouses, and other buildings.

Michels, Tim. *Solar Energy Utilization*. New York: Van Nostrand Reinhold Company, 1979.
This guide to residential use of solar energy discusses the cost-effectiveness of various solar system types for specific applications and how to maximize system efficiency.

Montgomery, Richard H. *The Solar Decision Book: A Guide to Heating Your Home with Solar Energy*. New York: John Wiley and Sons, 1978.
Offers an overview of the energy crisis and discussion of solar as an investment, with chapters on technical and economic considerations.

Olgyay, Aladar and Olgyay, Victor. *Solar Control and Shading Devices*. Princeton, NJ: Princeton University Press, 1957.

Recently reissued classic text on sun control, with photographs and evaluations of many built projects.

Reif, Daniel K. *Solar Retrofit: Adding Solar to Your Home*. Andover, MA: Brick House Publishing Company, 1981.
Discusses different solar systems and their selection. Includes step-by-step, how-to instructions for four retrofit projects.

Scully, Dan; Prowler, Don; and Anderson, Bruce. *The Fuel Savers: A Kit of Solar Ideas for Existing Homes*. Harrisville, NH: Total Environmental Action, 1978.
A small booklet containing eighteen ideas for do-it-yourself solar projects for the home. Includes economic evaluations and cost-effectiveness ratings for each project.

Shurcliff, William A. *New Inventions in Low-Cost Solar Heating*. Andover, MA: Brick House Publishing Company, 1979.
Descriptions and critiques of a great number of solar heating ideas adaptable to many different buildings.

____. *Solar Heated Buildings: A Brief Summary*. 13th ed. Cambridge, MA: William A. Shurcliff, 1977.
Lists solar-heated buildings in the United States. Final edition of listings that includes many historic buildings.

____. *Solar Heated Buildings of North America: 120 Outstanding Examples.* Andover, MA: Brick House Publishing Co., 1978.
Selected solar buildings in the United States and Canada.

Smithsonian Institution. *Fire of Life: The Smithsonian Book of the Sun*. New York: W.W. Norton and Company, 1981.
Man's relationship to the sun; solar physics; sun as life-giver.

Solar Energy Research Institute, Solar Energy Information Data Bank. *National Solar Energy Education Directory*. Third edition. Washington, DC: United States Government Printing Office, 1981.
A listing of solar-related courses, curricula, and programs at postsecondary institutions throughout the United States.

Strickler, Darryl J. *Passive Solar Retrofit*. New York: Van Nostrand Reinhold Company, 1982.
Describes why and how to add natural heating and cooling features to an existing home. Includes step-by-step instructions for many retrofit projects.

Szokolay, Steven V. *Solar Energy and Building*. New York: John Wiley and Sons, Halsted Press, 1975.
Describes solar energy collection and application, with examples; discusses economics and planning implications and gives a design guide.

Thomason, Harry E. and Thomason, Harry Jack Lee, Jr. *Solar House Plans*. Barrington, NJ: Edmund Scientific Company, 1972.
Explanation plus foldout drawings covering construction and installation of collectors on the Thomason house.

United States Department of Energy. *Passive Solar Design Analysis*. Prepared by J. Douglas Balcomb. Passive Solar Design Handbook, Volume 2. Springfield, VA: National Technical Information Service, United States Department of Commerce, 1980.
Design-analysis methods are given for passive design performance prediction for buildings using direct gain and thermal-storage walls. Includes sections on rock-bed storage, solar savings calculations, and economics.

____. *Passive Solar Design Concepts*. Prepared by Total Environmental Action. Passive Solar Design Handbook, Volume 1. Springfield, VA: National Technical Information Service, United States Department of Commerce, 1980.
Basic design concepts are presented for passive-solar buildings.

United States Department of Housing and Urban Development. *Regional Guidelines for Building Passive Energy Conserving Homes.* Prepared by the American Institute of Architects. Washington, DC: United States Government Printing Office, 1978.
Describes thirteen climate regions of the United States and gives passive design recommendations.

United States Energy Research and Development Administration. *Solar Use Now, a Resource for People: Extended Abstracts, 1975 International Solar Energy Congress and Exposition*. Prepared by the International Solar Energy Society. Washington, DC: Energy Research and Development Administration, 1975.
Bound abstracts from the 1975 ISES meeting.

United States Energy Research and Development Administration, Division of Solar Energy, Heating and Cooling Branch. *Passive Solar Heating and Cooling Conference and Workshop Proceedings, Albuquerque, New Mexico*. Springfield, VA: National Technical Information Service, United States Department of Commerce, 1976.
Collected papers and presentations from the first national passive-solar conference, held in 1976.

United States National Science Foundation. *An Assessment of Solar Energy as a National Energy Resource.* Washington, DC: United States Government Printing Office, 1972.
From the NSF/NASA Solar Energy Panel. Covers solar energy applications and other forms of natural energy.

United States National Science Foundation, Research Applied to National Needs. *Workshop Proceedings: Solar Cooling for Buildings.* Washington, DC: United States Government Printing Office, 1974.
Abstracts and presentations on research and development of systems to cool buildings using solar energy.

Watson, Donald. *Designing and Building a Solar House: Your Place in the Sun.* Charlotte, VT: Garden Way Publishing Company, 1977.
Illustrates how to combine solar-heating technology with residential design.

____. *Innovation in Solar Thermal House Design*. Washington, DC: American Institute of Architects, 1975
Covers the development of concepts for solar-thermal design in a climate-responsive house.

Waugh, Albert E. *Sundials: Their Theory and Construction*. New York: Dover Publications, 1973.
A comprehensive look at the many different kinds of sundials, the theory of their design, and construction methods.

Williams, J. Richard. *Solar Energy: Technology and Applications.* Ann Arbor, MI: Ann Arbor Science Publications, 1974.
Includes techniques and methods for collecting solar energy.

Wright, David. *Natural Solar Architecture: A Passive Primer.* Revised edition. New York: Van Nostrand Reinhold Company, 1980.
Well-known environmental architect describes microclimatic design and passive-solar techniques.

Wright, Sydney; Wright, Rodney; Selby, Bob; and Dieckmann, Larry. *The Hawkweed Passive Solar House Book*. Chicago: Rand McNally and Company, 1980.
Basic facts of passive-solar-energy systems, climatic analysis, and solar retrofits.

Yanda, Bill and Fisher, Rick. *The Food and Heat-Producing Solar Greenhouse*. Santa Fe, NM: John Muir Publications, 1976.
Solid guide to the solar-greenhouse concept. Describes design, construction, and operation.

Wind

Burke, Barbara L. and Robert L. Meroney. *Energy from the Wind*. Fort Collins, CO: Solar Energy Applications Laboratory, Colorado State University, 1977.
A very comprehensive bibliography with over nineteen hundred references current to 1975.

Cheremisinoff, Nicholas P. *Fundamentals of Wind Energy*. Ann Arbor, MI: Ann Arbor Science Pubs., 1978.
Overview of wind energy and wind technology.

Coonley, Douglas R. *An Introduction to the Use of Wind*. Harrisville, NH: Total Environmental Action, 1975.
Presents the basic principles of wind energy.

Eldridge, Frank R. *Wind Machines*. Second edition. New York: Van Nostrand Reinhold Company, 1980.
Review of wind power and wind machines. Discusses history, viability, design, application, potential.

Flavin, Christopher. *Wind Power: A Turning Point*. Worldwatch Paper 45. Washington, DC: Worldwatch Institute, 1981.
Author discusses history, economics, and potential of wind energy, and describes wind-power technology for water-pumping, small-scale electricity generation, and "wind farms."

Golding, E.W. *The Generation of Electricity by Wind Power.* New York: John Wiley and Sons, Halsted Press, 1955.
Classic reference work on wind power. Updated material was added in a 1976, revised edition.

Hackelman, Michael. *The Homebuilt, Wind-Generated Electricity Handbook.* Culver City, CA: Peace Press, 1975.
Presents in a conversational style a great deal of information on wind-power systems and their installation, especially on renovation of older equipment.

Inglis, David Rittenhouse. *Wind Power and Other Energy Options*. Ann Arbor, MI: Univ. of Michigan Press, 1978.
Author argues the necessity of rapid introduction of alternative energy sources, especially wind power and other solar-related energies.

Kovarik, Tom; Pipher, Charles; and Hurst, John. *Wind Energy*. Chicago: Quality Books, Domus Books, 1979.
Introduction to wind energy and wind-power generation, storage, and conversions.

McGuigan, Dermot. *Harnessing the Wind for Home Energy*. Charlotte, VT: Garden Way Publishing Company, 1978.
Describes wind-power installations for residences. Includes a list of manufacturers and bibliography.

Marier, Donald. *Wind Power for the Homeowner: A Guide to Selecting, Siting, and Installing an Electricity-Generating Wind Power System*. Emmaus, PA: Rodale Press, 1981.
Step-by-step guide on how to install a wind system, with options for battery storage, utility hookup, and other backup systems. Includes tables, appendix, source list, and bibliography.

Park, Jack. *The Wind Power Book*. Palo Alto, CA: Cheshire Books, 1981.
Discusses many wind-energy applications and explores the evolution of windmill designs.

Putnam, Palmer Cosslett. *Power from the Wind*. New York: Van Nostrand Reinhold Company, 1948.
Classic account of the design and testing of a one-hundred-seventy-five-foot, twelve-hundred-fifty-kilowatt wind machine at Grandpa's Knob, in Vermont, in the early 1940s.

Reynolds, John. *Windmills and Watermills.* New York: Praeger Publishers, 1970.
A collection of descriptions, photographs, and excellent illustrations of windmills and watermills in Europe and the United States.

Torrey, Volta. *Wind-Catchers*. Brattleboro, VT: Stephen Greene Press, 1976.
A broad, historic look at wind-power progress in the United States.

United States National Science Foundation, Research Applied to National Needs and the National Aeronautics and Space Administration. *Wind Energy Conversion Systems: Workshop Proceedings*. Joseph M. Savino, ed. Washington, DC: United States Government Printing Office, 1974.
Articles and presentations at the 1973 NSF/NASA wind-energy conference.

Index